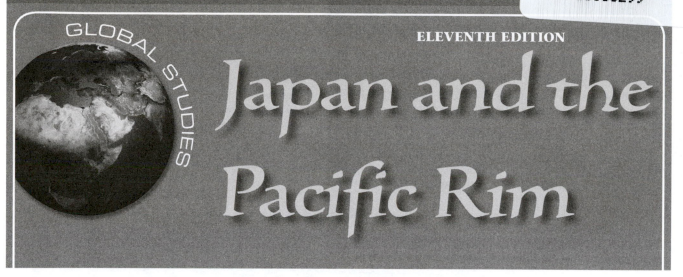

ELEVENTH EDITION

GLOBAL STUDIES

Japan and the Pacific Rim

Dean W. Collinwood

University of Utah

Series Consultant
Christopher J. Sutton
Western Illinois University

OTHER BOOKS IN THE GLOBAL STUDIES SERIES

- **Africa**
- **China**
- **Europe**
- **India and South Asia**
- **Latin America and The Caribbean**
- **The Middle East**
- **Russia and the Near Abroad**

Mc Graw Hill

*Connect
Learn
Succeed™*

GLOBAL STUDIES: JAPAN AND THE PACIFIC RIM, ELEVENTH EDITION

1 2 3 4 5 6 7 8 9 0 QDB/QDB 1 0 9 8 7 6 5 4 3 2

MHID: 0-07-802624-5
ISBN: 978-0-07-802624-9
ISSN: 1059-5988

Managing Editor: *Larry Loeppke*
Developmental Editor II: *Debra A. Henricks*
Permissions Coordinator: *Lenny J. Behnke*
Senior Marketing Communications Specialist: *Mary Klein*
Project Manager: *Erin Melloy*
Design Coordinator: *Brenda A. Rolwes*
Cover Graphics: *Rick D. Noel*
Buyer: *Nicole Baumgartner*
Media Project Manager: *Sridevi Palani*

Compositor: Laserwords Private Limited
Cover Image: ©David Welsh, 2011; Kinkaku-ji Temple (Golden Pavilion); Kyoto, Japan

Japan and the Pacific Rim

AUTHOR/EDITOR
Dr. Dean W. Collinwood

Dr. Dean W. Collinwood teaches international relations and comparative politics in the Political Science Department at the University of Utah in Salt Lake City, Utah, USA. He received his PhD from the University of Chicago where he studied in both the political science and education departments. His MSc in international relations is from the University of London, and his BA in political science with a minor in Japanese is from Brigham Young University. Dr. Collinwood was a Fulbright Scholar at Tokyo University and Tsuda College in Japan and has conducted research throughout Asia and the Pacific. He currently chairs the Asia Pacific Council and formerly directed the U.S. Japan Center's Japan Industry and Management Technology (JIMT) program and the U.S. China Center's China Professional Development Program for the U.S. government. He is past president of the Western Conference of the Association for Asian Studies and of the Salt Lake Committee on Foreign Relations. He is the author of several books on Japan, Korea, and other countries, and some of his books have been used by the U.S. State Department to train new Foreign Service officers.

Series Consultant
Dr. Christopher J. Sutton

Dr. Christopher J. Sutton is professor of geography at Western Illinois University. Born in Virginia and raised in Illinois, he received his bachelor's degree (1988) and master's degree (1991) in Geography from Western Illinois University. In 1995 he earned his PhD in Geography from the University of Denver. He is the author of numerous research articles and educational materials. A broadly trained geographer, his areas of interest include cartographic design, cultural geography, and urban transportation. After teaching at Northwestern State University of Louisiana for three years, Dr. Sutton returned to Western Illinois University in 1998, serving as chair of the Department of Geography from 2002 to 2007. Additionally, Dr. Sutton has served as president of the Illinois Geographical Society.

Academic Advisory Board

Members of the Academic Advisory Board are instrumental in the final selection of articles for each edition of Global Studies. Their review of articles for content, level, and appropriateness provides critical direction to the editors and staff. We think that you will find their careful consideration well reflected in this volume.

Global Studies: Japan and the Pacific Rim
Eleventh Edition

Academic Advisory Board Members

Acknowledgments

A special thank you to Stacy Banks who aided in the preparation of this manuscript. Stacy received her MEd from Seattle Pacific University and her BS from Brigham Young University. Her international experience includes living and working in the Gifu, Nagoya, Hokkaido, and Tokyo regions of Japan for nearly six years. For two of those years, she taught Japanese students as an Assistant Language Teacher for the Japan Exchange and Teaching Programme (JET). She recently finished two years of national service with AmeriCorps and currently volunteers with the American Red Cross. She lives with her family in Logan, Utah.

Contents

Global Studies: Japan and the Pacific Rim

Taiwan

Thailand

Vietnam

Preface

USING GLOBAL STUDIES: JAPAN AND THE PACIFIC RIM

The Global Studies Series

The Global Studies series was designed to provide readers with a basic knowledge and understanding of the regions and countries of the world. Each volume provides a foundation of information—geographic, cultural, economic, political, historical, and religious—that will allow readers to better assess the current and future problems within these countries and regions and to comprehend how events there might affect their own well-being. In short, these volumes present the background information necessary to respond to the realities of our global age.

Each of the volumes in the Global Studies series is crafted under the careful direction of an author/editor who is an expert in the area under study. The author/editors teach and conduct research and have traveled extensively through the regions about which they are writing.

Major Features of the Global Studies Series

The Global Studies volumes are organized to provide concise information on the regions and countries within those areas under study. The major sections and features of the books are described here.

Regional Essays

For *Global Studies: Japan and the Pacific Rim,* the author/editor has written two essays: "The Pacific Rim: Diversity and Interconnection," and "The Pacific Islands: Opportunities and Limits." Detailed maps accompany each essay.

Country Reports

Concise reports are written for each of the countries within the region under study. These reports are the heart of each Global Studies volume. *Global Studies: Japan and the Pacific Rim, Eleventh Edition,* contains 21 country reports, including a lengthy special report on Japan.

Each country report is comprised of: a detailed map visually positioning the country among its neighboring states; a current essay providing important historical, geographical, political, cultural, and economic information; a timeline of key historical events; and a summary of statistical information. "Did you Know?" boxes offer interesting facts to further engage student interest,

and "Further Investigation" boxes provide references to encourage students to explore a topic more deeply. Finally, Snapshot boxes summarize the country in terms of its development, freedom, health/welfare, and achievements.

A Note on the Statistical Reports

The statistical information provided for each country has been drawn from a wide range of sources. (The most frequently referenced are listed in the bibliography.) Every effort has been made to provide the most current and accurate information available. However, sometimes the information cited by these sources differs to some extent; and, all too often, the most current information available for some countries is somewhat dated. Aside from these occasional difficulties, the statistical summary of each country is generally quite complete and up to date. Care should be taken, however, in using these statistics (or, for that matter, any published statistics) in making hard comparisons among countries. Comparable statistics for the United States and Canada, can be found in Appendix A and B.

Articles From the World Press

A collection of carefully selected articles from a broad range of international periodicals and newspapers are reprinted in this volume. The articles have been chosen for currency, interest, and their differing perspectives on the subject countries. Learning Objectives and Challenge Questions accompany each article to enhance student learning and comprehension.

Internet References

An extensive annotated list of websites can be found in this edition of *Global Studies: Japan.* In addition, country-specific websites are provided at the end of most country reports. All of the website addresses were correct and operational at press time. Instructors and students alike are urged to refer to those sites often to enhance their understanding of the region and to keep up with current events.

Glossary, Bibliography, Index

At the back of each Global Studies volume, readers will find a glossary of terms and abbreviations, which provides a quick reference to the specialized vocabulary of the area under study and to the standard abbreviations used throughout the volume.

Following the glossary is a bibliography, which lists general works, national histories, and current-events publications and periodicals that provide regular coverage on Japan.

The index at the end of the volume is an accurate reference to the contents of the volume. Readers seeking specific information and citations should consult this standard index.

Currency and Usefulness

Global Studies: Japan, like the other Global Studies volumes, is intended to provide the most current and useful information available necessary to understand the events that are shaping the cultures of the region today.

This volume is revised on a regular basis. If you have an idea that you think will make the next edition more useful, an article or bit of information that will make it more current, or a general comment on its organization, content, or features, please send it in for serious consideration.

Guided Tour

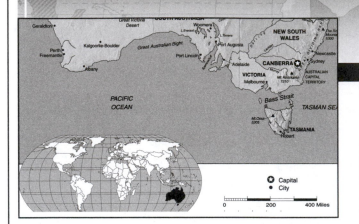

MAPS

Detailed maps accompany regional essays and country reports.

Australia
(Commonwealth of Australia)

Despite its out-of-the-way location far south of the main trading routes between Europe and Asia, seafarers from England, Spain, and the Netherlands began exploring parts of the continent of Australia in the seventeenth century. The French later made some forays along the coast, but it was the British who first found something to do with a land that others had disparaged as useless: they decided to send their prisoners there. The British had long believed that the easiest solution to prison overcrowding was expulsion from Britain. Convicts had been sent to the American colonies for many years, but after American independence was declared in 1776, Britain needed another place to send convicts, and Australia was it.

Australia seemed like the ideal spot for a penal colony: It was isolated from the centers of civilization; it had some good harbors; and, although much of the continent was a flat, dry, riverless desert with only sparse vegetation, the coastal fringes were well suited to human habitation. Indeed, although the British did not know it in the 1700s, they had come across a huge continent endowed with abundant natural resources. Along the northern coast (just south of present-day Indonesia and Papua New Guinea) was a tropical zone with heavy rainfall and tropical forests. The eastern coast was wooded with valuable pine trees, while the western coast was dotted with eucalyptus and acacia trees. Minerals, especially coal, gold, nickel, petroleum, iron, uranium, and bauxite, were plentiful, as

REGIONAL ESSAYS AND COUNTRY REPORTS

Concise regional essays and country reports include historical, cultural, religious, political, and economic information, along with detailed maps and statistics.

TIMELINE

1600s
European exploration of the Australian coastline begins
1688
British explorers first land in Australia
1788
The first shipment of English convicts arrives
1851
A gold rush lures thousands of immigrants
1901
Australia becomes independent within the Brit-

TIMELINE

A timeline of key historical events is included with each country report.

? DID YOU KNOW?

New South Wales passed a law in 2011 that requires anyone, including Muslim women wearing veils, to remove their face coverings when asked to do so by a police officer. The controversial new law was the result of a vehicle accident caused in part by a woman wearing a niqab. Penalties for those who refuse to remove the veil could include a year in prison and a fine of US$6,000. There are about 400,000 Muslims in Australia, and it is thought that about 2,000 of them wear veils.

DID YOU KNOW?

Did You Know? boxes offer interesting facts to engage student interest

FURTHER INVESTIGATION

Over the next five years, Australia will remove gender barriers in its military. Positions formerly reserved for men only, including front-line combat roles, will be available to both men and women. The move follows similar steps taken in New Zealand and Canada. To learn more, see: www.guardian.co.uk/world/2011/sep/27/australian-military-women-frontline-roles

FURTHER INVESTIGATION

Further Investigation boxes encourage students to explore a topic more deeply.

Statistical information from a wide range of sources is provided for each country.

Statistics

Geography

Area in Square Miles (Kilometers): 2,988,902 (7,741,220) (slightly smaller than the United States)

Capital (Population): Canberra (384,000)

Environmental Concerns: depletion of the ozone layer; pollution; soil erosion and excessive salinity; desertification; wildlife habitat loss; degradation of Great Barrier Reef; limited freshwater resources; drought

Geographical Features: mostly low plateau with deserts; fertile plain in southeast

Climate: generally arid to semiarid; temperate to tropical

People

Population

Total: 21,766,711

Annual Growth Rate: 1.2%

Rural/Urban Population (Ratio): 11/89

Major Languages: English; indigenous languages

Ethnic Makeup: 92% Caucasian; 7% Asian; 1% Aboriginal and others

Religions: 26% Roman Catholic; 19% Anglican; 19% other Christian; 2% Buddhist; 2% Muslim; 32% unaffiliated or other

Health

Life Expectancy at Birth: 79 years (male); 84 years

Transportation

Roadways in Miles (Kilometers): 505,157 (812,972)

Railroads in Miles (Kilometers): 23,889 (38,445)

Usable Airfields: 462

Government

Type: federal parliamentary democracy

Independence Date: January 1, 1901 (federation of U.K. colonies)

Head of State/Government: Queen Elizabeth II; Prime Minister Julia E. Gillard

Political Parties: Australian Greens; Australian Labor Party; Family First Party; Liberal Party; The Nationals

Suffrage: universal and compulsory at 18

Military

Military Expenditures (% of GDP): 3%

Current Disputes: territorial claim in Antarctica; boundary issues with East Timor and Indonesia

Economy

Currency ($ U.S. Equivalent): 1.09 Australian dollars = $1

Per Capita Income/GDP: 41,000/$882.4 billion

GDP Growth Rate: 2.7%

Inflation Rate: 2.9%

Unemployment Rate: 5.1%

Labor Force by Occupation: 75% services; 21% industry;

Snapshot: AUSTRALIA

Summarized below is a quick look at the country with regard to its development, freedom, health/welfare, and achievements.

Development

Mining of nickel, iron ore, uranium, and other metals continues to supply a substantial part of Australia's gross domestic product. The United States, Canada, South Korea, Taiwan, and China buy uranium, and Japan continues to buy beef and many other products. Seven out of 10 of Australia's largest export markets are now Asian countries. With 27 million head of cattle, Australia is the world's largest exporter of beef.

Freedom

Health/Welfare

Australia has developed a complex and comprehensive system of social welfare. Education is the province of the several states. Public education is compulsory. Australia boasts several world-renowned universities. The world's first voluntary-euthanasia law passed in Northern Territory in 1996, but legal challenges have thus far prevented its use. The government has proposed a nationwide filter to prevent pornographic images from being accessed on the Internet.

Achievements

The vastness and challenge of Australia's interior lands, called the "outback," have inspired a number of Australian writers to create outstanding poetry and fictional novels.

Snapshot boxes summarize each country with regard to its development, freedom, health/welfare, and achievements.

---

1 Deep Danger: Competing Claims in the South China Sea

Marvin C. Ott

"The South China Sea is a growing focus of concern in Washington, at the headquarters of the US Pacific Command in Honolulu, and in a number of Southeast Asian capitals."

Learning Objectives

After reading this article, you will more clearly understand the following:

• Mischief Reef

• South China Sea claims

• International waterways

• ASEAN Regional Forum (ARF)

• Freedom of navigation

• Political ambiguity

The waters of the South China Sea are dotted with hundreds of atolls, reefs, and small islands—only one of which has sufficient fresh water to qualify, under traditional international law, as capable of supporting (Hainan). The Philippines protested to China. Manila also attempted unsuccessfully to enlist US military support, but did succeed in persuading the Association of Southeast Asian Nations (ASEAN) to express strong collective concern to China.

Beijing responded with sustained outreach to Southeast Asia, and sought to strengthen ties in the region and burnish its image as a "good neighbor." A centerpiece of this was a Declaration on the Conduct of Parties in the South China Sea, signed by ASEAN and China in 2002, in which all parties pledged good behavior pending the resolution of conflicting claims. China also began to tout its willingness to engage in the "joint development" of presumed petroleum and mineral resources in the South China Sea while setting aside conflicting claims. China's efforts at reassurance did not, however, involve abandoning its claims to the South China Sea or its facility on Mischief Reef—where construction and upgrades proceeded apace.

A collection of carefully selected articles from the world press introduces students to relevant topics affecting the modern culture of the region.

Learning Objectives and ***Challenge Questions*** accompany each article to further enhance learning and comprehension.

Challenge Questions

The following questions will increase understanding of the contents of this article:

1. What is the status quo of territorial claims in the South China Sea, and what were the results of the Philippines finding a Chinese military outpost in the Spratly Islands?

2. Why is China a major part of foreign relations for Southeast Asian governments?

3. Describe some of the inconsistencies in China's reasons for claiming the South China Sea. How does China benefit from the resulting confusion?

4. Provide some examples of Chinese resistance to regional and international interference in its

---

Internet References

(Some websites continually change their structure and content, so the information listed here may not always be available.)

GENERAL SITES

CNN Online Page
www.cnn.com
A U.S. 24-hour video news channel. News is updated every few hours.

C-SPAN ONLINE
www.c-span.org
See international and archived C-SPAN (Cable-Satellite Public Affairs Network) programs, which broadcast unedited federal government and public affairs proceedings.

NationMaster.com
www.nationmaster.com/countries
A central data source that graphically compares nations by generating maps and graphs of statistics from the *CIA World Factbook*, UN, OECD, and others.

The New York Times
topics.nytimes.com/topics/news/international/

WWW Virtual Library Database
www.vlib.org
Easy search for country-specific sites that provide news, government, and other information is possible from this site.

JAPAN

Daily Yomiuri Online
www.yomiuri.co.jp/dy
Online edition of the *Daily Yomiuri* newspaper in Japan reporting top news of the day, including Japanese political, social, and general news, reports from foreign news services, and overseas English-language newspapers.

Japan Ministry of Foreign Affairs
www.mofa.go.jp
"What's New" lists events, policy statements, press releases on this website. The Foreign Policy section has speeches, archives, and information available in English and Japanese.

Internet References direct students to additional information.

This map provides a graphic picture of where the countries of the world are located, the relationship they have with their region and neighbors, and their positions relative to major economic and political power blocs. Certain areas have been focused on to illustrate these crowded regions more clearly. Japan and The Pacific Rim region is shaded for emphasis.

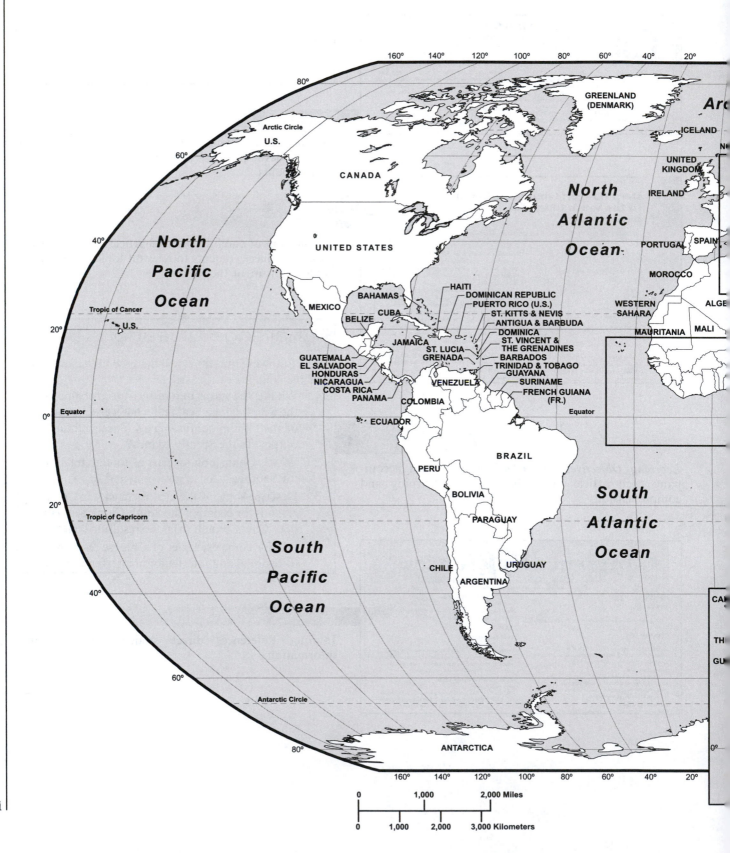

_an_

NORWAY
SWEDEN
DENMARK
NETHERLANDS
BELGIUM
LUX.
GERMANY
FRANCE
SWITZ.
SLOVENIA
SAN MARINO
MONACO
ANDORRA
VATICAN CITY
ITALY
ALBANIA
TUNISIA
MALTA
ALGERIA
LIBYA
ESTONIA
LATVIA
LITHUANIA
RUSSIA
BELARUS
POLAND
CZECHIA
SLOVAKIA
AUSTRIA HUNGARY
LIECHT.
CROATIA SERBIA
BOS.
MONT.
KOSOVO
MAC.
GREECE
UKRAINE
MOLDOVA
ROMANIA
BULGARIA
CYPRUS
LEBANON
ISRAEL
JORDAN
EGYPT
SYRIA
IRAQ
KUWAIT
SAUDI ARABIA
TURKEY
GEORGIA
AZERBAIJAN
ARMENIA
IRAN

0 300 600 Miles
0 300 600 Kilometers

60°
20°
40°
0°
40°
40°

NLAND

RUSSIA

KAZAKHSTAN
UZBEKISTAN
TURKMENISTAN
AFGHANISTAN
IRAN
BAHRAIN
SAUDI
ARABIA
QATAR U.A.E.
OMAN
YEMEN
DJIBOUTI
ERITREA
SUDAN
SOUTH
SUDAN
ETHIOPIA
SOMALIA
UGANDA
KENYA
EM. REP.
OF THE
CONGO
RWANDA
BURUNDI
TANZANIA
MALAWI
COMOROS
ZAMBIA
ZIMBABWE
TSWANA MOZAMBIQUE
OUTH
RICA
SWAZILAND
LESOTHO

EGYPT
A.R.
LA

MONGOLIA
TAJIKISTAN
KYRGYZSTAN
NEPAL
BHUTAN
CHINA
PAKISTAN
INDIA
BANGLADESH
MYANMAR
THAILAND
LAOS
VIETNAM
SRI
LANKA
CAMBODIA
BRUNEI
MALAYSIA
SINGAPORE
MALDIVES
SEYCHELLES
MADAGASCAR
MAURITIUS

JAPAN
TAIWAN
PHILIPPINES
PALAU
MICRONESIA
NAURU
KIRIBATI
PAPUA
NEW GUINEA
SOLOMON
ISLANDS
VANUATU
FIJI
TONGA
INDONESIA
TIMOR-LESTE

North
Pacific
Ocean

Indian
Ocean

AUSTRALIA

NEW ZEALAND

Arctic Circle
Tropic of Cancer
Equator
Tropic of Capricorn
Antarctic Circle

80°
60°
40°
20°
0°
20°
40°
60°
80°

MAURITANIA
MALI
NIGER
CHAD
BURKINA
FASO
BENIN
NIGERIA
GUINEA
COTE
D'IVOIRE
TOGO
GHANA
LIBERIA
CAMEROON
EQUATORIAL GUINEA
SÃO TOMÉ & PRÍNCIPE
GABON
CONGO REP.
NEGAL
NE

Miles
lometers

160°
180°
0°

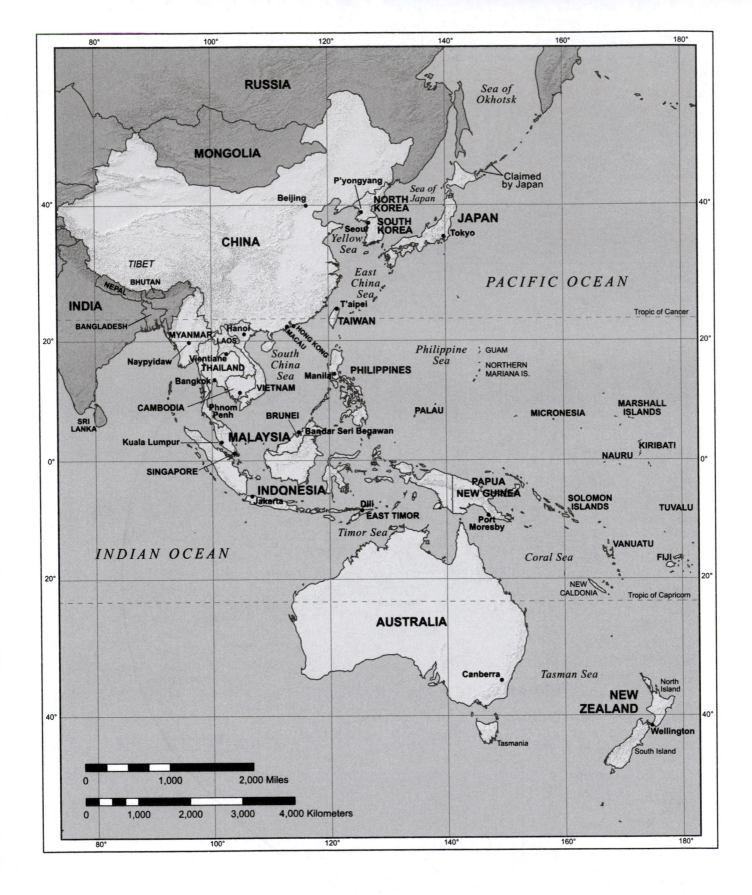

The Pacific Rim
Diversity and Interconnection

The term *Pacific Rim,* as used in this book, refers to 21 countries or administrative units along or near the Asian side of the Pacific Ocean, plus the numerous islands of the Pacific. Together, they are home to approximately 30 percent of the world's population and produce about 20 percent of the world's gross national product (GNP).

It is not a simple matter to decide which countries to include in a definition of the Pacific Rim. For instance, if we were thinking geographically, we might include Mexico, Chile, Canada, the United States, Russia, and numerous other countries that border the Pacific Ocean, while eliminating Myanmar (Burma) and Laos, since they are not technically on the rim of the Pacific. But our definition, and hence our selected inclusions, stem from fairly recent developments in economic and geopolitical power that have affected the countries of Asia and the Pacific in such a way that these formerly disparate regions are now being referred to by international corporate and political leaders as a single bloc.

People living in the region that we have thus defined do not think of themselves as "Pacific Rimmers." In addition, many social scientists, particularly cultural anthropologists and comparative sociologists, would prefer not to apply a single term to such a culturally, politically, and sociologically diverse region. Some countries, it is true, share certain cultural influences, such as Confucian family values and the hard work ethic in China, Japan, and Korea, and Theravada Buddhism and the culture of rice cultivation in Southeast Asia. But these commonalities have not prevented the region from fracturing into dozens of societies, often very antagonistic toward one another.

On the other hand, for more than 30 years, certain powerful forces have been operating in the Rim that have had the effect of pulling all of the countries toward a common philosophy. Indeed, it appears to be the case that most Pacific Rim countries are attempting to implement some version of free-market capitalism (as opposed to communism, socialism, asceticism, and other practices) and are rapidly acquiring the related values of materialism and consumerism. For most, although not all, of the Pacific Rim countries, a common awareness appears to be developing of the value of peaceful interdependence, rather than military aggression, for the improvement and maintenance of a high standard of living.

What are the powerful forces that launched the trend toward lifestyle convergence in the Rim? There are many, including nationalism and rapid advances in global communications due to the Internet and other new technologies. But one of the most important has been the Japanese yen. For years, Japanese money, in the form of direct investment, loans, and development aid, saturated the region. Using the infusion of capital, many Asia Pacific countries copied Japan's export-oriented market strategy and started producing economic growth beyond anything they had ever dreamed of. This has been particularly true in Southeast Asia where Japan was once called the "spark plug" of the region.

By the time Japan's economy had fallen into a prolonged stagnation, beginning in the 1990s, many of the surrounding economies were strong enough to move ahead on their own. One of those was China which had benefited for years from investments from Japan, Europe, and the United States. Japanese investment in the region has by no means ended, but emerging economies are now

(Courtesy of Amanda Campos)

A woman of the Yao ethnic group in southern China. China is home to more than 50 minority groups.

also looking to China, South Korea, Taiwan, and Hong Kong for large investments.

By the mid-1990s, the Japanese government, worried about the loss of jobs at home, began to urge private businesses to moderate their investments in other countries in the region. The Japanese government itself cut overseas development assistance (ODA) by 10 percent in 2002, and has made more cuts since. Before that time, Japan's annual ODA was US$11 billion compared to only US$8.4 billion from the United States. Despite its sluggish economy, Japan is not withdrawing from active involvement in the economies of the Rim, for the Rim countries provide much of the raw materials and inexpensive labor that Japanese companies need to compete successfully in the global market. Economic aid to Rim countries—such as the US$1.56 billion aid package granted to Indonesia in 2000, and the US$1.5 billion program for Mekong River projects in Thailand, Vietnam, Cambodia, Laos, and Myanmar—will continue. Thanks to the Japanese model and Japanese assistance, countries like South Korea, Taiwan, and China are now in a position to invest in other regional economies. South Korea, for example, entered the year 2010 with a US$42 billion trade surplus, surpassing Japan for the first time. It is also aggressively pursuing free trade agreements with Australia, Japan, China, and other countries. Thus, the integration and homogenization of the economies— and thus the cultures—of the Pacific Rim is continuing apace.

In the 1960s, when the Japanese economy had completely recovered from the devastation of World War II, the Japanese looked to North America and Europe for markets for their increasingly high-quality products. Japanese business continues to seek out markets and resources globally; but, in the 1980s, in response to the movement toward a truly common European economic community as well as in response to free trade agreements among North American countries, Japan began to invest more heavily in countries nearer its own borders. The Japanese hoped to guarantee themselves market and resource access should they find their products frozen out of the emerging European and North American economic blocs. Now China, Taiwan, Singapore, and others are doing the same thing. The unintended, but not unwelcome, consequences of this policy were the revitalization of many Asia–Pacific economies and the solidification of lines of communication between governments and private citizens within the region. Recognizing this interconnection has prompted many people to refer to the countries we treat in this book as a single unit, the Pacific Rim. Though set aside for awhile, the idea of a Pacific Rim free trade zone gains supporters each time the Asia Pacific Economic Cooperation forum meets, and some regional trade groupings have now been inaugurated, including the ASEAN-China Free Trade Area, which was formed in 2010. The agreement reduces tariffs to zero on 90 percent of imported goods within the region.

(Courtesy of Lisa Clyde Nielsen)

The small island of Macau was acknowledged by China as a Portuguese settlement in 1557. The Portuguese influence is evident in the architecture of the downtown plaza pictured above. Today Macau is a gambling mecca, drawing an enormous number of avid fans from Hong Kong. This last outpost of European colonial power returned to Chinese control on December 20, 1999, after 442 years under Portugal.

■ TROUBLES IN THE RIM

Twenty-five years ago, television images of billionaire Japanese businesspeople buying up priceless artworks at auction houses, and filthy-rich Hong Kong Chinese driving around in Rolls-Royces overshadowed the reality of the region: a place, where, for the most part, life is hard. For most of recorded history, most Pacific Rim countries have not met the needs of their peoples. Whether it is

? DID YOU KNOW?

New findings from the U.S. Census Bureau show that 14.6 million people of Asian descent live in the United States, with the Chinese being the largest bloc (23 percent). Filipinos represent 17 percent of the total.

the desire of rice farmers in Myanmar (formerly called Burma) for the right to sell their grain for personal profit, or of Chinese students to search uncensored Google sites or speak their minds without repression—in these and many other ways, the Pacific Rim has failed its peoples. In Vietnam, Myanmar, Laos, and Cambodia, for example, life is so difficult that thousands of families have risked their lives to leave their homelands. Some have swum across the wide Mekong River on moonless nights to avoid detection by guards, while others have sailed into the South China Sea on creaky and overcrowded boats (hence the name commonly given such refugees: "boat people"), hoping that people of goodwill, rather than marauding pirates, will find them and transport them to a land of safety. That does not always happen; in 2004, for example, Indonesia reported some 70 pirate attacks in its waters, and Malaysia reported 30. With the cut-off of refugee-support funds from the United Nations (UN), thousands of refugees remain unrepatriated, languishing in camps in Thailand, Malaysia, and other countries. Likewise, thousands of Karen and other ethnic villagers driven from their homes by the Myanmar army await return. Meanwhile, the number of defectors from North Korea continues to increase, as highlighted in 2009 when two U.S. journalists were captured while collecting data for a story on the exodus (they were released when former U.S. President Bill Clinton made a personal visit to North Korea).

Between 1975 and 1994, almost 14,000 refugees from various Asian countries reached Japan by boat, along with 3,500 Chinese nationals who posed as refugees in hopes of being allowed to live outside China. Many others have sought asylum in Australia and other countries. In 1998, the Malaysian government, citing its own economic problems, added to the dislocation of many people when it began large-scale deportations of foreign workers, many from Indonesia. Many of these individuals had lived in Malaysia for years. This "Operation Get Out" was expected to affect at least 850,000 people. These examples, and many others not mentioned here, stand as tragic evidence of the social and political instability of many Pacific Rim nations and of the intense ethnic rivalries that divide the people of the Rim.

Warfare

Historically, of all the Rim's troubles, warfare has been the most devastating. In Japan and China alone, an estimated 15.6 million people died as a result of World War II. Not only have there been wars in which foreign powers like Britain, the United States, France, and the former Soviet Union have been involved, but there have been and continue to be numerous battles between local peoples of different tribes, races, and religions.

The potential for serious conflict remains in most regions of the Pacific Rim. Despite international pressure, the military dictators of Myanmar continued for years to wage war against the Karens and other ethnic groups within its borders. Japan remains locked in a dispute with Russia over ownership of islands to the north

(© IMS Communications Ltd./Capstone Design/FlatEarth Images RF)

The number of elderly people in China will triple by the year 2025. Even though—and, ironically, because—it limits most couples to only one child, the Chinese government, rather than families, will need to care for retirement-age citizens.

of Hokkaido, and with South Korea over some small islets in the Japan Sea. Taiwan and China still lay claim to each other's territory, and North Korea has become even more aggressive than in the past toward South Korea, against which it has thousands of troops and hundreds of missiles positioned. Taiwan and Japan lay claim to the Senkaku Island chain. The Spratly Islands are claimed by half a dozen Pacific Rim countries. The list of disputed borders, lands, islands, and waters in the Pacific Rim is very long; indeed, there are some 30 unresolved disputes involving almost every country of Asia and some of the Pacific Islands.

Let us look more closely at the Spratly (sometimes spelled Spratley) Islands dispute. When the likelihood of large oil and cobalt deposits in the 340,000-square-mile ocean near the Spratlys was announced in the 1970s, China, Taiwan, Vietnam, the Philippines, Malaysia, and Brunei instantly laid claim to the area. By 1974, the Chinese Air Force and Navy were bombing a South Vietnamese settlement on the islands; by 1988, Chinese warships were attacking Vietnamese transport ships in the area. And by 1999, the Philippine Navy was sinking Chinese fishing boats off Mischief Reef. Both China and Vietnam have granted nearby oil-drilling concessions to different U.S. oil companies, so the situation remains tense, especially because China claims sovereignty, or at least hegemony, over almost the entire South China Sea and has been flexing its muscles in the area by stopping, boarding, and sometimes confiscating other nations' ships in the area. Some dispute management agreements among the

? DID YOU KNOW?

"The United States is expected to remain the third most populous country in the world through 2050, ranked behind China and India, while India's population is projected to exceed China's by 2025." Source: Newsday, June 28, 2011.

claimants were reached in 2001, but as long as each—especially China—claims ownership, the dispute will continue. In fact, in 2011, Vietnam and China engaged in live fire exercises in the region, and many Vietnamese staged protests against China's aggressive posture.

In addition to these cross-border disputes, internal ethnic tensions are sometimes severe. Most Pacific Rim nations are composed of scores of different ethnic groups with their own languages, religions, and world-views. In Fiji, it is the locals versus the immigrant Indians; in Southeast Asia, it is the locals versus the Chinese or the Muslims versus the Christians; in China, it is the Tibetans, the Uighurs, and many of the other 50 or so ethnic groups against the dominant Han Chinese.

Jockeying for the upper hand seems to be a way of life in Asia. Especially with the end of the cold war in the late 1980s and early 1990s, many Asian nations found it necessary to seek new military, political, and economic alliances. For example, South Korea made a trade pact with Russia, a nation that, during the Soviet era, would have dealt only with North Korea; and, forced to withdraw from its large naval base in the Philippines, the United States increased its military presence in Singapore and, in 2011, made an agreement with Australia to substantially increase its military presence there. Australia and New Zealand have created robust trade agreements with many other Pacific Rim countries—something they had not done in the past.

In the mid-1990s, the United States also began encouraging its ally Japan to assume a larger military role in the region. However, the thought of Japan re-arming itself causes considerable fear among Pacific Rim nations, almost all of which suffered defeat at the hands of the Japanese military only six decades ago. Nevertheless, Japan has acted to increase its military preparedness, even going so far as to propose changes in its constitution that would allow it to build a fully capable offensive military. In 2006, it upgraded its defense agency to a full Ministry of Defense. It now has the third-largest military budget in the world (after the United States and China), although it spends only about 1 percent of its budget on defense.

North Korea, always the wild card, has snubbed international opinion and continued to develop the capacity to create nuclear weapons and long-range ballistic missiles. China, with its booming economy, now has the ability to create a formidable military machine. It has been substantially increasing its purchases or development of sophisticated military hardware, including an aircraft carrier and stealth jet fighters. As a result, whereas the arms industry is in decline in some other world regions, it is big business in Asia. Four of the nine largest armies in the world are in the Pacific Rim. Adding tension to the entire region, North Korea, in 2006 and again in 2009, successfully tested a nuclear bomb, after having previously fired a missile across Japan's sovereign airspace. Thus, the tragedy of warfare, which has characterized the region for so many centuries, could continue unless governments manage conflict very carefully and come to understand the need for mutual cooperation.

Types Of Government In Selected Pacific Rim Countries

Parliamentary Democracies
Australia*
Fiji
New Zealand*
Papua New Guinea

Constitutional Monarchies
Brunei
Japan
Malaysia
Thailand

Republics
Indonesia
The Philippines
Singapore
South Korea
Taiwan

Socialist Republics
China
Laos
Myanmar (Burma)
North Korea
Vietnam

Overseas Territories/Colonies
French Polynesia
New Caledonia

*Some Australian and New Zealand leaders have declared their intentions of transforming their countries into completely independent republics with no governmental ties to England. However, the voters have not yet fully supported such plans.

In some cases, mutual cooperation is replacing animosity. Thailand and Vietnam are engaged in sincere efforts to resolve fishing-rights disputes in the Gulf of Thailand and water-rights disputes on the Mekong River; North and South Korea have agreed to allow some cross-border visitation and are cooperating on a mammoth industrial park just inside North Korea; and even Taiwan and China have amicably settled issues relating to fisheries, immigration, and hijackings. Yet greed and ethnic and national pride are far too often just below the surface; left unchecked, they could catalyze a major confrontation in the region.

Population Imbalances

Another serious problem in the Pacific Rim is population imbalance. In some cases, there are too many people; in others, there will be too few in the future. At the moment, there are well over 2 billion people living in the region. Of those, approximately 1.3 billion are Chinese. Even though China's government has implemented the strictest family-planning policies in world history, the country's annual growth rate is such that more than 1 million inhabitants are added *every month*. This means that more new Chinese are born each year than make up the entire populations of such countries as Portugal and Greece. China's efforts to reduce its out-of-control population have produced some unwanted results: a huge imbalance in the ratio of boys to girls. With births limited to just one for most families, and with the traditional Chinese preference for boys, many couples resort to selective abortions to make sure their only child will be a boy. As a result, there are now 120 boys born for every 100 girls, and as the boys reach marriage-able age, they will find there are simply not enough females to go around; they will be, as the Chinese call them, "bare branches" with no offspring. By 2020, it is estimated that China will have 13 million "excess" males.

Despite the unintended consequences of China's population control policy, most Pacific Rim countries continue to promote family planning, and it is working. The World Health Organization (WHO) reports that about 217 million people in East Asia use contraceptives, as compared to only 18 million in 1965. Couples in some countries, including Japan, Taiwan, and South Korea, have been voluntarily limiting family size. Other states, such as China and Singapore, have promoted family planning through government incentives and punishments. The effort is paying off. The United Nations now estimates that the proportion of the global population living in Asia will remain relatively unchanged between now and the year 2025, and China's share will decline. A drop in birth rates is characteristic of almost the entire region: Birth rates started to drop in Japan and Singapore in the 1950s; Hong Kong, South Korea, the Philippines, Brunei, Taiwan, Malaysia, Thailand, and China in the 1960s; and Indonesia and Myanmar in the 1970s. In fact, in some countries, especially Japan, South Korea, and Thailand, single-child families and an aging population are creating problems in their own right as the ratio of productive workers to the overall population declines. The birth rate in Tokyo, for example, is 1.01 children per couple—far below replacement level. At this rate, Japan's population will be half its current size by the year 2100. Some experts are calling the situation a "demographic collapse." The low births are caused, in part, by the number of Japanese who do not marry—approximately 14 percent of all males over age 40 and 6 percent of females over 40 are not married. Many of these unmarrieds continue to live in their parents' homes, coining the phrase "parasite singles." It is likely that these downward trends in Japan will soon be commonplace all over the Pacific Rim, some countries (e.g., the Philippines) excepted. China's demographic structure is slowly starting to mirror Japan's.

The *lack* of young workers will be a problem for the next generation, but for now, so many children have already been born that Pacific Rim governments simply cannot meet their needs. For these new Asians, schools must be built, health facilities provided, houses constructed, and jobs created. This is not an easy challenge for many Rim countries. Moreover, as the population density increases, the quality of life decreases. By way of comparison, think of New York City, where the population is about 1,100 per square mile. Residents there, finding the crowding to be too uncomfortable, frequently seek more relaxed lifestyles in the suburbs. Yet in Tokyo, the density is approximately 2,400 per square mile; and in Macau, it is 57,576! Today, many of the world's largest cities are in the Pacific Rim: Shanghai, China, has about 17 million people (in the wider metropolitan area); Jakarta, Indonesia, has close to 10 million; Manila, the Philippines, is home to over 10 million; and Bangkok, Thailand, has about 9 million inhabitants. And migration to the cities continues despite miserable living conditions for many (in some Asian cities, 50 percent of the population lives in slum housing). One incredibly rapid-growth country is the Philippines; home to only about 7 million in 1898, when it was acquired by the United States, it is projected to have 130 million people in the year 2020.

Absolute numbers alone do not tell the whole story. In many Rim countries, 40 percent or more of the population is under age 15. Governments must provide schooling and medical care as well as plan for future jobs and housing for all these children. Moreover, as these young people age, they will require increased medical and social care. Scholars point out that between 1985 and 2025, the numbers of old people will double in Japan, triple in China, and quadruple in Korea. In Japan, where replacement-level fertility was achieved in the 1960s, government officials are already concerned about the ability of the nation to care for the growing number of retirement-age people while paying the higher wages that the increasingly scarce younger workers are demanding.

It is important to remember that conditions can vary dramatically across an area as large as the Pacific Rim. While Japan is wrestling with de-population, and while the pace of population growth is slowing in some countries, many other countries are straining to find ways to feed their citizens. China's prosperity, for example, is pushing up the demand for traditional meats such as pork. But despite being the second largest corn-growing nation on earth, China cannot produce enough feed for its farm animals, and the cost of food for humans is rising rapidly as well. China's population will hit 1.5 billion by around the year 2030, and many experts predict the country will not be able to produce enough grain. Rice production is expected to decline by 3 percent in China and by 4 percent in Southeast Asia in the next few years.

DID YOU KNOW?

In some cities in China, there are more than 150 boys born to every 100 girls.

Political Instability

One consequence of the overwhelming problems of population imbalances, urbanization, and continual military or ethnic conflict is disillusionment with government. In many countries of the Pacific Rim, as with many countries in the Middle East, people are questioning the very right of their governments to rule or are demanding a complete change in the political philosophy that undergirds government.

For instance, despite the risk of death, torture, or imprisonment, many college students in Myanmar demonstrated for years against the military dictatorship, and rioting students and workers in Indonesia were successful in bringing down the corrupt government of President Suharto. In some Rim countries, opposition groups armed with sophisticated weapons obtained from foreign nations roam the countryside, capturing towns and military installations. In less than a decade, the government of the Philippines endured six coup attempts; elite military dissidents have wanted to impose a patronage-style government like that of former president Ferdinand Marcos, while armed rural insurgents have wanted to install a Communist government. Thousands of students have been injured or killed protesting the governments of South Korea and China. China's increasingly repressive tactics against Tibetans and Uighurs have generated a steady drip of violent protests which have tarnished the image of the Chinese government and caused provincial leaders to be sacked. Thailand has been beset by no fewer than 18 coups as well as by an insurgency against the government from Muslims in the southern part of the country. Similarly, the former British colony of Fiji has endured three coups and remains very unstable. Half a million residents of Hong Kong took to the streets to oppose Great Britain's decision to turn over the territory to China in 1997, and protests against China continue regularly in Hong Kong. Military takeovers, political assassinations, and repressive policies have been the norm in most of the countries in the region. Millions have spent their entire lives under governments they have never agreed with, and unrest is bound to continue, because people are showing less and less patience with imposed government.

Identity Confusion

A related problem is that of confusion about personal and national identity. Many nation-states in the Pacific Rim were created in response to Western pressure. Before Western influences came to be felt, many Asians, particularly Southeast Asians, did not identify themselves with a nation but, rather, with a tribe or an ethnic group. National unity has been difficult in many cases because of the archipelagic nature of some countries or because political boundaries have changed over the years, leaving ethnic groups from adjacent countries inside the neighbor's territory. The impact of colonialism has left many people, especially those in places like Singapore, Hong Kong, and the Pacific islands, unsure as to their roots; are they European or Asian/Pacific, or something else entirely?

Indonesia illustrates this problem. People think of it as an Islamic country, as overall, its people are 87 percent Muslim. But in regions like North Sumatra, 30 percent are Protestant; in Bali, 94 percent are Hindu. In the former Indonesian territory and now independent country of East Timor, about 95 percent are Catholic, although about half of these believers mix Catholicism with various forms of animism. Such stark differences in religious worldview, combined with the tendency of people of similar faith to live near each other and separate themselves geographically from others, make it difficult for a clear national identity to form. It is not surprising that Indonesia has had to frequently grapple with separationist movements. The Philippines is another example. With 88 different languages spoken, its people spread out over 12 large islands, and a population explosion (the average age is just 16), the Philippines is a classic case of psychological (and economic and political) fragmentation. Coups and countercoups rather than peaceful political transitions seem to be the norm, as people have not yet developed a sense of unified nationalism.

Uneven Economic Development

While millionaires in Singapore, Hong Kong, and Japan wrestle with how best to invest their wealth, others worry about how they will obtain their next meal. Such disparity illustrates another major problem afflicting the Pacific Rim: uneven economic development. The United Nations' Food and Agricultural Organization (FAO) estimates that 20 percent of the people in the Pacific Rim (the data exclude China) are undernourished. And while some people in, say, Tokyo and Singapore have a lifestyle similar to people in Europe and North America, others, for instance in Papua New Guinea, have a lifestyle not that different from that believed to have existed in the Stone Age. (It is also true that Asia overall consumes far less energy per person—0.99 ton of oil—than do Canada and the United States—8.67 and 8.77 tons, respectively. If energy consumption is equated to quality of life, then the gap between Asia and the developed world must stand as painful evidence of general worldwide inequality.)

Many Asians, especially those in the Northeast Asian countries of Japan, Korea, and China, are finding that rapid economic change seems to render the traditions of the past meaningless. Moreover, economic success has produced a growing Japanese interest in maximizing investment returns, with the result that Japan (and, increasingly, South Korea, Taiwan, Singapore, and Hong Kong) is successfully searching out more ways to make money—developing the *anime* industry, for example—while resource-poor regions like the Pacific islands lag behind. Indeed, with China receiving "normal trade relations" status from the United States in late 2000 and acceding to membership in the World Trade Organization (WTO) in 2001, many smaller countries, such as the Philippines, worry that business will move away from them and toward China, where labor costs are considerably lower. Indeed, the so-called "China Price," or the cost to make and sell goods made by low-wage

Chinese laborers, is putting downward pressure on customary revenue streams in all Rim countries. Investors who formerly looked at Southeast Asia now cast their eyes toward China (and India). In fact, so much investment money is flowing into China, that less development is taking place elsewhere in Asia, reinforcing the area's general inequality.

The *developed nations* are characterized by political stability and long-term industrial success. Their per capita incomes are comparable to those of Canada, Northern Europe, and the United States, and they have achieved a level of economic sustainability. These countries are closely linked to North America economically. Japan, for instance, exports almost 20 percent of its products to the United States.

The *newly industrializing countries* (NICs) are currently capturing world attention because of their rapid growth. Hong Kong, for example, has exported more manufactured products per year for the past decade than did the former Soviet Union and Central/Eastern Europe combined. Taiwan, famous for cameras and calculators, has had the highest average economic growth in the world for the past 20 years. South Korea is tops in shipbuilding and steel manufacturing and is the eleventh-largest trading nation in the world.

The *resource-rich developing nations* have tremendous natural resources but have been held back economically by political and cultural instability and by insufficient capital to develop a sound economy. Malaysia is an example of a country attempting to overcome these drawbacks. Ruled by a coalition government representing nearly a dozen ethnic-based parties, Malaysia is richly endowed with tropical forests and large oil and natural-gas reserves. Developing these resources has taken years (the oil and gas fields began production as recently as 1978) and has required massive infusions of investment monies from Japan and other countries. By the mid-2000s, nearly 3,000 foreign companies were doing business in Malaysia, and the country was moving into the ranks of the world's large exporters.

Command economies lag far behind the rest, not only because of the endemic inefficiency of the system but because military dictatorships and continual warfare have sapped the strength of the people, as in Laos, for example. Yet significant changes in some of these countries are now emerging. China and Vietnam, in particular, are eager to modernize their economies and institute market-based reforms. Vietnam and even Laos now have their own stock exchanges, for example. It is estimated that about half of China's economy is now privatized, although substantial government regulation remains throughout the system. Historically having directed its trade to North America and Europe, Japan is now finding its Asian/Pacific neighbors—especially the socialist-turning-capitalist ones—to be convenient recipients of its powerful economic and cultural influence.

Many of the *less developed countries* (LDCs) are the small micro-states of the Pacific with limited resources and tiny internal markets. Others, like Papua New Guinea,

have only recently achieved political independence and are searching for a comfortable role in the world economy.

Environmental Destruction and Health Concerns

Environmental destruction in the Pacific Rim is a problem of mammoth proportions. For more than 20 years, the former Soviet Union dumped nuclear waste into the Sea of Japan; China's use of coal in industrial development has produced acid rain over Korea and Japan; deforestation in Thailand, Myanmar, and other parts of Southeast Asia and China has destroyed many thousands of acres of watershed and wildlife habitat. On the Malaysian island of Sarawak, for example, loggers work through the night, using floodlights, to cut timber to satisfy the demands of international customers, especially the Japanese. The forests there are disappearing at a rate of 3 percent a year. Highway and dam construction in many countries in Asia has seriously altered the natural environment. But environmental damage is perhaps most noticeable in the cities. Mercury pollution in Jakarta Bay has led to brain disorders among children in Indonesia's capital city; air pollution in Manila and Beijing ranks among the world's worst, while not far behind are Bangkok and Seoul; water pollution in Hong Kong has forced the closure of many beaches. The World Resources Institute of the U.S. Environmental Protection Agency (EPA) reports that greenhouse-gas emissions (annual emissions in billions of tons) are 5.7 for the United States, 6.2 for Canada, but in excess of 8 for Asia (of which most is from China, making it the single largest greenhouse gas polluter in the world).

An environmentalist's nightmare came true in 1997 and 1998 in Asia, when thousands of acres of timber went up in a cloud of smoke. Fueled by the worst El Niño–produced drought in 30 years and started by farmers seeking an easy way to clear land for farming, wildfires in Malaysia, Indonesia, and Brunei covered much of Southeast Asia in a thick blanket of smoke for months. Singapore reported its worst pollution-index record ever. All countries in the region complained that Indonesia, in particular, was not doing enough to put out the fires. Foreign-embassy personnel, citing serious health risks, left the region until rains—or the lack of anything more to burn—extinguished the flames. Airports had to close, hundreds of people complaining of respiratory problems sought help at hospitals, and many pedestrians and even those inside buildings donned face masks. With valuable timber becoming more scarce all the time, many people around the world reacted with anger at the callous disregard for the Earth's natural resources. Some environmental problems cannot be prevented. The massive tsunami in 2004 that devastated parts of Indonesia, Thailand, Malaysia, and Burma was not preventable. Other problems, however, such as the mudslides that now frequently sweep away villages in the Philippines and Southeast Asia, are the result of careless harvesting of the forests by companies that are more concerned about profit than about sustainable development.

While conservationists are raising the alarm about the region's polluted air and declining green spaces, medical professionals are expressing dismay at the speed with which serious diseases such as AIDS are spreading in Asia. The Thai government, despite impressive gains in the fight against HIV/AIDS, now believes that more than 2.4 million Thais are HIV-positive. World Health Organization data show that the AIDS epidemic is growing faster in Asia and Africa than anywhere else in the world.

In late 2002, a new and deadly coronavirus emerged among the population of Guangdong province in southern China. Called SARS (severe acute respiratory syndrome), the flu-like illness, for which there was no cure, quickly spread to Hong Kong, Singapore, Vietnam, Taiwan, Canada, and other countries, infecting thousands and killing hundreds. Not only did SARS create a serious health threat, it also damaged the reputation of China's leaders, who did not act quickly to handle the crisis and then tried to downplay the severity of the problem. Initially, China even refused to allow World Health Organization officials to visit Taiwan, claiming that Taiwan was not an independent country and could not invite such organizations inside its borders without China's permission. Although China relented on this point, the damage to its reputation had already been done. Various health officials issued advisories against travel to infected regions, and the tourism sector of the region had already suffered losses in excess of US$11 billion by early 2003. Hotels and restaurants reduced their staff, and airlines, already impacted by the fallout of the September 11, 2001, terrorist attacks on the New York City World Trade Center, were saddled with yet more layoffs and more red ink.

The region had just started to recover from the SARS problem (which had mercifully caused fewer deaths than expected) when another virus emerged: the bird flu. Affecting ducks, chickens, and even some non-fowl animals, the disease can only be eradicated by "culling" or killing all flocks where an infected bird is found. Millions of birds have been thus killed in Vietnam, Thailand, Malaysia, Laos, Indonesia, and other countries, with a resulting impact on people's incomes. Apparently, humans can also contract the virus, and with some 120 people felled by the disease in Asia between 2003 and 2006, some epidemiologists were predicting a world pandemic in which the bird flu would rival or exceed the human deaths attributed to the worldwide Spanish Flu epidemic in 1918. As of 2012, that had not happened, nor had a third threat, the H1N1 or "swine flu," caused much of a problem in Asia, but coping with these and the other health emergencies has produced one unexpected positive outcome—more communication and cooperation across borders.

Natural Disasters

In December 2004, the world watched in horror as a mammoth tsunami or tidal wave generated by a 9.0 earthquake off the coast of Indonesia devastated the shores of 11 countries and killed some 230,000 people around the Indian Ocean. Billions of dollars of aid were sent to Indonesia, Malaysia, Myanmar, Thailand, and other affected countries to help the survivors. While this

Selected Natural Disasters in the Pacific Rim 1920–2004

Year	Place	Event	Deaths
1920	Gansu, China	8.6 earthquake	200,000
1923	Kanto, Japan	7.9 earthquake	143,000
1932	Gansu, China	7.6 earthquake	70,000
1927	Tsinghai, China	7.9 earthquake	200,000
1976	Tangshan, China	7.5 earthquake	255,000
2004	Southeast Asia	9.0 earthquake and tsunami	230,000
2008	Sichuan, China	7.9 earthquake	69,000
2011	Tohoku, Japan	9.0 earthquake and tsunami	23,000

Source: U.S. Geological Survey

event was truly cataclysmic, it must be remembered that similar events have always been part of the Pacific Rim. The chart above shows just a few of the natural disasters that have afflicted the region over the years.

In the year 2004 alone (and one could compose a similar list for any year), the following natural disasters (a selected sample only) hit the region, causing loss of life and severe economic hardship for both individuals and governments: a tropical storm followed by a typhoon in the Philippines killed 566 people and left hundreds more missing or homeless; flooding in New Zealand killed 2 and caused US$200 million in damage; landslides in the southern Philippines left 13 dead and 150 missing; a 6.4-magnitude earthquake killed 22 people and caused dozens of buildings to collapse in southern Indonesia; the Mount Awu volcano in Indonesia erupted and forced thousands from their homes; 6 people were killed by flash floods in Vietnam; a cyclone with 105 mph (168 kmh) winds hit Myanmar killing 140 people and leaving 18,000 people homeless. The coastline of the Asian continent is one of the world's most seismically active zones and is known as the "ring of fire." Earthquakes and volcanic eruptions are commonplace. But the combination of seismic activity with tsunamis, cyclones, droughts, floods, and fires produces significant destabilization in the region.

Of course, the world continues to watch with horror as the drama of the March 2011 earthquake and tsunami in Japan unfolds. Besides the devastation caused by the shaking and waves, the residents of Tohoku, Japan have had to deal with radiation from a severely damaged nuclear reactor. Nine months after the devastation, thousands of Japanese were still living in schools and other temporary shelters.

■ GUARDED OPTIMISM

Warfare, population imbalances, political instability, identity confusion, uneven development, and environmental health, and natural disasters would seem to be an irresolvable set of problems for the peoples of the Pacific Rim, but there is reason for guarded optimism about the future of the region. Unification talks between North and South Korea have finally resulted in real breakthroughs. For instance, in 2000, a railway between the demilitarized zone (DMZ) separating the two antagonists was reconnected after years of disuse. President Vladimir Putin of Russia agreed to reopen discussion with Japan over the decades-long North Territories dispute, and Cambodia settled a land claims dispute with Vietnam in 2006. Other important issues, for example the complex Spratly Islands problem, are also under discussion all over the region, and the UN peacekeeping effort in Cambodia seems to have paid off—at least there is a legally elected government in place, and most belligerents have put down their arms. Similarly, in 2010, Myanmar held elections for the first time in two decades and released its most famous political prisoner, Aung San Suu Kyi.

Until the Asian financial and currency crises of 1998–1999, the world media carried glowing reports on the burgeoning economic strength of many Pacific Rim countries. Typical was the *CIA World Factbook 1996–1997,* which reported high growth in gross national product per capita for most Rim countries: South Korea, 9.0 percent; Hong Kong, 5.0 percent; Indonesia, 7.5 percent; Japan (due to recession), 0.3 percent; Malaysia, 9.5 percent; Singapore, 8.9 percent; and Thailand, 8.6 percent. By comparison, the U.S. GNP growth rate was 2.1 percent; Great Britain, 2.7 percent; and Canada, 2.1 percent. Other reports on the Rim compared 1990s investment and savings percentages with those of 20 years earlier; in almost every case, there had been tremendous improvements in the economic capacity of these countries.

Throughout the 1980s and most of the 1990s, the rate of economic growth in the Pacific Rim was indeed astonishing. In 1987, for example, the rate of real gross domestic product growth in the United States was 3.5 percent over the previous year. By contrast, in Hong Kong, the rate was 13.5 percent; in Taiwan, 12.4 percent; in Thailand, 10.4 percent; and in South Korea, 11.1 percent. In 1992, GDP growth throughout Asia averaged 7 percent, as compared to only 4.8 percent for the rest of the world. But the collapse of currencies in Southeast Asia in 1998 had a ripple effect on all economies in the Rim. Furthermore, the lingering recession in Japan greatly reduced the flow of Japanese ODA into the region, thus putting development projects on hold all over Asia. Indonesia was particularly hard hit. The Rim's growth rate had started to return to its formerly rapid pace only to be significantly stalled again by the world economic meltdown of 2009 which started in the United States and spread worldwide. Notwithstanding that new blow to the region, most Rim countries continued to grow faster than the developed nations of the West. Most countries in the region rebounded quickly and strongly from these financial crises, and some hardly noticed the downturn. Indeed, China's growth rate at the height of the recession in 2009 was over 8 percent, and it has now overtaken Japan in the size of its gross domestic product (although it ranks 127th in the world in per capita GDP).

Ranking of Pacific Rim Countries by Per Capita GDP in 2010 (Purchasing Power Parity; US Dollar)

Rank	Country/Region	Per Capita GDP
1	Singapore	$62,100
2	Brunei	$51,600
3	Hong Kong	$45,900
4	Australia	$41,000
5	Taiwan	$35,700
6	Japan	$34,000
7	Macau	$33,000
8	South Korea	$30,000
9	New Zealand	$27,700
10	Malaysia	$14,700
11	Thailand	$8,700
12	China	$7,600
13	Indonesia	$4,200
14	Philippines	$3,500
15	Vietnam	$3,100
16	East Timor	$2,600
17	Laos	$2,500
18	Papua New Guinea	$2,500
19	Cambodia	$2,100
20	North Korea	$1,800
21	Myanmar	$1,400

Comparisons: United States: $47,200; Canada: $39,400
Sources: *CIA World Fact Book 2010* and International Monetary Fund 2010

The significance of high growth rates, in addition to improvements in the standard of living, is the shift in the source of development capital, from North America to Asia. Historically, the economies of North America (and their European counterparts) were regarded as the engine behind Pacific Rim growth; and yet today, growth in the United States and Canada trails many of the Rim economies. This anomaly can be explained, in part, by the hard work and savings ethics of Pacific Rim peoples and by their external-market–oriented development strategies. But, without venture capital and foreign aid, hard work and clever strategies would not have produced the rapid economic improvement that Asia has experienced over the past several decades. Japan's contribution to this improvement, through investments, loans, and donations, and often in much larger amounts than other investor nations such as the United States, cannot be overstated.

Some subregions are also emerging. There is, of course, the Association of Southeast Asian Nations (ASEAN) regional trading unit. In 2001, the ASEAN countries and China agreed to create a free trade area similar to the North American Free Trade Association (NAFTA). The new bloc was launched in 2010, much sooner than expected.

But the grouping that is gaining world attention is the informal region that people are calling "Greater China," consisting of the emerging capitalist enclaves of the People's Republic of China, Hong Kong, and Taiwan. Copying Japanese strategy and aided by a common written language and culture, this region has the potential of exceeding even the mammoth U.S. economy in the future. While the world's attention is increasingly drawn to China, one must remember that the engine behind the Pacific Rim's growth in the past 50 years has been Japan. Indeed, as recently as 2005, Japan's economy was bigger than the combined economies of South Korea, Indonesia, Thailand, Malaysia, Philippines, Hong Kong and half a dozen other Asian countries. Let us briefly review Japan's contribution to the region.

Japan has been investing in the Asia/Pacific region for several decades. However, growing protectionism in its traditional markets as well as changes in the value of the yen and the need to find cheaper sources of labor (labor costs are 75 percent less in Singapore and 95 percent less in Indonesia) raised Japan's level of involvement so high as to have given it the upper hand in determining the course of development and political stability for the entire region. This heightened level of investment started to gain momentum in the mid-1980s. Between 1984 and 1989, Japan's overseas development assistance to the ASEAN countries amounted to $6.1 billion. In some cases, this assistance translated to more than 4 percent of a nation's annual national budget and nearly 1 percent of GDP. Private Japanese investment in ASEAN countries plus Hong Kong, Taiwan, and South Korea was $8.9 billion between 1987 and 1988. In more recent years, the Japanese government or Japanese business invested $582 million in an auto-assembly plant in Taiwan, $5 billion in an iron and steel complex in China, $2.3 billion in a bullet-train plan for Malaysia, and $530 million in a tunnel under the harbor in Sydney, Australia. Japan is certainly not the only player in Asian development (Japan has "only" about 20 projects under way in Vietnam, for example, as compared to 80 for Hong Kong and 39 for Taiwan), but the volume of Japanese investment both historically and in the present is staggering. In Australia alone, nearly 900 Japanese companies are now doing business. Throughout Asia, Japanese has become a major language of business (along with Chinese, which is also in demand as the Chinese economy grows in influence).

Although Japan works very hard at globalizing its markets and its resource suppliers, it has also developed closer ties with its nearby Rim neighbors. In a recent year, out of 20 Rim countries, 13 listed Japan as their first- or second-most-important trading partner, and several more put Japan third. Japan receives 42 percent of Indonesia's exports and 26 percent of Australia's; in return, 23 percent of South Korea's imports, 29 percent of Taiwan's, 30 percent of Thailand's, 24 percent of Malaysia's, and 23 percent of Indonesia's come from Japan. But Pacific Rim countries are now developing to the point that they need less of Japan's financial assistance. For example, Malaysia has a goal to be listed as a "developed nation" by 2020, and Laos has established a goal to be off the "least developed" list by the same year. China's economy is so strong that Japan, in 2006, stopped its loan program to China several years earlier than planned.

■ JAPANESE INFLUENCE, PAST AND PRESENT

The past few decades are certainly not the first time in modern history that Japanese influence has swept over Asia and the Pacific. A major thrust began in 1895, when Japan, like the European powers, started to acquire bits and pieces of the region. By 1942, the Japanese were in control of Taiwan, Korea, Manchuria and most of the populated parts of China and Hong Kong; what are now Myanmar, Vietnam, Laos, and Cambodia; Thailand; Malaysia; Indonesia; the Philippines; part of New Guinea; and dozens of Pacific islands. In effect, by the 1940s, the Japanese were the dominant force in precisely the area that they have been influencing recently and that we are calling the Pacific Rim.

The similarities do not end there, for, while many Asians of the 1940s were apprehensive about or openly resistant to Japanese rule, many others welcomed the Japanese invaders and even helped them to take over their countries. This was because they believed that Western influence was out of place in Asia and that Asia should be for Asians. They hoped that the Japanese military would rid them of Western rule—and it did: After the war, very few Western powers were able to regain control of their Asian and Pacific colonies.

Today, many Asians and Pacific islanders are concerned about Japanese financial and industrial influence in their countries, but they welcome Japanese investment anyway because they believe that it is the best and cheapest way to rid their countries of poverty and underdevelopment. So far, they are right—by copying the Japanese model of economic development, and thanks to Japanese trade, foreign aid, and investment, the entire region— some countries excepted—has increased its wealth and positioned itself to be a major player in the world economy for the foreseeable future.

It is important to note that many Rim countries, such as China, Taiwan, Hong Kong, and South Korea, are strong challengers to Japan's economic dominance; in addition, Japan has not always felt comfortable about its position as head of the pack, for fear of a backlash. For example, Japan's higher regional profile has prompted complaints against the Japanese military's World War II treatment of civilians in Korea and China and forced Japan to pledge $1 billion to various Asian countries as a

Economic Development in Selected Pacific Rim Countries

Economists have divided the Rim into five zones, based on the level of development, as follows:

Developed Nations

Australia

Japan

New Zealand

Newly Industrializing Countries (NICS)

Hong Kong

Singapore

South Korea

Taiwan

Resource-Rich Developing Economies

Brunei

Indonesia

Malaysia

Philippines

Thailand

Command Economies*

Cambodia

China

Laos

Myanmar (Burma)

North Korea

Vietnam

Less Developed Countries (LDCS)

Papua New Guinea

Pacific Islands

*China, Vietnam, and, to a much lesser degree, North Korea are moving toward free-market economies.

---

symbolic act of apology. China, in particular, is very sensitive about Japan's heretofore dominant role and tries to restrain Japan at every opportunity. For example, most nations agree that Japan should have a permanent seat in the UN Security Council, but China refuses to vote for Japan's admission, and this is not likely to change despite the Japanese prime minister's 2009 near-apology for the Pacific War.

Why have the Japanese re-created a modern version of the old Greater East Asian Co-Prosperity Sphere of the imperialistic 1940s? We cannot find the answer in the propaganda of wartime Japan, because fierce devotion to the emperor and the nation, and belief in the superiority of Asians over all other races are no longer the propellants in the Japanese economic engine. Rather, Japan courts Asia and the Pacific today to acquire resources to sustain its civilization. Japan is about the size of California, but it has five times as many people and not nearly as much arable land. Much of Japan is mountainous; many other parts are off limits because of active volcanoes (one-tenth of all the active volcanoes in the world are in Japan); and, after 2,000-plus years of intensive and uninterrupted habitation, the natural forests have long since been consumed (though they have been replanted with new varieties), as are most of the other natural resources—most of which were scarce to begin with.

In short, Japan continues to extract resources from the rest of Asia and the Pacific because it is the same Japan as before—environmentally speaking, that is. Take oil, for example. In the early 1940s, Japan needed oil to keep its industries (as well as its military machine) operating, but the United States wanted to punish Japan for its military expansion in Asia, so it shut off all shipments to Japan of any kind, including oil. That may have seemed politically right to the policymakers of the day, but it did not change Japan's resource environment; Japan still did not have its own oil, and it still needed as much oil as before. So Japan decided to capture a nearby nation that did have natural reserves of oil; in 1941, it attacked Indonesia and obtained by force the resource it had been denied through trade.

Japan has no more domestic resources now than it did half a century ago, and yet its needs—for food, minerals, lumber, paper—are greater. Except for fish, you name it—Japan does not have it. A realistic comparison is to imagine trying to feed half the population of the United States solely from the natural output of the U.S. state of Montana. As it happens, however, Japan sits next to the continent of Asia, which is rich in almost all the materials it needs. For lumber, there are the forests of Malaysia; for food, there are the farms and ranches of New Zealand and Australia; and for oil, there are Indonesia and Brunei, the latter of which sells about half of its exports to Japan. In sum, the quest for resources to maintain its quality of life is the reason Japan flooded its neighbors with Japanese yen in recent decades and why it will continue to maintain an active engagement with all Pacific Rim countries well into the future.

Catalyst for Development

In addition to the need for resources, Japan has turned to the Pacific Rim in an attempt to offset the anti-Japanese import or protectionist policies of its historic trading partners. Because so many import tariffs are imposed on products sold directly from Japan, Japanese companies find that they can avoid or minimize tariffs if they cooperate on joint ventures in Rim countries and have products shipped from there. The result is that both Japan and its host countries have prospered beyond expectation. For years, Sony Corporation assembled parts made in both Japan and Singapore to

(Courtesy of Bryan Ischo and Nancy Lau. www.ischo.com/gallery/albums.php)

Tokyo fish market: The early morning tuna auction.

construct videocassette recorders at its Malaysian factory, for export to North America, Europe, and other Rim countries. Toyota Corporation assembles its automobile transmissions in the Philippines and its steering-wheel gears in Malaysia, and then assembles the final product in whichever country intends to buy its cars.

So helpful has Japanese investment been in spawning indigenous economic powerhouses that many other Rim countries are now reinvesting in the region. In particular, Hong Kong, Singapore, Taiwan, and South Korea are now in a position to seek cheaper labor markets in Indonesia, Malaysia, the Philippines, and Thailand. In recent years, they have invested billions of dollars in the resource- and labor-rich economies of Southeast Asia, increasing living standards and adding to the growing interconnectivity of the region. An example is a Taiwanese company that has built the largest steel-production facility in the world—in Malaysia—and ships its entire product to Korea and Japan.

Eyed as a big consumer as well as a bottomless source of cheap labor is the People's Republic of China. Many Rim countries, such as South Korea, Taiwan, Hong Kong, and Japan, are working hard to increase

their trade with China. For over a decade, annual trade between Taiwan and China and between Hong Kong and China has been in the billions of dollars, despite political tensions. Japan was especially eager to resume economic aid to China in 1990 after temporarily withholding aid to China because of the Tiananmen Square massacre in Beijing. For its part, China is establishing free-enterprise zones that will enable it to participate more fully in the regional and world economy. China's membership in the World Trade Organization will force it to be even more engaged in world trade. Already the Bank of China is the second-largest bank in Hong Kong.

Japan and a handful of other economic power-houses of the Rim are not the only big players in regional economic development. The United States and Canada are major investors in the Pacific Rim (in computers and automobiles, for example), and Europe maintains its historical linkages with the region (such as oil). But there is no question that Japan has been the main catalyst for development. As a result, Japan itself has become wealthy. The Japanese stock market rivals the New York Stock Exchange, and, despite years of a sluggish economy, there is a long list of Japanese billionaires:

Masyoshi Son, worth US$7.1 billion (during the dot.com boom he was briefly the world's richest person); Tadashi Yanai, worth US$7.6 billion; Akira Mori, worth US$6.8 billion; and many others. Loans secured now-deflated land prices have damaged the Japanese banking industry in recent years, but it appears that the banks are recovering and that Japan will continue to play an active role in the economic development of Asia, albeit at a reduced level compared to the past.

Not everyone is pleased with the way Japan has been giving aid and making loans in the region. Money invested by the Japan International Development Organization (JIDO) has usually been closely connected to the commercial interests of Japanese companies. For instance, commercial-loan agreements have often required that the recipient of low-interest loans purchase Japanese products. Nevertheless, it is clear that many countries would be a lot less developed without Japanese aid. In a recent year, JIDO aid around the world was $10 billion. Japan is the dominant supplier of foreign aid to the Philippines and a major investor; in Thailand, where U.S. aid recently amounted to $20 million, while Japanese aid was close to $100 million. Some of this aid, moreover, gets recycled from one country to another within the Rim. Thailand, for example, receives more aid from Japan than any other country, but in turn, it supplies major amounts of aid to other nearby countries. Thus we can see the growing interconnectivity of the region, a reality now recognized formally by the establishment of the Asia-Pacific Economic Cooperation forum (APEC). This economic planning organization includes most Pacific Rim countries (as well as others outside the Rim) and is increasingly interested in creating a Rim-wide free trade zone.

During the militaristic 1940s, Japanese dominance in the region produced antagonism and resistance. However, it also gave subjugated countries new highways, railways, and other infrastructural improvements. Today, while host countries continue to benefit from infrastructural advances, they also receive quality manufactured products. After World War II, Japan was considered a pariah nation in Asia; now most countries no longer believe that Japan has military aspirations against them (despite the rhetoric from China), and they regard Japanese investment as a first step toward becoming economically strong themselves. Many people are eager to learn the Japanese language; in some cities, such as Seoul, Japanese has displaced English as the most valuable business language.

■ ASIAN FINANCIAL CRISIS

All over Asia, but especially in Thailand, Indonesia, Malaysia, South Korea, and the Philippines, business leaders and government economists found themselves scrambling in 1998 and 1999 to minimize the damage from the worst financial crisis in decades, a crisis that exploded in late 1997 with currency devaluations in Southeast Asia. With their banks and major corporations in deep trouble, governments began shutting down some of their expensive overseas consular offices, canceling costly public-works projects, and enduring abuse heaped on them by suddenly unemployed citizens.

For years, Southeast Asian countries copied the Japanese model: They stressed exports, and they allowed their governments to decide which industries to develop. This economic-development approach worked very well for a while, but governments were not eager to let natural markets guide production. So, even when profits from government-supported industries were down, the governments believed that they should continue to maintain these industries through loans from Japan and other sources. But banks can loan only so much, especially if it is "risky" money, and eventually the banks' creditworthiness was called into question. Money became tighter. Currencies were devalued, making it harder still to pay off loans and forcing many companies and banks into bankruptcy. Stock markets nose-dived. Thousands of workers were laid off, and many of the once-booming Asian economies hit hard times.

Japan's own economic sluggishness meant that it was not able to carry the same weight in solving the crisis as it might have done in the 1980s. One after another, the affected countries requested bailout assistance from the International Monetary Fund (IMF), a pool of money donated by some 183 nations. IMF funding halted the flow of red ink, and many countries were on the road to recovery when the United States' banking and real estate crisis hit the world in 2008 and 2009. This once again weakened Japan's ability to resume the powerful role it had played in igniting economic growth in the region. With the rise of China, it is doubtful that Japan will ever again be the undisputed leader of the economic pack in the region, but it's role in making Asia what it is today cannot be discounted.

■ POLITICAL AND CULTURAL CHANGES

Although economic issues are important to an understanding of the Pacific Rim, political and cultural changes are also crucial. The new, noncombative relationship between the United States and the former Soviet bloc means that special-interest groups and governments in the Rim will be less able to rely on the strength and power of those nations to help advance or uphold their positions. Communist North Korea, for instance, can no longer rely on the former Soviet bloc for trade and ideological support. North Korea may begin to look for new ideological neighbors or, more significantly, to consider major modifications in its own approach to organizing society.

In the case of Hong Kong, the British government shied away from extreme political issues and agreed to the peaceful annexation in 1997 of a capitalist bastion by a Communist nation, China. It is highly unlikely that such a decision would have been made had the issue of Hong Kong's political status arisen during the anti-Communist years of the cold war. One must not get the

Pacific Rim Billionaires 2006

Name	Country	Age	Worth in Billions $
Li Ka-shing	Hong Kong	83	26.0
Raymond, Thomas & Walter Kwok	Hong Kong	NA	20.0
Lee Shau Kee	Hong Kong	83	19.0
Robert Kuok	Malaysia	87	12.5
Ananda Krishnan	Malaysia	73	9.5
Robin Li	China	42	9.4
Georgina Rinehart	Australia	57	9.0
Cheng Yu-tung	Hong Kong	85	9.0
Kun-Hee & family	South Korea	69	8.6
Masayoshi Son	Japan	53	8.1
Liang Wengen	China	54	8.0
Stanley Ho	Hong Kong	84	6.5
Yasuo Takei & family	Japan	76	5.4
YC Wang	Taiwan	89	5.4
Kunio Busujima & family	Japan	80	5.2
James Packer	Australia	38	5.0
Nobutada Saji & family	Japan	60	4.7
Akira Mori & family	Japan	68	4.5
Shin Kyuk-Ho & family	South Korea	83	4.5
Terry Gou	Taiwan	55	4.3
Eitaro Itoyama	Japan	63	4.2
Nina Wang	Hong Kong	NA	4.2
Tadashi Yanai & family	Japan	57	4.2

Source: *Forbes*

impression, however, that suddenly peace has arrived in the Pacific Rim. But outside support for extreme ideological positions seems to be giving way to a pragmatic search for peaceful solutions. This should have a salutary effect throughout the region.

The growing pragmatism in the political sphere is yielding changes in the cultural sphere. Whereas the Chinese formerly looked upon Western dress and music as decadent, most Chinese now openly seek out these cultural commodities and are finding ways to merge these things with the repressive human rights conditions under which they live. It is also increasingly clear to most leaders in the Pacific Rim that international mercantilism has allowed at least one regional country, Japan, to rise to the highest ranks of world society, first economically and now culturally and educationally. The fact that one Asian nation has accomplished this impressive achievement fosters hope that others can do so also.

Rim leaders also see, however, that Japan achieved its position of prominence because it was willing to change traditional mores and customs and accept outside modes of thinking and acting. Religion, family life,

gender relations, recreation, and many other facets of Japanese life have altered during Japan's rapid rise to the top. Many other Pacific Rim nations—including Thailand, Singapore, and South Korea—seem determined to follow Japan's lead in this regard. Therefore, we are witnessing in certain high-growth Rim economies significant cultural changes: a reduction in family size, a secularization of religious impulses, a desire for more leisure time and better education, and a move toward acquisition rather than "being" as a determinant of one's worth. That is, more and more people are likely to judge others' value by what they own rather than by what they believe or do. Buddhist values of self-denial, Shinto values of respect for nature, and Confucian values of family loyalty are giving way slowly to Western-style individualism and the drive for personal comfort and monetary success. Formerly close-knit communities, such as those in the South Pacific, are finding themselves struggling with drug abuse, AIDS, and gang-related violence, just as in the more metropolitan countries. These changes in political and cultural values are at least as important as economic growth in projecting the future of the Pacific Rim.

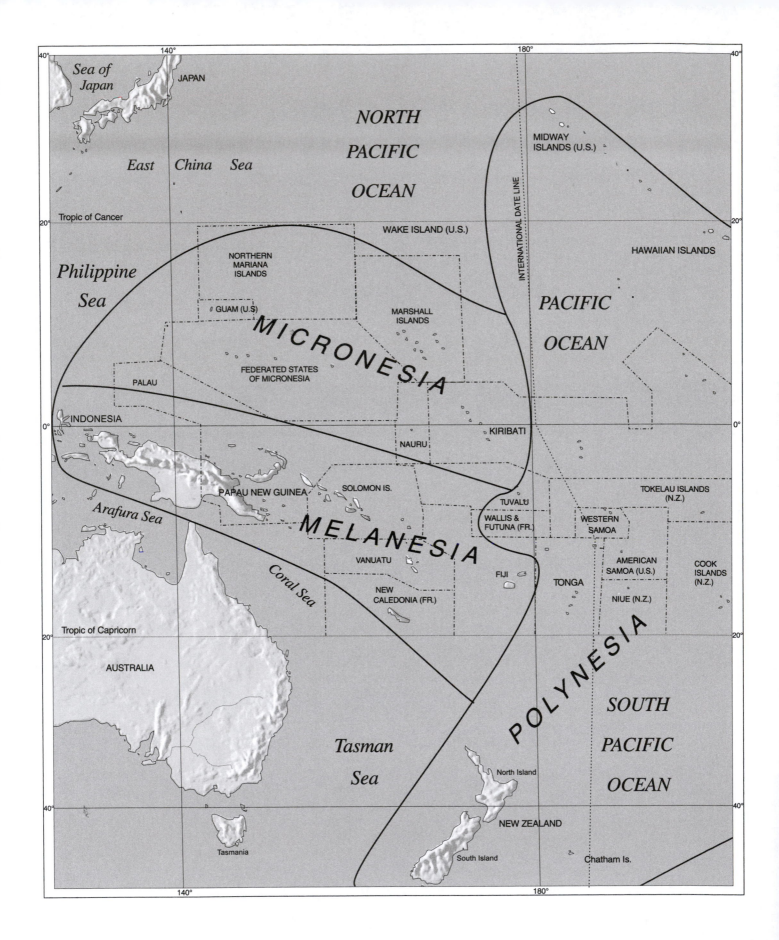

The Pacific Islands
Opportunities and Limits

There are about 30,000 islands in the Pacific Ocean. Most of them are found in the South Pacific and have been classified into three mammoth regions: *Micronesia,* composed of some 2,000 islands with such names as Palau, Nauru, and Guam; *Melanesia,* where 200 different languages are spoken on such islands as Fiji and the Solomon Islands; and *Polynesia,* comprised of Hawaii, Samoa, Tahiti, and other islands.

Straddling both sides of the equator and divided by the International Dateline, these territories are characterized as much by what is *not* there as by what *is*—that is, between any two tiny islands there might lie hundreds or even thousands of miles of open ocean. A case in point is the Cook Islands in Polynesia. Associated with New Zealand, this 15-island group contains only 92 square miles of land but is spread over 714,000 square miles of open sea. So expansive is the space between islands that early explorers from Europe and the Spanish lands of South America often unknowingly bypassed dozens of islands that lay just beyond view in the vastness of the 64 million square miles of the Pacific—the world's largest ocean.

However, once the Europeans found and set foot on the islands, they inaugurated a process that irreversibly changed the history of island life. Their goals in exploring the Pacific were to convert islanders to Christianity and to increase the power and prestige of their homelands (and themselves) by obtaining resources and acquiring territory. They thought of themselves and European civilization as superior to others and often treated the "discovered" peoples with contempt. An example is the "discovery" of the Marquesas Islands (from whence came some of the Hawaiian people) by the Peruvian Spaniard Alvaro de Mendana. Mendana landed in the Marquesas in 1595 with some women and children—and, significantly, 378 soldiers. Within weeks, his entourage had planted three Christian crosses, declared the islands to be the possession of the king of Spain, and killed 200 islanders. Historian Ernest S. Dodge describes the brutality of the first contacts:

> The Spaniards opened fire on the surrounding canoes for no reason at all. To prove himself a good marksman one soldier killed both a Marquesan and the child in his arms with one shot as the man desperately swam for safety. . . . The persistent Marquesans again attempted to be friendly by bringing fruit and water, but again they were shot down

when they attempted to take four Spanish water jars. Magnanimously the Spaniards allowed the Marquesans to stand around and watch while mass was celebrated. . . . When [the islanders] attempted to take two canoe loads of . . . coconuts to the ships half the unarmed natives were killed and three of the bodies hung in the rigging in grim warning. The Spaniards were not only killing under orders, they were killing for target practice.

> —*Islands and Empires; Western Impact on the Pacific and East Asia*

All over the Pacific, islanders were "pacified" through violence or deception inflicted on them by the conquering nations of France, England, Spain, and others. Rivalries among the European nations were often acted out in the Pacific. For example, the Cook Islands, inhabited by a mixture of Polynesian and Maori peoples, were partly controlled by the Protestant Mission of the London Missionary Society until the threat of incursions by French Catholics from Tahiti persuaded the British to declare the islands a protectorate of Britain. New Zealand eventually annexed the islands, and it controlled them until 1965.

Business interests frequently took precedence over islanders' sovereignty. In Hawaii, for instance, when Queen Liliuokalani proposed to limit the influence of the business community in island governance, a few dozen American business leaders—without the knowledge of the U.S. president or Congress, yet with the unauthorized help of 160 U.S. Marines—overthrew the Hawaiian monarch, installed Sanford Dole (of Dole Pineapple fame) as president, and petitioned Congress for Hawaii's annexation as a U.S. territory.

Whatever the method of acquisition, once the islands were under European or American control, the colonizing nations insisted that the islanders learn Western languages, wear Western clothing, convert to Christianity, and pay homage to faraway rulers whom they had never seen.

This blatant Eurocentrism ignored the obvious—that the islanders already had rich cultural traditions that both predated European culture and constituted substantial accomplishments in technology, the arts, and social structure. Islanders were skilled in the construction of boats suitable for navigation on the high seas and of homes and religious buildings of varied

(© 2002 Eric Guinther, GNU Free Documentation License)

In the South Pacific area of Micronesia, some 2,000 islands are spread over an ocean area of 3 million square miles. There remain many relics of the diverse cultures found on these islands; these highly prized stone discs were used as money on the islands of the Yap District.

architecture; they had perfected the arts of weaving and cloth-making, tattooing (the word itself is Tahitian), and dancing. Some cultures organized their political affairs much as did early New Englanders, with village meetings to decide issues by consensus. Others developed strong chieftainships and kingships with an elaborate variety of rituals and taboos (a Tongan word) associated with the ruling elite. Island trade involving vast distances brought otherwise disparate people together. And although reading and writing were not known on most islands, some evidence of an ancient writing system has been found on Easter Island.

Despite these cultural attributes and a long history of skill in interisland or intertribal warfare, the islanders could not withstand the superior force of European firearms. Within just a few generations, the entire Pacific had been conquered and colonized by Britain, France, the Netherlands, Germany, the United States, and other nations.

■ CONTEMPORARY GROUPINGS

The Pacific islands today are classified into three racial/cultural groupings. The first, Micronesia, with a population of approximately 533,000 people, contains seven political entities, four of which are politically independent and three of which are affiliated with the United States. Guam is perhaps the best known of these islands. Micronesians share much in common genetically and culturally with Asians. The term *Micronesia* refers to the small size of the islands in this group.

The second grouping, Melanesia, with a population of some 7.4 million (if Papua New Guinea is included), contains six political entities, four of which are independent and two of which are affiliated with colonial powers.

Types of Pacific Island Governments

The official names of some of the Pacific Island nations indicate the diversity of government structures found there:

Republic of Fiji
Federated States of Micronesia
Independent State of Papua New Guinea
Kingdom of Tonga

The best known of these islands is probably Fiji, which has a population of nearly 840,000. The term *Melanesia* refers to the dark skin of the inhabitants, who, despite appearances, apparently have no direct ties with Africa.

Polynesia, the third grouping, with a population of approximately 636,000, contains 12 political entities, 3 of which are independent, while the remaining are affiliated with colonial powers. *Polynesia* means "many islands," the most prominent grouping of which is probably Hawaii. Most of the cultures in Polynesia have some ancient connections with the Marquesas Islands or Tahiti.

Subtracting the atypically large population of the island of Papua New Guinea and that of New Zealand leaves about 2.3 million inhabitants in the region that we generally think of as the Pacific Islands. That is not many people, but annual population growth is 2.2 percent, which means that the regional population could double in just 30 years. A handful of islands have actually experienced population decline, but not because of decreasing fertility, but rather because of out-migration. The overall regional population does not decline, however, because most emigrants simply take up residence in nearby more economically or politically stable islands. The population growth rate of some islands is extremely high. For example, the Solomon Islands is growing at 6 percent annually. Probably never in history have the islands been peopled so heavily, and many islands are near the breaking point. Unable to make a living in the countryside, many islanders have migrated to the cities, living marginal lives on the fringes of towns.

The origins of the peoples of the Pacific are not clear. It is possible that some of the islands were peopled by or had contact with ancient civilizations of South America, but the overwhelming weight of scholarship places their roots in Southeast Asia, Indonesia, and Australia.

Geologically, the islands may be categorized into the tall, volcanic islands, which have abundant water, flora, and fauna, and are suitable for agriculture; and the dry, flat, coral islands, which have fewer resources. Two of those islands, Tuvalu and Tokelau, declared a state of emergency in 2011 when, due to a prolonged drought from a strong La Niña system, the drinking water supplies ran dangerously low. Some islands are rich in phosphate, but many have little by way of

The Case of the Disappearing Island

It wasn't much to begin with, but the way things are going, it won't be *anything* very soon. Nauru, a tiny, 8 1/2-square-mile dot of phosphate dirt in the Pacific, is being gobbled up by the Nauru Phosphate Corporation. Made of bird droppings (guano) mixed with marine sediment, Nauru's high-quality phosphate has a ready market in Australia, New Zealand, Japan, and other Pacific Rim countries, where it is used in industry, medicine, and agriculture.

Until recently, most Pacific Islanders would have envied the 12,809 Nauruans. With lots of phosphate to sell to willing buyers, Nauru earned so much income that its people had to pay no taxes and were provided with free health and dental care, bus transportation, and even their daily newspaper. Schooling (including higher education in Australia) was also paid for. Rent for government-built homes, supplied with telephones and electricity, was only US$5 a month. Nor did Nauruans have to work particularly hard for their living, since most phosphate pit laborers, about 3000 of them, were imported from other islands, while most managers and other professionals came from Australia (which once controlled the island), New Zealand, and Great Britain.

It sounds too good to be true, and it is, for Nauru's phosphate has been its only export, and now, the pits are nearly empty. Already there is only just a little fringe of green left along the shore, where everyone lives, and the government (Nauru is considered to be the smallest independent republic in the world) is debating what will happen when even the ground under people's homes is mined and shipped away. Some think that topsoil should be brought in to see if the moon-like surface of the excavated areas could be revitalized. Others think that moving away makes sense; that is, the government has been encouraged (by Australia and others) to just buy another island somewhere and move everyone there.

But all these dreams are now on hold while the tiny island deals with an even more urgent problem: near bankruptcy. As it turns out, Nauruans knew the phosphate was running out, so they put large amounts of their earnings into investments and trusts. They purchased several Boeing 737s, a number of hotels on other islands, and the tallest skyscraper in Melbourne, Australia. But many of their investments went sour, their former properties were taken over by GE Capital Corporation (which now says Nauru owes it US$176 million), and they began to drain their trust funds. Today, on the verge of bankruptcy, the government has frozen wages, eliminated many of the benefits people once enjoyed, and has lobbied Australia for more and more assistance. So now, along with having little land left, Nauru also has little of its economy left. Whether there will even be a Nauru in the near future is a very serious question.

---

natural resources, save the fish in the sea around them. It also appears that the farther away an island is from the Asian and Australian continental landmasses, the less varied and plentiful are the flora and fauna. In 2006 and 2009 former U.S. President George W. Bush named a total of 558,280 sq. miles in the U.S.-managed areas of the Pacific as marine monuments. The designation limits fishing and oil exploration in the Mariana Trench—the world's deepest sea canyon—as well as around uninhabited islands in the Northern Marianas, the Rose Atoll in American Samoa, and the northwestern Hawaiian islands.

■ THE PACIFIC COMMUNITY

During the early years of Western contact and colonization, maltreatment of the indigenous peoples, along with Western diseases such as measles and influenza, greatly reduced their numbers and their cultural strength. Moreover, the carving up of the Pacific by different Western powers superimposed a cultural fragmentation on the region that added to the separateness resulting naturally from distance. Today, however, the withdrawal or realignment of European and American political power under the post–World War II United Nations policy of decolonization has permitted the growth of regional organizations.

First among the postwar regional groups was the South Pacific Commission. Established in 1947, when Western powers were still largely in control, many of its functions have since been augmented or superseded by indigenously created organizations such as the South Pacific Forum (now called the Pacific Islands Forum), which was organized in 1971 and has since spawned numerous other associations, including the South Pacific Regional Trade and Economic Agency and the South Pacific Islands Fisheries Development Agency. Through an executive body (the South Pacific Bureau for Economic Cooperation), these associations handle such issues as relief funding, the environment, fisheries, trade, and regional shipping. The organizations have produced a variety of duty-free agreements among countries, and have yielded joint decisions about regional transportation and cultural exchanges. As a result, regional art festivals and sports competitions are now a regular feature of island life, and a regional university in New Zealand attracts several thousand island students a year, as do colleges in Hawaii. A planned Australian Pacific Islands Technical College is envisioned to help train students in technical subjects and in the hospitality industry. Some 16 nations (as well as several observer countries) met under the auspices of the Pacific Islands Forum Secretariat in late 2005 to establish the "Pacific Plan." The Plan, updated in 2006, includes steps for regional cooperation and integration in the areas of trade, transportation, digital communication, energy and sustainable development, health (including HIV/AIDS policies), and many others.

Guam: This is Liberation?

In 1994, the people of the U.S. Territory of Guam celebrated the 50th anniversary of their liberation by U.S. Marines and Army Infantry from the occupying troops of the Japanese Army. During the three years that they controlled the tiny, 30-mile-long island, the Japanese massacred some of the Guamanians and subjected many others to forced labor and internment in concentration camps.

Their liberation, therefore, was indeed a cause for celebration. But the United States quickly transformed the island into its military headquarters for the continuing battle against the Japanese. The entire northern part of the island was turned into a base for B-29 bombers, and the Pacific submarine fleet took up residence in the harbor. Admiral Chester W. Nimitz, commander-in-chief of the Pacific, made Guam his headquarters. By 1946, the U.S. military government in Guam had laid claim to nearly 80 percent of the island, displacing entire villages and hundreds of individual property owners.

Since then, some of the land has been returned, and large acreages have been handed over to the local civilian government—which was to have distributed most of it, but has not yet done so. The local government still controls about one third of the land, and the U.S. military controls another third, meaning that only one third of the island is available to the residents for private ownership. Litigation to recover the land has been bitter and costly, involving tens of millions of dollars in legal expenses since 1975. The controversy has prompted some local residents to demand a different kind of relationship with the United States, one that would allow for more autonomy. It has also spurred the growth of nativist organizations such as the Chamorru Nation, which promotes the Chamorru language of the original Malayo–Polynesian inhabitants (spelled Chamorro by the Spanish) and organizes acts of civil disobedience against both civilian and military authorities.

Guam was first overtaken by Spain in 1565. It has been controlled by the United States since 1898, except for the brief Japanese interlude. Whether the local islanders, who now constitute a fascinating mix of Chamorro, Spanish, Japanese, and American cultures, will be able to gain a larger measure of autonomy after hundreds of years of colonization by outsiders is difficult to predict, but the ever-present island motto, *Tano Y Chamorro* ("Land of the Chamorros"), certainly spells out the objective of many of those who call Guam home.

Some regional associations have been able to deal forcefully with much more powerful countries. For instance, when the regional fisheries association set higher licensing fees for foreign fishing fleets (most fleets are foreign-owned, because island fishermen usually cannot provide capital for such large enterprises), the Japanese protested vehemently. Nevertheless, the association held firm, and many islands terminated their contracts with the Japanese rather than lower their fees. In 1994, the Cook Islands, the Federated States of Micronesia, Fiji, Kiribati, the Marshall Islands, Nauru, Niue, Papua New Guinea, the Solomon Islands, Tonga, Tuvalu, Vanuatu, and Western Samoa signed an agreement with the United States to establish a joint commercial commission to foster private-sector businesses and to open opportunities for trade, investment, and training. Through this agreement, the people of the islands hoped to increase the attractiveness of their products to the U.S. market.

Increasingly important issues in the Pacific are the testing of nuclear weapons and the disposal of toxic waste. Island leaders, with the occasional support of Australia and the strong support of New Zealand, have spoken out vehemently against the continuation of nuclear testing in the Pacific by the French government (Great Britain and the United States tested hydrogen bombs on coral atolls for years, but have now stopped) and against the burning of nerve-gas stockpiles by the United States on Johnston Atoll. In 1985, the 13 independent or self-governing countries of the South Pacific adopted their first collective agreement on regional security, the South Pacific Nuclear Free Zone Treaty. Encouraged by New Zealand and Australia, the group declared the Pacific a nuclear-free zone and issued a communique criticizing the dumping of nuclear waste in the region. Some island leaders, however, see the storage of nuclear waste as a way of earning income to compensate those who were affected by the nuclear testing on Bikini and Enewetak Islands. The Marshall Islands, for example, are interested in storing nuclear waste on already-contaminated islands; however, the nearby Federated States of Micronesia, which were observers at the Nuclear Free Zone Treaty talks, oppose the idea and have asked the Marshalls not to proceed. Caring for those whose health has been damaged by the nuclear testing is becoming a heavy burden on other islands. Those residents of Palau, the Marshall Islands, and Micronesia who have higher rates of cancer and kidney disease as a result of the testing, have to leave their islands to seek medical treatment elsewhere because there are no facilities to handle such illnesses on their islands. Many of them go to Guam and Hawaii for treatment. Hawaii's government spent more than US$52 million on such care in 2010, and government leaders have been searching for ways to handle the increasing demands on their services.

World leaders met in Jamaica in 1982 to sign into international law the Law of the Sea. This law, developed under the auspices of the United Nations, gave added power to the tiny Pacific Island nations because it extended the territory under their exclusive economic control to 12 miles beyond their shores or 200 miles of undersea continental shelf. This put many islands in undisputed control of large deposits of nickel, copper, magnesium, and other valuable metals. The seabed areas away from continents and islands were declared the world's common heritage, to be mined by an international company, with profits channeled to developing countries. The United States has negotiated for years to increase the role of industrialized nations in mining the seabed areas; if modifications are made to the treaty allowing a larger role for private enterprise to mine the seabeds, the United States would be more willing to sign it, as it stands, the United States has so far refused to sign.

■ COMING OF AGE?

If the peoples of the Pacific Islands are finding more reasons to cooperate economically and politically, they are still individually limited by the heritage of cultural fragmentation left them by their colonial pasts. Western Samoa, for example, was first annexed by Germany in 1900, only to be given to New Zealand after Germany's defeat in World War I. Today, the tiny nation of mostly Christian Polynesians, independent since 1962, uses both English and Samoan as official languages and embraces a formal governmental structure copied from Western parliamentary practice. Yet the structure of its hundreds of small villages remains decidedly traditional, with clan chiefs ruling over large extended families, who make their not particularly profitable livings by farming breadfruit, taro, yams, bananas, and copra.

Political independence has not been easy for those islands that have embraced it, nor for those colonial powers that continue to deny it. Anticolonial unrest continues on many islands, especially the French ones. However, concern over economic viability has led most islands to remain in some sort of loose association with their former colonial overseers. After the defeat of Japan in World War II, the Marshall Islands, the Marianas, and the Carolines (now Federated States of Micronesia) were assigned by the United Nations to the United States as trust territories. The French Polynesian islands have remained overseas "departments" (similar to U.S. states) of France.

In such places as New Caledonia, however, there has been a growing desire for autonomy, which France has attempted to meet in various ways while still retaining sovereignty. The UN decolonization policy has made it possible for most Pacific islands to achieve independence if they wish, but many are so small that true economic independence in the modern world will never be possible. In French Polynesia, the independence issue continues to flare up. Some 15,000 people marched against the removal of pro-independence president Oscar Temaru after parliament voted to censure him in 2004. A more conservative former president who did not advocate separation from France replaced Temaru. Financial assistance keeps many of the islands connected to Western powers even though they might prefer it otherwise. For instance, the Pacific Islands Trade and Investment Commission is fully funded by New Zealand. In 2006, voters in Tokelau, a dependency of New Zealand, voted against independence. Eighty percent of Tokelau's budget comes from New Zealand. Australia has given US$600 million to the Solomon Islands since 2000.

In addition to relations with their former colonial masters, the Pacific islands have found themselves in the middle of many serious domestic political problems. For example, Fiji, a former British colony, weathered two military coups in 1987, a bizarre coup in 2000, and another coup in late 2006. After the 2006 elections, indigenous islanders attacked and burned homes and businesses in Suva's Chinatown, claiming that the Chinese had unduly influenced the election. Some 240 Chinese fled the island. In the 2000 episode, native Fijian businessman George Speight captured the prime minister and other government leaders and held them hostage until they resigned their posts. The prime minister, Mahendra Chaudhry, was not a native Fijian but, rather, an Indian. So many immigrants from India had settled in Fiji over the past few decades that they now constituted some 44 percent of the population, enough to wield political power on the island. This situation had frightened the native Fijians into stripping Indians of various civil rights (restored in 1997), but the election of the first Indian prime minister seemed to push some of the native Fijians past their tolerance limit.

While not all native Fijians feel threatened by the rise of the Indian population, when a Suva television station criticized the latest coup and called for racial tolerance, the station was ransacked by Fijian rioters, who also destroyed some 20 other buildings in the capital city. Tensions continued in 2001 when the new prime minister, Laisenia Qarase, refused to seat Indo-Fijians in his cabinet until a court ruling ordered him to do so. In 2002, George Speight was sentenced to life in prison for his takeover of the Parliament, a development that produced a foiled plot by his supporters to kidnap government leaders and hold them hostage until Speight was released. Finally, in 2004, all coup leaders were convicted and sent to jail. In 2007, Commodore Frank Bainimarama took office as prime minister after deposing the previous government (and banning the former prime minister to his home on another island) in a bloodless military coup. This coup was triggered, in part, by the government's plan to pardon those involved in the 2000 coup and by plans to give land to indigenous Fijians at the expense of the Indian minority. Coup leaders' plans were mollified somewhat by the traditional Fijian Council of Chiefs, which continues to hold influence in island affairs. Bainimarama restored the president to his post and agreed to select a cabinet. But then in 2009, the president, claiming that the government was not legal, repealed the constitution and voided the appointments of all judges, while reappointing Bainimarama as prime minister. Elections were not promised until 2014. This violation of democratic principles caused the Pacific Islands Forum (whose headquarters is in Suva, Fiji) to suspend Fiji in 2009. In 2010, the government convicted eight people of the 2007 plot against the Prime Minister.

Similarly, tribal tensions have wracked the Solomon Islands. In the past few years, residents of the island of Malaita have migrated in large numbers to the larger, main island of Guadalcanal. The locals have complained that the newcomers have taken over their land and the government. Tribal violence related to this issue has killed some 50 islanders in the recent past; but in 2000, Malaita's prime minister, a Malaitan, was kidnapped in Honiara, the capital city, and forced to resign. Violence forced 20,000 residents to flee their homes, and foreign nationals were evacuated to avoid the fierce fighting by opposing paramilitary groups. It is clear that in Fiji, the Solomon Islands, and other places, Pacific islanders will have to find creative ways to allow all members of society to participate in the political process. Without political stability, other pressing issues will be neglected.

No amount of political realignment can overcome the economic dilemma of most of the islands. Japan, the single largest purchaser of island products, as well as the United States and others, are good markets for the Pacific economies, but exports are primarily of mineral and agricultural products (coffee, tea, cocoa, sugar, tapioca, coconuts, mother-of-pearl) rather than of the more profitable manufactured or "value-added" items produced by industrial nations. In addition, there will always be the cost of moving products from the vastness of the Pacific to the various mainland markets.

Another problem is that many of the profits from the island's resources do not redound to the benefit of the islanders. Tuna, for example, is an important and profitable fish catch, but most of the profits go to the Taiwanese, Korean, Japanese, and American fleets that ply the Pacific. Similarly, tourism profits largely end up in the hands of the multinational hotel owners. About 80 percent of visitors to the island of Guam since 1982 have been Japanese (more than half a million people annually)—seemingly a gold mine for local Guamanians, since each traveler typically spends more than US$2,000. However, because the tourists tend to purchase their tickets on Japanese airlines and book rooms in Japanese-owned or -managed hotels, much of the money that they spend for their vacations in Guam never reaches the Guamanians. Some enterprising Fijians, together with Chinese, Malaysian, and Hong Kong gangs, have been attempting to make money their own way: in Fiji's capital of Suva, they created the largest methamphetamine lab in the Southern Hemisphere. Located in a three-building complex stocked with drums of chemicals, the lab was capable of producing US$540 million worth of "meth." Police shut down the lab in 2004, but the case illustrates how desperate some islanders are for income.

The poor economies, especially in the outer islands, have prompted many islanders to move to larger cities (about 1 million islanders now live in the Pacific's larger cities) to find work. Indeed, there is currently a tremendous mixing of all of the islands' peoples. Hawaii, for example, has long been peopled by Samoans, Filipinos, and many other islanders; pure Hawaiians are a minority, and despite efforts to preserve the Hawaiian language, it is used less and less. Similarly, Fiji, as we have seen, is now populated by nearly as many immigrants from India as by native Fijians. New Caledonians are outnumbered by Indonesians, Vietnamese, French, and others. Guam is peopled with islanders from all of Micronesia as well as from Samoa and other islands. And, of course, whites have long outnumbered the Maoris on New Zealand.

In addition to interisland migration, many islanders emigrate to Australia, New Zealand, the United States, or other countries and then send money back home to sustain their families. Those remittances are important to the economies of the islands, but the absence of parents or adult children for long periods of time does considerable damage to the social fabric. In a few cases, such as in the Cook Islands and American Samoa, more islanders live abroad than remain on the islands. For example, whereas American Samoa has about 58,000 residents, more than 130,000 Samoans live on the U.S. mainland, in such places as Los Angeles and Salt Lake City. Those who leave often find life abroad quite a shock, for the island culture, influenced over the decades by the missionary efforts of Mormons, Methodists, Seventh-day Adventists, and especially the London Missionary Society, is conservative, cautious, and personal. Metropolitan life, by contrast, is considered by some islanders to be wild and impersonal. Some young emigrants respond to the "cold" environment and marginality of big-city life by engaging in deviant behavior themselves, such as selling drugs and joining gangs.

Island society itself, moreover, is not immune from the social problems that plague larger societies. Many islands report an increasing number of crimes and suicides. Young Samoans, for example, are afflicted with many of the same problems—gangs, drugs, and unemployment—as are their U.S. inner-city counterparts. Samoan authorities have reported increases in incidences of rape, robbery, and other socially dysfunctional behaviors. In addition, the South Pacific Commission and the World Health Organization are now reporting an alarming increase in HIV/AIDS and other sexually transmitted diseases.

Pacific islanders are trying to cope with a variety of other problems as well. Tuvalu, for example, has a very threatening problem: its nine coral atolls are sinking into the sea. With sea levels rising due to global warming, the country is already subjected to monthly flooding. The 11,000 inhabitants are pleading with both Australia and New Zealand to allow or assist with relocation to those countries or to the island of Niue. The highest point on the islands is only 16 feet above sea level, and drinking water is already becoming brackish as the ocean penetrates groundwater reserves. In 50 years or less, the land will be entirely submerged. New Zealand has already accepted some islanders, but Australia has denied both access and relocation assistance. Other problems caused by nature also make life difficult for islanders. Low-lying islands often suffer serious damage from cyclones, such as the category 5 cyclone that hit Tokelau in 2004 and completely leveled the capital city. An 8 foot tsunami in 2010 left 1,000 people homeless in the Solomon Islands.

For decades, and notwithstanding the imposition of foreign ways, islanders have shared a common culture; many people know how to raise bananas, coconuts, and yams, how to roast pigs and fish, and how to make breadfruit, tapioca, and poi. But much of island culture has depended on an identity shaped and preserved by isolation from the rest of the world. Whether the essence of island life—and especially the identity of the people—can be maintained in the face of increasing integration into a much larger world remains to be seen.

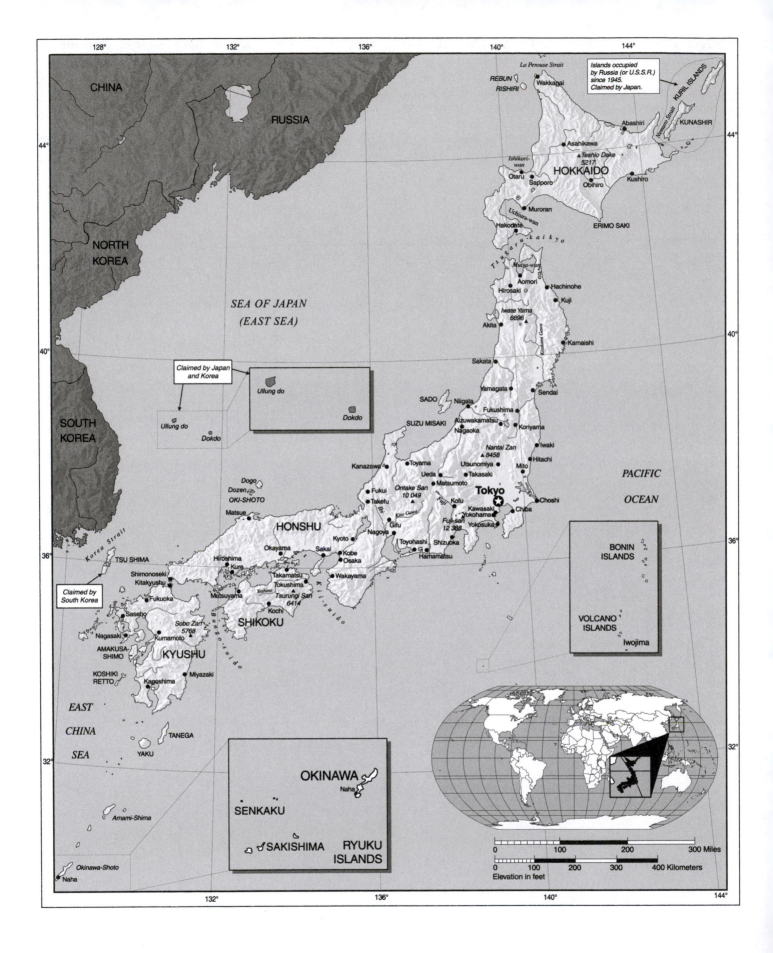

CHINA

RUSSIA

La Perouse Strait

REBUN
RISHIRI
Wakkanai

Islands occupied
by Russia (or U.S.S.R.)
since 1945.
Claimed by Japan.

KURIL ISLANDS

NORTH
KOREA

Abashiri

KUNASHIR

Asahikawa

Nemuro Strait

SEA OF JAPAN
(EAST SEA)

Ishikari-wan

▲ Teshio Dake
5217

HOKKAIDO

Kushiro

Otaru

Obihiro

Sapporo

Muroran

Uchiura-wan

ERIMO SAKI

SOUTH
KOREA

Hakodate

Tsugaru-kaikyo

Claimed by Japan
and Korea

Ullung do

Dokdo

Mutsu-wan

Aomori

Hachinohe

Hirosaki

Kuji

Ullung do

Dokdo

Iwate Yama
6696 ▲

Akita

Kitakami Gawa

Kamaishi

Sakata

Yamagata

Sendai

SADO

Niigata

SUZU MISAKI

Fukushima

Aizuwakamatsu

Koriyama

Nagaoka

Iwaki

Nantai Zan
▲ 8458

Hitachi

Kanazawa

Toyama

Mito

Ueda

Takasaki

Fukui

Ontake San
10 049

Matsumoto

Utsunomiya

Dogo
Dozen
OKI-SHOTO

Takefu

▲

Tokyo

Kofu

Chiba

Fuji

Kawasaki
Yokohama

Choshi

PACIFIC

OCEAN

Matsue

Iti

Kiso Gawa

Gifu

Fuji-san
12 388

Yokosuka

HONSHU

Kyoto

Nagoya

Toyohashi

Shizuoka

Okayama

Sakai

Kobe

Hamamatsu

Hiroshima

Osaka

TSU SHIMA

Kure

Takamatsu

Wakayama

BONIN
ISLANDS

Shimonoseki

Tokushima

Claimed by
South Korea

Kitakyushu

Matsuyama

Yoshino

Tsurugi San
6414

Korea Strait

Fukuoka

Kochi

Kii-suido

Sasebo

SHIKOKU

VOLCANO
ISLANDS

Sobo Zan
5768 ▲

Nagasaki

Kumamoto

Iwojima

Bungo-suido

AMAKUSA-
SHIMO

KYUSHU

KOSHIKI
RETTO

Kageshima

Miyazaki

EAST

CHINA

TANEGA

SEA

YAKU

OKINAWA

Naha

Amami-Shima

SENKAKU

SAKISHIMA

RYUKU
ISLANDS

Okinawa-Shoto

Naha

0   100   200   300 Miles

0   100   200   300   400 Kilometers

Elevation in feet

26

Japan

On Friday afternoon, March 11, 2011, a 133 foot (40 meter) wave of roiling ocean water (as high as a ten-story building) slammed into the Tohoku region of Japan, traveling inland up to 6 miles (9.5 kilometers). More mammoth waves followed. The tsunami was triggered by a 9.0 earthquake 43 miles (70 kilometers) east of Japan, the strongest earthquake ever to hit the country. It was so powerful that it actually moved the entire island of Honshu some 8 feet (26 meters) east, forced the seabed upward, caused the earth's axis to shift by 10 inches (.24 meters), and increased the speed of the earth's rotation, shortening the length of a day by 1.8 microseconds. The earthquake may have also been related to a volcanic eruption that occurred in Kyushu two days later. Liquefaction in many areas, including as far away as Tokyo, caused serious damage to thousands of buildings. Many towns in the Tohoku region were completely wiped out along with all their inhabitants. Never had the Japanese experienced such a massive natural disaster, and the Prime Minster, Naoto Kan, declared it to be Japan's most serious crisis since World War II.

But there was much more pain to come. The earthquake and tsunami damaged several nuclear power plants, causing some to explode and go into meltdown. Residents who had survived the earthquake and tsunami now had to flee their homes to avoid radiation emanating from the damaged plants. Some of the homes became so irradiated that the residents would never be allowed to return. Fires ignited in many places; a dam ruptured; 125,000 buildings, including 11 hospitals, completely collapsed; the supply of electricity and drinking water ceased; highways, bridges, train lines, ships, and cars were destroyed; and at least 23,000 died and 300,000 were left homeless. More than 230 children were orphaned.

Residents felt the shaking for 5 minutes, and then they had only about 8 minutes before the tsunami hit because even though the epicenter was miles away, the waves were traveling as fast as a jet airplane. Aftershocks from the earthquake, some 900 of them, and some of them over 7 on the Richter scale, continued for months and caused additional damage and deaths. The cost of rebuilding what was left of the area has been estimated to be in excess of US$300 billion. Eight months after the disaster, hundreds of residents, their homes and belongings totally obliterated, remained in schools and other temporary shelters.

The heartbreaking devastation alone was enough to cause people everywhere much reflection. But there was something else that emerged from the rubble—something that reveals what kind of people the Japanese are. When Hurricane Katrina struck New Orleans in the United States, residents were seen stealing and looting everything they could get their hands on. When there were gasoline shortages, people got into fights (including one man who killed another) rather than wait their turn to fill up. But in Japan? There was virtually no looting. Instead, vending machine owners were giving out free drinks to needy people, and Japanese from far and near were walking the littered beaches looking for photos or other items of personal value to return to their owners. A ration system was implemented, and people waited patiently in line to shop for their ten allowed items of food. Wallets with money intact were returned to police stations, and senior citizens, believing their lives were nearly over anyway, volunteered to

TIMELINE

20,000–4,500 B.C.
Prepottery, paleolithic culture

4,500–250 B.C.
Jomon culture with distinctive pottery

250 B.C.–A.D. 300
Yayoi culture with rice agriculture, Shinto religion, and Japanese language

A.D. 300–700
The Yamato period; warrior clans import Chinese culture

710–794
The Nara period; Chinese-style bureaucratic government at the capital at Nara

794–1185
The Heian period; the capital is at Kyoto

1185–1333
The Kamakura period; feudalism and shoguns; Buddhism is popularized

1333–1568
The Muromachi period; Western missionaries and traders arrive; feudal lords control their own domains

1568–1600
The Momoyama period; feudal lords become subject to one central leader; attempted invasion of Korea

1600–1868
The Tokugawa Era; self-imposed isolation from the West

1868–1912
The Meiji Restoration; modernization; Taiwan and Korea are under Japanese control

1912–1945
The Taisho and Showa periods; militarization leads to war and Japan's defeat

1945
Japan surrenders; the U.S. Occupation imposes major changes in the organization of society

1951
Sovereignty is returned to the Japanese people by treaty

1980s
The ruling party is hit by scandals but retains control of the government; Emperor Hirohito dies; Emperor Akihito succeeds

1990s
After years of a slow economy, Japan officially admits it is in a recession; a devastating earthquake in Kobe kills more than 6,000 people

(continued)

TIMELINE

2000s

A U.S. Navy submarine accidentally rams into and sinks a Japanese trawler; nine Japanese are killed; relations with the United States are strained

Experts predict that the worker/retiree ratio will drop from 6:1 to 2:1 by 2020, as the population ages

Japanese business undergoes a shift toward more openness and flexibility

Weaknesses in the Japanese economy are a concern all over the world

Japan's economy begins a slow rebound from a decade of stagnation

Japan is denied membership in the UN Security Council for the 14th time

The government launches a plan to alter the peace clause of the constitution to allow the development of a full, offensive-capable military

The Democratic Party of Japan soundly defeats the Liberal Democratic Party

Mandatory Saturday school attendance was eliminated

Massive earthquake and tsunami hit Tohoku region

(Japan National Tourist Organization)

The Japanese emperor has long been a figurehead in Japan. In 1926, Hirohito (above) became emperor and ushered in the era named *Showa*. He died on January 7, 1989, having seen Japan through World War II and witnessed its rise to the economic world power it is today. He was succeeded by his son, Akihito, who named his reign *Heisei*, meaning "Achieving Peace."

replace younger workers in the damaged nuclear plants who were trying to stop the radiation leaks.

In the worst possible conditions, the Japanese people acted with dignity and integrity. What produced a people of this sort? Perhaps this chapter will help answer that question.

The Japanese nation is thought to have begun about 250 B.C., when ancestors of today's Japanese people began cultivating rice, casting objects in bronze, and putting together the rudiments of the Shinto religion. However, humans are thought to have inhabited the Japanese islands as early as 20,000 B.C. Some speculate that remnants of these or other early peoples may be the non-Oriental Ainu people (now largely Japanized) who still occupy parts of the northern island of Hokkaido. Asiatic migrants from China and Korea and islanders from the South Pacific occupied the islands between 250 B.C. and A.D. 400, contributing to the population base of modern Japan.

Between A.D. 300 and 710, military aristocrats from some of the powerful clans into which Japanese society was divided established their rule over large parts of the country. Eventually, the Yamato clan leaders, claiming divine approval, became the most powerful. Under Yamato rule, the Japanese began to import ideas and technology from nearby China, including the Buddhist religion and the Chinese method of writing—which the elite somewhat awkwardly adapted to spoken Japanese, an entirely unrelated language. The Chinese bureaucratic style of government and architecture was also introduced; Japan's first permanent capital was constructed at the city of Nara between the years 710 and 794, and it mimicked the style of Chinese cities.

As Chinese influence waned in the period 794–1185, the capital was relocated to Kyoto, with the Fujiwara family wielding real power under the largely symbolic figurehead of the emperor. A warrior class controlled by shoguns, or generals, held power at Kamakura between 1185 and 1333 and successfully defended the country from invasion by the Mongols. Buddhism became the religion of the masses, although Shintoism was often embraced simultaneously. Between 1333 and 1568, a very rigid class structure developed, along with a feudalistic economy controlled by *daimyos,* feudal lords who reigned over their own mini-kingdoms.

In 1543, Portuguese sailors landed in Japan, followed a few years later by the Jesuit missionary Francis Xavier. An active trade with Portugal began, and many Japanese (perhaps half a million), including some feudal lords, converted to Christianity. The Portuguese introduced firearms to the Japanese and perhaps taught them Western-style techniques of building castles with moats and stone walls. Wealthier feudal lords were able to utilize these innovations to defeat weaker rivals; by 1600, the country was unified under a military bureaucracy, although feudal lords still retained substantial sovereignty over their fiefs. During this time, the general Hideyoshi attempted an unsuccessful invasion of nearby Korea.

The Tokugawa Era

In the period 1600 to 1868, called the Tokugawa Era, the social, political, and economic foundations of modern Japan were put in place. The capital was moved to Tokyo (although the emperor remained isolated in Kyoto), cities began to grow in size, and a merchant class arose that was powerful enough to challenge the hegemony of the centuries-old warrior class. Strict rules of dress and behavior for each of the four social classes (samurai, farmer, craftsman, and merchant) were imposed, and the

Japanese people learned to discipline themselves to these codes. Western ideas came to be seen as a threat to the established ruling class. The military elite expelled foreigners and put the nation into 2½ centuries of extreme isolation from the rest of the world. Christianity was banned, as was most trade with the West. Even Japanese living or traveling abroad were forbidden to return, for fear that they might have become contaminated with foreign ideas.

During the Tokugawa Era, indigenous culture expanded rapidly. Puppet plays and a new form of drama called *kabuki* became popular, as did *haiku* poetry and Japanese pottery and painting. The samurai code, called *bushido,* along with the concept of *giri,* or obligation to one's superiors, suffused Japanese society. Literacy among males rose to about 40 percent, higher than most European countries of the day. With the nation at peace, the samurai busied themselves with the education of the young, using teaching methods that included strict discipline, hard work, and self-denial.

During the decades of isolation, Japan grew culturally strong but militarily weak. In 1853, a U.S. naval squadron appeared in Tokyo Bay to insist that Japan open up its ports to foreign vessels needing supplies and desiring to trade. Similar requests had been denied in the past, but the sophistication of the U.S. ships and their advanced weaponry convinced the Japanese military rulers that they could no longer keep Japan isolated from the outside.

The Era of Modernization: The Meiji Restoration

Treaties with the United States and other Western nations followed, and the dislocations associated with the opening of the country to the world soon brought discredit to the ruling shoguns. Provincial samurai took control of the government. The emperor, long a figurehead in Kyoto, away from the center of power, was moved to Tokyo in 1868, beginning the period known as the Meiji Restoration.

Although the Meiji leaders came to power with the intention of ousting all the foreigners and returning Japan to its former state of domestic tranquillity, they quickly realized that the nations of the West were determined to defend their newly won access to the ports of Japan. To defeat the foreigners, they reasoned, Japan must first acquire Western knowledge and technology.

Thus, beginning in 1868, the Japanese leaders launched a major campaign to modernize the nation. Ambassadors and scholars were sent abroad to learn about Western-style government, education, and warfare. One such group, the Iwakura Embassy of 1872, spent months in San Francisco, Salt Lake City, and Washington, D.C., observing legislatures and discussing America's electoral system, farm policies, taxation policies, religious freedom, and other matters with government leaders and citizens. Implementing these ideas resulted in the abolition of the feudal system and the division of Japan into 43 prefectures, or states, and other administrative districts under the direct control of the Tokyo government. Legal codes that established the formal separation of society into social classes were abolished, and Western-style dress, music, and education were embraced. The old samurai class turned its attention from warfare to leadership in the government, in schools, and in business. Freedom of religion was established, and a national army, rather than clan-based military units, was created. Factories and railroads were constructed, and public education was expanded. By 1900, Japan's literacy rate was 90 percent, the highest in all of Asia. Parliamentary rule was established along the lines of the government in Prussia, agricultural techniques were imported from the United States, and banking methods were adopted from Great Britain.

Japan's rapid modernization soon convinced its leaders that the nation was strong enough to begin doing what other advanced nations were doing: acquiring empires. Japan went to war with China, acquiring the Chinese island of Taiwan in 1895. In 1904, Japan attacked Russia and successfully acquired Korea and access to Manchuria (both areas having been in the sphere of influence of Russia). Siding against Germany in World War I, Japan was able to acquire Germany's Pacific empire—the Marshall, Caroline, and Mariana Islands. Western nations were surprised at Japan's rapid empire-building but did little to stop it. Indeed, some Westerners viewed Japan's aggression against others as a sign of a progressive nation.

The Great Depression of the 1930s caused serious hardships in Japan because, being resource-poor yet heavily populated, the country had come to rely on international trade to supply its people's basic needs. Many Japanese advocated the forced annexation of Manchuria as a way of acquiring needed resources. This was accomplished easily, albeit with much brutality, in 1931. With militarism on the rise, the Japanese nation began moving away from democracy and toward a military dictatorship. Political parties were eventually banned, and opposition leaders were jailed and tortured.

WORLD WAR II AND THE JAPANESE EMPIRE

The battles of World War II in Europe, initially won by Germany, promised to substantially realign the colonial empires of France and other European powers in Asia. The military elite of Japan declared its intention of creating a Greater East Asia Co-Prosperity Sphere—in effect, a Japanese empire created out of the ashes of the European empires in Asia that were then dissolving. In 1941, under the guidance of General Hideki Tojo and with the tacit approval of the emperor, Japan captured the former French colony of Indochina (Vietnam, Laos, and Cambodia), bombed Pearl Harbor in Hawaii, and captured oil-rich Indonesia. These victories were followed by others: Japan captured all of Southeast Asia, including Burma (now called Myanmar), Thailand, Malaya, the Philippines, and parts of New Guinea; and expanded its hold in China and in the islands of the

South Pacific. Many of these conquered peoples, lured by the Japanese slogan of "Asia for the Asians," were initially supportive of the Japanese, believing that Japan would rid their countries of European colonial rule. It soon became apparent, however, that Japan had no intention of relinquishing control of these territories and would go to brutal lengths to subjugate the local peoples. Japan soon dominated a vast empire, the constituents of which were virtually the same as those making up what we call the Pacific Rim today.

In 1941, the United States launched a counteroffensive against the powerful Japanese military. (American history books refer to this offensive as the Pacific Theater of World War II, but the Japanese call it the *Pacific War.* We use the term *World War II* in this text, for reasons of clarity and consistency.) By 1944, the U.S. and other Allied troops, at the cost of tens of thousands of lives, had ousted the Japanese from most of their conquered lands and were beginning to attack the home islands themselves. Massive firebombing of Tokyo and other cities, combined with the dropping of two atomic bombs on Hiroshima and Nagasaki, convinced the Japanese military rulers that they had no choice but to surrender.

This was the first time in Japanese history that Japan had been conquered, and the Japanese were shocked to hear their emperor, Hirohito—whose voice had never been heard on radio—announce on August 14, 1945, that Japan was defeated. The emperor cited the suffering of the people—almost 2 million Japanese had been killed—as well as the devastation of Japan's cities brought about by the use of a "new and most cruel bomb," and the possibility that, without surrender, Japan as a nation might be completely "obliterated." Emperor Hirohito then encouraged his people to look to the future, to keep pace with progress, and to help build world peace by accepting the surrender ("enduring the unendurable and suffering what is insufferable").

This attitude smoothed the way for the American Occupation of Japan, led by General Douglas MacArthur. Defeat seemed to inspire the Japanese people to adopt the ways of their more powerful conquerors and to eschew militarism. Under the Occupation forces, the Japanese Constitution was rewritten in a form that mimicked that of the United States. Industry was restructured, labor unions encouraged, land reform accomplished, and the nation as a whole demilitarized.

FURTHER INVESTIGATION:

In 2010, the Japanese Prime Minister delivered a "heartfelt apology" to South Korea (but not North Korea) for the havoc the Japanese wreaked on Koreans during the years of Japanese colonial rule. To learn more about Japan-Korea relations, see: http://www.eastasiaforum.org/2010/08/24/japan-korea-relations-in-a-new-era/ and also http://www.cfr.org/japan/japans-relationship-south-korea/p9108

Economic aid from the United States, as well as the prosperity in Japan that was occasioned by the Korean War in 1953, allowed Japanese industry to begin to recover from the devastation of war. The United States returned the governance of Japan back to the Japanese people by treaty in 1951 (although some 35,000 troops still remain in Japan as part of an agreement to defend Japan from foreign attack).

By the late 1960s, the Japanese economy, stronger already than some European economies, was more than self-sustaining, and the United States was Japan's primary trading partner. Today, reflecting the diversification of Japan's markets, about 10 percent of the items Japan imports come from the United States, and the United States buys about 18 percent of Japan's exports. China now supplies more of Japan's imports than does the United States, but many of those imports are made by U.S. companies operating in China. For many years after World War II, Japan's trade with its former Asian empire, however, was minimal, because of lingering resentment against Japan for its wartime brutalities. As recently as the late 1970s, for example, anti-Japanese riots and demonstrations occurred in Indonesia upon the visit of the Japanese prime minister to that country, and the Chinese government raises the alarm each time Japan effects a modernization of its military.

Between the 1960s and early 1990s, Japan experienced an era of unprecedented economic prosperity. Annual economic growth was three times as much as in other industrialized nations. Japanese couples voluntarily limited their family size so that each child born could enjoy the best of medical care and social and educational opportunities. The fascination with the West continued, but eventually, rather than "modernization" or "Americanization," the Japanese began to speak of "internationalization," reflecting both their capacity for and their actual membership in the world community, politically, culturally, and economically. Militarily, however, Japan was forbidden by its U.S.-drafted Constitution to develop a war capacity. In recent years, Japanese troops have engaged in UN peacekeeping operations in Cambodia and have participated in noncombat duties in the Iraq War, but even that level of military involvement has caused tremendous domestic and international comment. Japan was severely denounced by China and South Korea in 2007 when it announced the upgrading of its military bureaucracy from that of an "agency" to a full "ministry of defense." More alarming to some Asian countries is the government's decision to amend the constitution to allow Japan to possess offensive military capabilities. The United States is supportive of these changes, because it wants Japan to shoulder more of the responsibility of defending allies in the face of such threats as the newly nuclearized North Korea and the rising military strength of China. With North Korea firing missiles over Japan's airspace in 2008 and 2009, some of the Japanese public's opposition to the military-related Constitutional changes softened, although the majority still prefer the so-called "peace Constitution."

(U.S. Navy)

On December 7, 1941, Japan entered World War II as a result of its bombing of Pearl Harbor in Hawaii. This photograph, taken from an attacking Japanese plane, shows Pearl Harbor and a line of American battleships.

The Japanese government as well as private industry began to accelerate the drive for diversified markets and resources in the mid-1980s. This was partly in response to protectionist trends in countries in North America and Europe with which Japan had accumulated huge trade surpluses, but it was also due to changes in Japan's own internal social and economic conditions. Japan's recent resurgence of interest in its neighboring countries and the resulting rise of a bloc of rapidly developing countries we are calling the Pacific Rim can be explained by both external protectionism and internal changes. This time, however, Japanese influence is generally welcomed by nations in Asia because it provides them with substantial amounts of capital to invest in developing their own economies. Indeed, Japan has been the "spark plug" of economic growth in Southeast Asia in particular. Even China, which sees Japan as its number one competitor, has benefited greatly from Japanese investment.

■ DOMESTIC CHANGE

What internal conditions caused Japan's renewed interest in Asia and the Pacific? One change involves wage structure. For several decades, Japanese exports were less expensive than competitors' because Japanese workers were not paid as well as workers in North America and Europe. Today, however, the situation is reversed: Average manufacturing wages in Japan are higher than those paid to workers in most other developed nations, including the United States. Schoolteachers, college professors, and many white-collar workers are also better off in Japan. These wage differentials are the result of successful union activity and demographic changes (although the high cost of living reduces the actual domestic buying power of Japanese wages to about 70 percent that of the United States).

Whereas prewar Japanese families—especially those in the rural areas—were large, today's modern household typically consists of a couple and only one or two children. As Japan's low birth rate began to affect the supply of labor, companies were forced to entice workers with higher wages. The cost of land, homes, food—even Japanese-grown rice—is so much higher in Japan than in most of its neighbor countries that employees in Japan expect high wages.

Given conditions like these, many Japanese companies have found that they cannot be competitive in world markets unless they move their operations to countries like the Philippines, Singapore, or China, where an abundance of laborers keeps wage costs 75 to 95 percent lower than in Japan. Abundant, cheap labor (as well as a desire to avoid import tariffs) is also the reason why so many Japanese companies have been constructed in the economically depressed areas of the U.S. Midwest and South.

Another internal condition that is spurring Japanese interest in the Pacific Rim is a growing public concern for the domestic environment. Beginning in the 1970s, the Japanese courts handed down several landmark decisions in which Japanese companies were held liable for damages to people caused by chemical and other industrial wastes. Japanese industry, realizing that it no longer had a carte blanche to make profits at the expense of the environment, began moving some of its smokestack industries to new locations in developing-world countries, just as other industrialized nations had done. This has turned out to be a wise move economically (although obviously not environmentally for the host countries) for many companies, as it has put their operations closer to their raw materials. This, in combination with cheaper labor costs, has allowed them to remain globally competitive. It also has been a tremendous benefit to the host countries, although environmental groups in many Rim countries are also now becoming active, and industry in the future may be forced to effect actual improvements in their operations rather than just move polluting technologies to "safe" areas. In 2009, Japan's former prime minister, Yukio Hatoyama, announced an ambitious plan to reduce greenhouse gas emissions by 25 percent below their 1990 levels by the year 2020—a plan lauded around the world but resisted by Japan's manufacturing sector.

Attitudes toward work are also changing in Japan. Although the average Japanese worker still works about six hours more per week than the typical North American, the new generation of workers—those born and raised since World War II—are not so eager to sacrifice as much for their companies as were their parents. Recent policies have eliminated weekend work in virtually all government offices and many industries, and sports and other recreational activities are becoming increasingly popular. Given these conditions, Japanese corporate leaders are finding it more cost effective to move operations abroad to countries like South Korea, where labor legislation is weaker and long work hours remain the norm.

■ MYTH AND REALITY OF THE ECONOMIC MIRACLE

The Japanese economy, like any other economy, must respond to market as well as social and political changes to stay vibrant. It just so happened that for several decades, Japan's attempt to keep its economic boom alive created the conditions that, in turn, furthered the economies of all the countries in the Asia/Pacific region. That a regional "Yen Bloc" (so called because of the dominance of the Japanese currency, the yen) had been created was revealed in the late 1990s when sluggishness in the Japanese economy contributed to dramatic downturns in the economies of surrounding countries.

For many years, world business leaders were of the impression that whatever Japan did—whether targeting a certain market or reorienting its economy toward regional trade—turned to gold, as if the Japanese possessed some secret that no one else understood. But when other countries in Asia began to copy the Japanese model, their economies also improved—until, that is, lack of moderation and an inflexible application of the model produced a major correction in the late 1990s, which slowed economic growth throughout the entire region. Rather than possessing a magical management style or other great business secret, the Japanese achieved their phenomenal success thanks to hard work, advanced planning, persistence, and, initially, large doses of outside financial help.

However, even with those ingredients in place, Japanese enterprises often fall short. In many industries, for example, Japanese workers are less efficient than are workers in other countries. Japan's national railway system was once found to have 277,000 more employees on its payroll than it needed. At one point, investigators revealed that the system had been so poorly managed for so many years that it had accumulated a public debt of $257 billion. Multimillion-dollar train stations had been built in out-of-the-way towns, for no other reason than that a member of the *Diet* (the Japanese Parliament) happened to live there and had pork-barreled the project. Both government and industry have been plagued by bribery and corruption, as occurred in the Recruit Scandal of the late 1980s, which caused many implicated government leaders, including the prime minister, to resign.

Nor is the Japanese economy impervious to global market conditions. Values of stocks traded on the Tokyo Stock Exchange took a serious drop in 1992; investors lost millions of dollars, and many had to declare bankruptcy. Moreover, the tenacious 20-year recession that hit Japan in the early 1990s and that is likely to continue beyond 2010 has seriously damaged many corporations. Large companies such as Yamaichi Securities and Sogo (a department store chain) had to declare bankruptcy, as did many banks. Indeed, at one time, the majority of the top 10 banks in the world were Japanese, but by 2009, only 1 Japanese bank was on that prestigious list, whereas 3 Chinese banks were. In 2010, one of Japan's most famous and powerful companies, Toyota, was suffering from allegations that faulty construction of its cars had caused the deaths of many people. Clearly, Japan's economy has taken a beating. Still the rise of Japan from utter devastation in 1945 to the second-largest economy in the world (a position it held for some 30 years until China displaced it in 2010) has been nothing short of phenomenal. It will be helpful to review in detail some of the reasons for that success. We might call these the 10 commandments of Japan's economic success.

■ THE 10 COMMANDMENTS OF JAPAN'S ECONOMIC SUCCESS

1. Some of Japan's entrenched business conglomerates, called *zaibatsu,* were broken up by order of the U.S. Occupation commander after World War II; this allowed competing businesses to get a start. Similarly, the physical infrastructure—roads, factories—was destroyed during the war. This was a blessing in disguise, for it paved the way for newer equipment and technologies to be put in place quickly.

2. The United States, seeing the need for an economically strong Japan in order to offset the growing attraction of Communist ideology in Asia, provided substantial reconstruction aid. For instance, Sony Corporation got started with help from the Agency for International Development (AID)—an organization to which the United States is a major contributor. Mazda Motors got its start by making Jeeps for U.S. forces during the Korean War. (Other Rim countries that are now doing well can also thank U.S. generosity: Taiwan received $5.6 billion and South Korea received $13 billion in aid during the period 1945–1978.)

3. Japanese industry looked upon government as a facilitator and received useful economic advice as well as political and financial assistance from government planners. (In this regard, it is important to note that many of Japan's civil servants are the best graduates of Japan's colleges and universities—government service always having pride of place among career choices.) Also, the advice and help coming from the government were fairly consistent over time, because the same political party, the Liberal Democratic Party, remained in power for almost the entire postwar period.

4. Japanese businesses selected an export-oriented strategy that stressed building market share over immediate profit.

5. Except in certain professions, such as teaching, labor unions in Japan were not as powerful as in Europe and the United States. This is not to suggest that unions were not effective in gaining benefits for workers, but the structure of the union movement—individual company unions rather than industry-wide unions—moderated the demands for improved wages and benefits.

6. Company managers stressed employee teamwork and group spirit and implemented policies such as "lifetime employment" and quality-control circles, which contributed to group morale. In this they were aided by the

FURTHER INVESTIGATION

With Japan's economy in stagnation for the past 20 years, many other countries are fearful that their own economies could go the way of Japan, a process people are calling economic "Japanization." To learn more, see: www.ft.com/cms/s/0/c86470b2-ca7b-11e0-94d0-00144feabdc0.html

tendency of Japanese workers to grant to the company some of the same level of loyalty traditionally reserved for families. In certain ways, the gap between workers and management was minimized. (Because of changing internal and external conditions, many Japanese firms are now abandoning the lifetime employment model in favor of "employment by performance" models).

7. Companies benefited from the Japanese ethic of working hard and saving much. For most of Japan's postwar history, workers labored six days a week, arriving early and leaving late. The paychecks were carefully managed to include a substantial savings component—generally between 15 and 25 percent. This guaranteed that there were always enough cash reserves for banks to offer company expansion loans at low interest.

8. The government spent relatively little of its tax revenues on social-welfare programs or military defense, preferring instead to invest public funds in private industry.

9. A relatively stable family structure (i.e., few divorces and substantial family support for young people, many of whom remained at home until marriage at about age 27), produced employees who were reliable and psychologically stable.

10. The government as well as private individuals invested enormous amounts of money and energy into education, on the assumption that in a resource-poor country, the mental energies of the people would need to be exploited to their fullest.

Some of these conditions for success are now part of immutable history; but others are open to change as the conditions of Japanese life change. A relevant example is the practice of lifetime employment. Useful as a management tool when companies were small and *skilled* laborers were difficult to find, it is now giving way to a freer labor-market system. In some Japanese industries, as many as 30 percent of new hires quit after two years on the job. In other words, the aforementioned conditions for success were relevant to one particular era of Japanese and world history and may not be as effective in other countries or other times. Selecting the right strategy for the right era has perhaps been the single most important condition for Japanese economic success.

■ CULTURAL CHARACTERISTICS

All these conditions notwithstanding, Japan would never have achieved economic success without its people possessing certain social and psychological characteristics, many of which can be traced to the various religious/ethical philosophies that have suffused Japan's 2,000-year

(DigitalVision/Getty Images RF)

Daigo-ji Temple reflected in a pond, Kyoto

history. Shintoism, Buddhism, Confucianism, Christianity, and other philosophies of living have shaped the modern Japanese mind. This is not to suggest that Japanese are tradition-bound; nothing could be further from the truth. Even though many Westerners think "tradition" when they think Japan, it is more accurate to think of Japanese people as imitative, preventive, pragmatic, obligative, and inquisitive rather than traditional. These characteristics are discussed in this section.

Imitative

The capacity to imitate one's superiors is a strength of the Japanese people; rather than representing an inability to think creatively, it constitutes one reason for Japan's legendary success. It makes sense to the Japanese to copy success, whether it is a successful boss, a company in the West, or an educational curriculum in Europe. It is true that imitation can produce conformity; but, in Japan's case, it is often conformity based on respect for the superior qualities of someone or something rather than simple, blind mimicry.

Once Japanese people have mastered the skills of their superiors, they believe that they have the moral right to a style of their own. Misunderstandings on this point arise often when East meets West. One American schoolteacher, for example, was sent to Japan to teach Western art to elementary-school children. Considering her an expert, the children did their best to copy her work to the smallest detail. Misunderstanding that this was at once a compliment and the first step toward creativity, the teacher removed all of her art samples from the classroom in order to force the students to paint something from their own imaginations. Because the students found this to be a violation of their approach to creativity, they did not perform well, and the teacher left Japan believing that Japanese education teaches

conformity and compliance rather than creativity and spontaneity.

Japan's approach to imitation helps us predict the future role of Japan vis-à-vis the West. After decades of imitating the West, Japanese people are becoming more assertive; they now feel that they have the skills and the moral right to create styles of their own. We can expect to see, therefore, an explosion of Japanese creativity in the near future. Some observers have noted, for example, that the global fashion industry seems to be gaining more inspiration from designers in Tokyo than from those in Milan, Paris, or New York. And the Japanese have often registered more new patents with the U.S. Patent Office than any other nation except the United States. The Japanese are also now winning more Nobel Prizes than in the past, including the prize for literature in 1994, the prizes for chemistry in 2000, 2001, and 2002, and the prize for physics in 2002. Four Japanese scientists were awarded Nobel Prizes in 2008, and two were awarded the prize in 2010. Government officials have set a goal of having Japanese intellectuals and scientists receive 50 Nobel Prizes in the next 30 years, and it is quite possible that

? DID YOU KNOW?

It's possible to ascertain cultural differences between Japan and the West just by dropping by a Kentucky Fried Chicken store. Instead of dumping fried chicken into buckets the way it is done in America, for example, the Japanese arrange chicken in a single layer and then place the chicken in boxes with ribbed, plastic bottoms to minimize grease absorption. Japanese KFC store managers say that the product would never have been popular in Japan if the American way of serving it had been maintained.

they could achieve that goal, especially as they already have received 40 Nobel awards.

Preventive

Japanese individuals, families, companies, and the government generally prefer long-range over short-range planning, and they greatly prefer foreknowledge over postmortem analysis. Assembly-line workers test and retest every product to prevent customers from receiving defective products. Some store clerks plug in and check electronic devices in front of a customer in order to prevent bad merchandise from sullying the good reputation of the store, and commuter trains in Japan have three times as many "Watch your step" and similar notices as do trains in the United States and Europe. Insurance companies do a brisk business in Japan; even though all Japanese citizens are covered by the government's national health plan, many people buy additional coverage—for example, cancer insurance—just to be safe.

This concern with prevention trickles down to the smallest details. At train stations, multiple recorded warnings are given of an approaching train to commuters standing on the platform. Parent–teacher associations send teams of mothers around the neighborhood to determine which streets are the safest for the children. They then post signs designating certain roads as "school passage roads" and instruct children to take those routes even if it takes longer to walk to school. The Japanese think that it is better to avoid an accident than to have an emergency team ready when a child is hurt. Whereas Americans say, "If it ain't broke, don't fix it," the Japanese say, "Why wait 'til it breaks to fix it?"

Pragmatic

Rather than pursue a plan because it ideologically fits some preordained philosophy, the Japanese try to be pragmatic on most points. Take drugs as an example. Many nations say that drug abuse is an insurmountable problem that will, at best, be contained but probably never eradicated, because to do so would violate civil liberties. But, as a headline in the *Asahi Evening News* proclaimed a few years ago, "Japan Doesn't Have a Drug Problem and Means to Keep It That Way." Reliable statistics support this claim, but that is not the whole story. In 1954, Japan had a serious drug problem, with 53,000 drug arrests in one year. At the time, the authorities concluded that they had a big problem on their hands and must do whatever was required to solve it. The government passed a series of tough laws restricting the production, use, exchange, and possession of all manner of drugs, and it gave the police the power to arrest all violators. Users were arrested as well as dealers: It was reasoned that if the addicts were not left to buy the drugs, the dealers would be out of business. Their goal at the time was to arrest all addicts, even if it meant that certain liberties were briefly circumscribed. The plan, based on a do-what-it-takes pragmatism, worked; today, Japan is the only industrialized country without a widespread drug problem. In this case, to pragmatism was added the

Japanese tendency to work for the common rather than the individual good.

This approach to life is so much a part of the Japanese mind-set that many Japanese cannot understand why the United States and other industrialized nations have so many unresolved social and economic problems. For instance, when it comes to availability of money for loans to start up businesses or purchase homes, it is clear that one of the West's most serious problems is a low personal savings rate. For several decades, Americans saved only about 3 to 5 percent of their incomes, while the Japanese saved 11 to 15 percent. As a result of the prolonged recession in Japan and the resulting decline in incomes, the Japanese in recent years have been saving less and less (now less than 3 percent). But for many years, the Japanese were able to get inexpensive loans from banks because banks had plenty of people's saved earnings to loan out for business or other purposes.

Obligative

The Japanese have a great sense of duty toward those around them. Thousands of Japanese workers work late without pay to improve their job skills so that they will not let their fellow workers down. Good deeds done by one generation are remembered and repaid by the next, and lifelong friendships are maintained by exchanging appropriate gifts and letters. North Americans and Europeans are often considered untrustworthy friends because they do not keep up the level of close, personal communications that the Japanese expect of their own friends; nor do the Westerners have as strong a sense of place, station, or position.

Duty to the group is closely linked to respect for superior authority. Every group—indeed, every relationship—is seen as a mixture of people with inferior and superior resources. These differences must be acknowledged, and in the ideal situation, no one is disparaged for bringing less to the group than someone else. However, equality is assumed when it comes to basic commitment to or effort expended for a task. Slackers are not welcome. Obligation to the group along with respect for superiors motivated Japanese pilots to fly suicide missions during World War II, and it now causes workers to go the extra mile for the good of the company.

That said, it is also true that changes in the intensity of commitment are becoming increasingly apparent. More Japanese than ever before are beginning to feel that their own personal goals are more important than those of their companies or extended families. This is no doubt a result of the Westernization of the culture since the Meiji Restoration era of the late 1800s, and especially of the experiences of the growing number of Japanese—approximately half a million in a given year—who live abroad and then take their newly acquired values back to Japan. (About half of these "away Japanese" live in North America and Western Europe.)

There is no doubt that the pace of "individualization" of the Japanese psyche is increasing and that, more and more, the Japanese attitude toward work is approaching that of the West. Many Japanese companies are now allowing employees to set their own "flex-time" work schedules, and some companies have even asked employees to stop addressing superiors with their hierarchical titles and instead refer to everyone as *san,* or Mr. or Ms.

The lengthening of the time after college and before marriage for young people has caused the growth of a new phenomenon: "parasite singles." These are singles who work full-time and continue to live at home but who contribute little to the family. They spend what they earn at work on personal items such as expensive clothing and electronic gadgets, but do not help with their parents' mortgages or other expenses. This period of almost complete personal indulgence is not welcomed by the older generation and is evidence that the traditional sense of familial duty is, while not gone, definitely on the wane in modern Japan. It may soon be the case that we shall have to add another category to this list of cultural characteristics, namely, "hedonistic."

Inquisitive

The image of dozens of Japanese business-people struggling to read a book or newspaper or intently listening to their iPods while standing inside a packed commuter train is one not easily forgotten, symbolizing as it does the intense desire among the Japanese for knowledge, especially knowledge of foreign cultures. Nearly 6 million Japanese travel abroad each year (many to pursue higher education), and for those who do not, the government and private radio and television stations provide a continuous stream of programming about everything from Caribbean cuisine to French ballet. The Japanese have a yen for foreign styles of dress, foreign cooking, and foreign languages. The Japanese study languages with great intensity. Every student is required to study English; many also study Chinese, Greek, Latin, Russian, Arabic, and other languages, with French being the most popular after English—although it is evident that, on the whole, the Japanese do not have a gift for languages and struggle in vain to gain fluency.

Observers inside and outside of Japan are beginning to comment that the Japanese are recklessly discarding Japanese culture in favor of foreign ideas and habits, even when they make no sense in the Japanese context. A tremendous intellectual debate, called *Nihonjinron,* is now taking place in Japan over the meaning of being Japanese and the Japanese role in the world. There is certainly value in these concerns, but, as was noted previously, the secret about Japanese traditions is that they are not traditional. That is, the Japanese seem to know that, in order to succeed, they must learn what they need to know for the era in which they live, even if it means modifying or eliminating the past. This is probably the reason why the Japanese nation has endured for more than 2,000 years while many other empires have fallen. In this sense, the Japanese are very forward-looking people and, in their thirst for new modes of thinking and acting, they are, perhaps, revealing their most basic and useful national personality

characteristic: inquisitiveness. Given this attitude toward learning, it should come as no surprise that formal schooling in Japan is a very serious business to the government and to families. It is to that topic that we now turn.

■ SCHOOLING

Probably most of the things that the West has heard about Japanese schools are distortions or outright falsehoods. We hear that Japanese children are highly disciplined, for example; yet in reality, Japanese schools at the elementary and junior high levels are rather noisy, unstructured places, with children racing around the halls during breaks and getting into fights with classmates on the way home. Japan actually has a far lower percentage of its college-age population enrolled in higher education than is the case in the United States—47 percent as compared to 66 percent. Moreover, the Japanese government does not require young people to attend high school (they must attend only until age 15), although 94 percent do anyway. Given these and other realities of school life in Japan, how can we explain the consistently high scores of Japanese on international tests and the general agreement that Japanese high school graduates know almost as much as college graduates in North America?

Structurally, schools in Japan are similar to those in many other countries: There are kindergartens, elementary schools, junior high schools, and high schools. Passage into elementary and junior high is automatic, regardless of student performance level. But admission to high school and college is based on test scores from entrance examinations. Preparing for these examinations occupies the full attention of students in their final year of both junior high and high school, respectively. Both parents and school authorities insist that studying for the tests be the primary focus of a student's life at those times. For instance, members of a junior high soccer team may be allowed to play on the team only for their first two years; during their last year, they are expected to be studying for their high school entrance examinations. School policy reminds students that they are in school to learn and to graduate to the next level, not to play sports. Many students even attend after-hours "cram schools" (*juku*) several nights a week to prepare for the exams.

Time for recreational and other nonschool activities is restricted, because Japanese students attend school more days per year than students in North America. When mandatory school attendance on Saturdays was eliminated in 2002 as part of a major school reform program, many parents protested vigorously. They worried that their children would fall behind. School leaders agreed to start voluntary Saturday schools, and when they did, as many as 75 percent of their students attended. Summer vacation is only about six weeks long, and students often attend school activities during most of that period. Japanese youths are expected to treat schooling as their top priority over part-time jobs (usually prohibited by school policy during the school year, except for the needy), sports, dating, and even family time.

Children who do well in school are generally thought to be fulfilling their obligations to the family, even if they do not keep their rooms clean or help with the dishes. The reason for this focus is that parents realize that only through education can Japanese youths find their place in society. Joining Japan's relatively small military is generally not an option, opportunities for farming are limited because of land scarcity, and most major companies will not hire a new employee who has not graduated from college or a respectable high school. Thus, the Japanese find it important to focus on education—to do one thing and do it well.

Teachers are held in high regard in Japan, partly because when mass education was introduced, many of the high-status samurai took up teaching to replace their martial activities. In addition, in modern times, the Japan Teacher's Union has been active in agitating for higher pay for teachers. As a group, teachers are the highest-paid civil servants in Japan. They take their jobs very seriously. Public-school teachers, for example, visit the home of each student each year to merge the authority of the home with that of the school, and they insist that parents (usually mothers) play active supporting roles in the school.

Some Japanese youths dislike the system, and discussions are currently under way among Japanese educators on how to improve the quality of life for students. Occasionally the pressure of taking examinations (called "exam hell") produces such stress that a desperate student will commit suicide rather than try and fail. Stress also appears to be the cause of *ijime,* or bullying of weaker students by stronger peers. In recent years, the Ministry of Education has worked hard to help students deal with school stress, with the result that Japan's youth suicide rate has dropped dramatically, far lower than the youth rate in the United States (although suicide among adults varies with the ups and downs of the economy). Despite these and other problems, most Japanese youths enjoy school and value the time they have to be with their friends, whether in class, walking home, or attending cram school. Some of those who fail their college entrance exams continue to study privately, some for many years, and take the exam each year until they pass. Others travel abroad and enroll in foreign universities that do not have such rigid entrance requirements. Still others enroll in vocational training schools. But everyone in Japan realizes that education—not money, name, or luck—is the key to success.

Parents whose children are admitted to the prestigious national universities—such as Tokyo and Kyoto Universities—consider that they have much to brag about. Other parents are willing to pay as much as US$45,000 on average for four years of college at the private (but usually not as prestigious) universities. Once admitted, students find that life slows down a bit. For one thing, parents typically pay more than 65 percent of the costs, and approximately 3 percent is covered by scholarships. This leaves only about 30 percent to be earned by the students; this usually comes from tutoring high school students who are studying for the entrance exams. Contemporary parents are also willing to pay the cost of a son's or daughter's traveling to and spending a few months in North America or Europe either before college begins or during summer breaks—a practice that is becoming de rigueur for Japanese students, much as taking a "grand tour" of Europe was once expected of young, upper-class Americans and Canadians.

College students may take 15 or 16 courses at a time, but classes usually meet only once or twice a week, and sporadic attendance is the norm. Straight lecturing rather than class discussion is the typical learning format, and there is very little homework beyond studying for the final exam. Students generally do not challenge the professors' statements in class, but some students develop rather close, avuncular-type relationships with their professors outside of class. Hobbies, sports, and club activities (things the students did not have time to do while in public school) occupy the center of life for many college students. Japanese professors visiting universities in North America and Europe are often surprised at how diligently students in those places study during their college years. By contrast, Japanese students spend a lot of time making friendships that will last a lifetime and be useful in one's career and private life.

■ THE JAPANESE BUSINESS WORLD

Successful college graduates begin their work careers in April, when most large companies do their hiring (although this practice is slowly giving way to individual hiring throughout the year). They may have to take an examination to determine how much they know about electronics or stocks and bonds, and they may have to complete a detailed personality profile. Finally, they will have to submit to a very serious interview with company management. During interviews, the managers will watch their every move; the applicants will be careful to avoid saying anything that will give them "minus points."

Once hired, individuals attend training sessions in which they learn the company song and other rituals as well as company policy on numerous matters. They may be housed in company apartments (or may continue to live at home), permitted to use a company car or van, and advised to shop at company grocery stores. Almost never are employees married at this time, and so they are expected to live a rather spartan life for the first few years.

Employees are expected to show considerable deference to their section bosses, even though, on the surface, bosses do not appear to be very different from other employees. Bosses' desks are out in the open, near the employees; they wear the same uniform; they socialize with the employees after work; even in a factory, they are often on the shop floor rather than sequestered away in private offices. Long-term employees often come to see the section leader as an uncle figure (bosses are usually male) who will give them advice about life, be the best man at their weddings, and provide informal marital and family counseling as needed.

Although there are cases of abuse or unfair treatment of employees, Japanese company life can generally be described as somewhat like a large family rather than a military squad; employees (sometimes called associates) often obey their superiors out of genuine respect rather than forced compliance. Moreover, competition between workers is reduced because everyone hired at the same time receives more or less the same pay and most workers receive promotions at about the same time. Only later in one's career are individualistic promotions given. That said, it is important to note that, under pressure to develop a more inventive workforce to compete against new ideas and products from the West, many Japanese companies are now experimenting with new pay and promotion systems that reward performance, not longevity.

Employees are expected to work hard, for not only are Japanese companies in competition with foreign businesses, but they also must survive the fiercely competitive business climate at home. Indeed, the Japanese skill in international business was developed at home. There are, for example, hundreds of electronics companies and thousands of textile enterprises competing for customers in Japan. And whereas the United States has only four automobile-manufacturing companies, Japan has nine. All these companies entice customers with deep price cuts or unusual services, hoping to edge out unprepared or weak competitors. Many companies fail. There were once, for instance, almost 40 companies in Japan that manufactured calculators, but today only half a dozen remain, the rest victims of tough internal Japanese competition.

At about age 30, after several years of working and saving money for an apartment, a car, and a honeymoon (wedding and reception costs of approximately US$27,000 on average are shared by the bride and groom's parents), the typical Japanese male worker marries (although there is an increasing number of Japanese males and females who opt to never marry). The average bride, about age 27, may have taken classes in college typical to academic institutions, or she may have taken private lessons in flower arranging, the tea ceremony, sewing, cooking, and perhaps a musical instrument like the *koto,* the Japanese harp. She probably will not have graduated from college, although more and more Japanese women are obtaining college degrees. If she is working, she likely is paid much less than her husband, even if she has an identical position (despite equal-pay laws enacted in 1986). Although an increasing number of female office workers are given meaningful assignments, many still spend their workday in the company preparing and serving tea for clients and employees, dusting the office, running errands, and answering telephones. When she has a baby, she will be expected to quit—although more women today are choosing to remain on the job, and some are advancing into management or are leaving to start their own companies.

Because the wife is expected to serve as the primary caregiver for the children, the husband is expected always to make his time available for the company. He may be asked to work weekends, to stay out late most of the week (about four out of seven nights), or even to be transferred to another branch in Japan or abroad without his family. This loyalty is rewarded in numerous ways: Unless the company goes bankrupt or the employee is unusually inept, he may be permitted to work for the company until he retires, usually at about age 55 or 60, even if the company no longer really needs his services; he and his wife will be taken on company sightseeing trips; the company will pay most of his health-insurance costs (the government pays the rest); and he will have the peace of mind that comes from being surrounded by lifelong friends and workmates. His association with company employees will be his main social outlet, even after retirement; upon his death, it will be his former workmates who organize and direct his funeral services. There are many younger employees who fear the strictures of the traditional company and prefer to work in small and medium-sized businesses with a more individualistic work environment. Polls show that workers entering companies today are less willing to accept long hours, more likely to change jobs, and more determined to make a separation between their private lives and their lives on the job.

■ THE FAMILY

The loyalty once given to the traditional Japanese extended family, called the *ie,* has been transferred to the modern company. This is logical from a historical perspective, since the modern company once began as a family business and was gradually expanded to include more workers, or "siblings." Thus, whereas the family is seen as the backbone of most societies, it might be more accurate to argue that the *kaisha,* or company, is the basis of modern Japanese society. As one Japanese commentator explained, "In the West, the home is the cornerstone of people's lives. In Tokyo, home is just a place to sleep at night. . . . Each family member—husband, wife, and children—has his own community centered outside the home."

Thus, the common image that Westerners hold of the centrality of the family to Japanese culture may be inaccurate. For instance, father absence is epidemic in Japan. It is an unusual father who eats more than one meal a day with his family. He may go shopping or to a park with his family when he has free time from work, but he is more likely to go golfing with a workmate. Schooling occupies the bulk of the children's time, even on weekends. And with far fewer children (Japan has one of the lowest fertility rates in the world, with an average of 1.37 children per woman) than in earlier generations and with appliance-equipped apartments, many Japanese women rejoin the workforce after their children are self-maintaining.

Japan's divorce rate, while rising for awhile, is considerably lower than in other industrialized nations. It is now 1.9 divorces per thousand in Japan compared to 4.0 per thousand in the United States), a fact that may seem incongruent with the conditions just described. Yet, as explained by one Japanese sociologist, Japanese couples "do not expect much emotional closeness; there is less pressure on us to meet each other's emotional needs. If we become close, that is a nice dividend, but if we do not, it is not a problem because we did not expect it in the first place."

Despite these modifications to the common Western image of the Japanese family, Japanese families have significant roles to play in society. Support for education is one of the most important. Families, especially mothers, support the schools by being actively involved in the parent–teacher association, by insisting that children be given plenty of homework, and by saving for college so that the money for tuition is available without the college student having to work.

Another important function of the family is mate selection. Somewhat fewer than half of current Japanese marriages are arranged by the family or have occurred as a result of far more family involvement than in North America. Families sometimes ask a go-between (an uncle, a boss, or another trusted person) to compile a list of marriageable candidates. Criteria such as social class, blood type, and occupation are considered. Photos of prospective candidates are presented to the unmarried son or daughter, who has the option to veto any of them or to date those he or she finds acceptable. Young people, however, increasingly select their mates with little or no input from parents.

Finally, families in Japan, even those in which the children are married and living away from home, continue to gather for the purpose of honoring the memory of deceased family members or to enjoy one another's company for New Year's Day, Children's Day, and other celebrations.

■ WOMEN IN JAPAN

Ancient Confucian values held that women were legally and socially inferior to men. This produced a culture in feudal Japan in which the woman was expected to walk several steps behind her husband when in public, to eat meals only after the husband had eaten, to forgo formal education, and to serve the husband and male members of the family whenever possible. A "good woman" was said to be one who would endure these conditions without complaint. This pronounced gender difference (though minimized substantially over the centuries since Confucius) can still be seen today in myriad ways, including in the preponderance of males in positions of leadership in business and politics, in the smaller percentage of women college graduates, and in the pay differential between women and men.

Given the Confucian values just noted, one would expect that all top leaders would be males. However, women's roles are also subject to the complex interplay of both ancient and modern cultures. Between A.D. 592 and 770, for instance, of the 12 reigning emperors, half were women. The debate on whether or not to allow a woman to become empress became quite heated recently because the crown prince's first and only child was a girl. When another royal family member gave birth to a male baby,

(© Imageworks/Getty Images RF)

In Japan, not unlike in many other parts of the world, economic well-being often requires two incomes. Still, there is strong social pressure on women to stop working once they have a baby.

the proposal to alter the system was almost immediately dropped. Change in the imperial system will come slowly if at all. But in rural areas today, women take an active decision-making role in farm associations. In the urban workplace, some women occupy typically pink-collar positions (nurses, clerks, and so on), but many women are also doctors and business executives; over 40 percent of Japan's workforce is comprised of women, but only about 14 percent hold positions of leadership, but that figure includes over 30,000 female company presidents—a major change from past years.

Thus, it is clear that within the general framework of gender inequality imposed by Confucian values, Japanese culture, especially at certain times, has been rather lenient in its application of those values. The declining population in Japan is forcing changes in the traditional preference for women to quit working once they marry or have a baby. Some companies are now encouraging mothers to stay on the job, and women are responding in kind. Nearly 60 percent of the female workforce is married. An equal-pay law was enacted in 1986 that made it illegal to pay women less for doing comparable work

? DID YOU KNOW?

Japanese soccer players beat the Americans to win the women's World Cup in 2011 with a score of 3–1. Certainly, the Americans wanted to win, but the comments most often heard after the game—by Americans—was that it had been nice for Japan to win and have something to cheer about after the devastating earthquake and tsunami four months before.

(although it may take years for companies to comply fully). And the Ministry of Education has mandated that home economics and shop classes now be required for both boys and girls; that is, both girls and boys will learn to cook and sew as well as construct things out of wood and metal.

In certain respects, and contrary to the West's image of Japanese gender roles, some Japanese women seem more assertive than women in the West. For example, in a recent national election, a wife challenged her husband for his seat in the House of Representatives (something that has not been done in the United States, where male candidates usually expect their wives to stump for them). Significantly, too, the former head of the Japan Socialist Party was an unmarried woman, and in 1999 Osaka voters elected a woman as mayor for the first time. Women have been elected to the powerful Tokyo Metropolitan Council and awarded professorships at prestigious universities such as Tokyo University. And, while women continue to be used as sexual objects in pornography and prostitution, certain kinds of misogynistic behavior, such as rape and serial killing, are less frequent in Japan than in Western societies. New laws against child pornography may reduce the abuse of young girls, but a spate of bizarre killings by youths and mafia in recent years is causing the sense of personal safety in Japan to dissipate. Signs in train stations warn of pickpockets, infrequently traveled paths sometimes have signs warning of molesters, and, to avoid the growing problem of women being groped in packed trains, some commuter trains now have cars reserved exclusively for females.

Recent studies show that many Japanese women believe that their lives are easier than those of most Westerners. With their husbands working long hours and their one or two children in school all day, Japanese women find they have more leisure time than Western women. Gender-based social divisions remain apparent throughout Japanese culture, but modern Japanese women have learned to blend these divisions with the realities and opportunities of the contemporary workplace and home.

■ RELIGION/ETHICS

There are many holidays in Japan, most of which have a religious origin. This fact, as well as the existence of numerous shrines and temples, may leave the impression that Japan is a rather religious country. This is not true, however. Most Japanese people do not claim any active religious affiliation, but many will stop by a shrine occasionally to ask for divine help in passing an exam, finding a mate, or recovering from an illness. When the economy is in trouble, the number of people going to Shinto and Buddhist temples to pray goes up.

Nevertheless, modern Japanese culture sprang from a rich religious heritage. The first influence on Japanese culture came from the animistic Shinto religion, from whence modern Japanese acquired their respect for the beauty of nature. Confucianism brought a respect for hierarchy and education. Taoism stressed introspection, and Buddhism taught the need for good behavior now in order to acquire a better life in the future.

Shinto was selected in the 1930s as the state religion and was used as a divine justification for Japan's military exploits of that era, but most Japanese today will say that Japan is, culturally, a Buddhist nation. Some new Buddhist denominations have attracted thousands of followers. The rudiments of Christianity are also a part of the modern Japanese consciousness, but few Japanese have actually joined Christian churches. Sociologically, Japan, with its social divisions and hierarchy, is probably more of a Confucian society than it is Buddhist or any other philosophy, although Confucianism is so deeply woven into the fabric of Japanese life that few would recognize it as a distinct philosophy.

Most Japanese regard morality as springing from within the group rather than pronounced from above. That is, a Japanese person may refrain from stealing so as not to offend the owner of an object or bring shame upon the family, rather than because of a divine prohibition against stealing. Thus we find in Japan a relatively small rate of violent—that is, public—crimes, and a much larger rate of white-collar crimes such as embezzlement, in which offenders believe that they can get away with something without creating a public scandal for their families.

■ THE GOVERNMENT

The Constitution of postwar Japan became effective in 1947 and firmly established the Japanese people as the ultimate source of sovereignty, with the emperor as the symbol of the nation. The national Parliament, or *Diet*, is empowered to pass legislation. The Diet is divided into two houses: the House of Representatives, with 480 members elected for four-year terms; and the House of Councillors, with 252 members elected for six-year terms from each of the 47 prefectures (states) of Japan as well as nationally. The prime minister, assisted by a cabinet, is also the leader of the party with the most seats in the Diet. Prefectures are governed by an elected governor and an assembly, and cities and towns are governed by elected mayors and town councils. The Supreme Court, consisting of a chief judge and 14 other judges, is independent of the legislative branch of government.

Japan's Constitution forbids Japan from engaging in war or from having military capability that would allow it to attack another country. Japan does maintain a well-equipped self-defense force, but it relies on a security treaty with the United States in case of serious aggression against it. In recent years, the United States has been encouraging Japan to assume more of the burden of the military security of the Asian region. In response (and as an outcome of a slow increase in Japanese nationalism within the general population), Japan has taken significant steps toward building a fully equipped, offensive-capable military. The first step was to propose the alteration of Article 9, or the so-called "peace clause" of the Constitution (along with other parts of the U.S.-written document). The revision process is now underway, and if completed as some people hope, the prohibition against a fully developed military will be eliminated. But the Democratic Party of Japan, which came to power in 2009, has not been in favor of altering the Constitution, so the process's future is in doubt. Other steps taken to date include the upgrading in 2007 of the Self-Defense Force to a Ministry of Defense with cabinet-level status, and the sending of troops into noncombat positions overseas. Some 1,000 troops provided rear-guard support in Iraq and Afghanistan during those U.S.-led wars, and nearly 2,000 troops were sent to Cambodia during UN-supervised peacekeeping operations there (Japan has 240,000 troops total). In 2009, the Japanese navy was sent to protect Japanese ships from pirates in international waters. As early as the 1990s, Japan had started building its own version of the F-16 fighter jet.

Generally speaking, the Japanese public is not pleased with the efforts of its government to move away from the "peace clause" mentality that has been in force since the end of World War II. They remember the devastation of the atomic bombs dropped on Hiroshima and Nagasaki and have become firmly committed to a pacifist foreign policy. Protests have marked each acceleration of Japan's military status. But, in recent years, developments in nearby countries have caused a softening of the Japanese public's resistance.

In 1998, North Korea fired a test missile across Japan's sovereign airspace, and then it detonated a small test nuclear bomb in 2006. At the same time, the Japanese public has become increasingly aware of the growing military might of China, a country that has now surpassed Japan in the size of its military budget to have the second largest military budget in the world and that is acquiring aircraft carriers and other military hardware. In 2007, China exploded a satellite in space with a missile, further exacerbating the general public's feeling that the United States may not be able to fully defend Japan in the event of war. Thus, with its neighbors becoming more aggressive militarily, the Japanese are coming to feel that it is time for Japan to become a "normal nation" and possess a fully developed military. Furthermore, they are tired of being criticized as a nation for sending only financial support instead of actual military support when aggressor nations are being restrained. Such was the case in 1990 when a coalition of countries drove Saddam Hussein from Kuwait but no Japanese soldiers participated (although Japan contributed $9 billion to the effort). The

tenure of Japanese prime ministers is extremely short. In the six years from 2006 to 2011, there have been five prime ministers, a tenure of less than one year per office holder. Scandal brought down some of them as did public dissatisfaction and other issues. Former Prime Minister Hatoyama stepped down after he switched sides on the Okinawa bases issue. At first he opposed some of the U.S. plans, but then he changed his mind, much to the consternation of the public who voted for him based on his original position. The current Prime Minister, Yoshihiko Noda of the Democratic Party of Japan took up his post in late 2011. But whoever the person is, he (there has, as yet, been no females) will be confronted with many of the same nagging issues: the sluggish economy, the disagreement with Russia over islands north of Hokkaido, fishing-rights and ownership disputes with South Korea over Takeshima (Dokdo) Island, the reconstruction of the Tohoku region after the devastating earthquake and tsunami, and disputes with China and other nations over the Spratly islands.

The Japanese have formed numerous political parties to represent their views in government. Among these have been the Japan Communist Party, the Social Democratic Party, and the New Frontier Party. For nearly 40 years, however, the most powerful party was the Liberal Democratic Party (LDP). Formed in 1955, it guided Japan to its current position in the top category of developed nations, but a series of sex and bribery scandals caused it to lose control of the government in 1993. A shaky coalition of eight parties took control for about a year but was replaced by an even more unlikely coalition of the LDP and the Japan Socialists—historic enemies who were unable to agree on most policies. Eventually, the LDP was able to regain some of its lost political clout; but with some half a dozen changes in the prime ministership in the 1990s, increasing frustration over the sluggish economy, and embarrassing episodes such as a brawl on the floor of the Diet in which members of the upper house physically assaulted each other, the LDP was finally brought down in 2009 by the Democratic Party of Japan (DPJ), a party considerably more left of center than the LDP. DPJ policy is to slim down the size of government by eliminating as many as 80 seats in the Diet, to give more power to local governments, and to provide monetary "child allowances" for all children until they finish junior high school.

Part of the reason for Japan's political instability can be explained by Japan's party faction system. Party politics in Japan has always been a mixture of Western-style democratic practice and feudalistic personal relationships. Japanese parties are really several parties rolled into one. That is, parties are divided into several factions, each comprised of a group of loyal younger members headed by a powerful member of the Diet. The senior member has a duty to pave the way for the younger members politically, but they, in turn, are obligated to support the senior member in votes and in other ways. The faction leader's role in gathering financial support for faction members is particularly important, because Diet

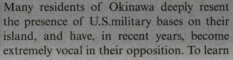

FURTHER INVESTIGATION

Many residents of Okinawa deeply resent the presence of U.S. military bases on their island, and have, in recent years, become extremely vocal in their opposition. To learn more about this long-standing controversy, see www.globalsecurity.org/military/facility/okinawa.htm

members are expected by the electorate to be patrons of numerous causes, from charity drives to the opening of a constituent's fast-food business. Because parliamentary salaries are inadequate to the task, outside funds, and thus the faction, are crucial. The size and power of the various factions are often the critical elements in deciding who will assume the office of prime minister and who will occupy which cabinet seats. The role of these intraparty factions is so central to Japanese politics that attempts to ban them have never been successful.

The factional nature of Japanese party politics means that cabinet and other political positions are frequently rotated. This would yield considerable instability in governance were it not for the stabilizing influence of the Japanese bureaucracy. Large and powerful, the career bureaucracy is responsible for drafting more than 80 percent of the bills submitted to the Diet. Many of the bureaucrats are graduates from the finest universities in Japan, particularly Tokyo University, which provides some 80 percent of the senior officials in the more than 20 national ministries. Many of them consider their role in long-range forecasting, drafting legislation, and implementing policies to be superior to that of the elected officials under whom they work. They reason that, whereas the politicians are bound to the whims of the people they represent, bureaucrats are committed to the nation of Japan—to, as it were, the *idea* of Japan. So superior did bureaucrats feel to their elected officials (and bosses) that until recently, elected officials were not questioned directly in the Diet; career bureaucrats represented their bosses to the people. Generally speaking, government service is considered a higher calling than are careers in private business, law, and other fields. However, in recent years, a number of scandals involving both corrupt and incompetent bureaucrats in the Ministry of Finance and other ministries have tarnished the image of this once-unassailable category of professionals.

In addition to the bureaucracy, Japanese politicians have leaned heavily on big business to support their policies of postwar reconstruction, economic growth, and social reform. Business has accepted heavy taxation so that social-welfare programs such as the national health plan are feasible, and they have provided political candidates with substantial financial help. In turn, the government has seen its role as that of facilitating the growth of private industry (some critics claim that the relationship between government and business is so close that Japan is best described not as a nation but as "Japan, Inc."). Consider, for example, the powerful Ministry of Economy, Trade, and Industry (METI, formerly MITI). Over the years, it has worked closely with business, particularly the Federation of Economic Organizations (Keidanren) to forecast potential market shifts, develop strategies for market control, and generally pave the way for Japanese businesses to succeed in the international marketplace. The close working relationship between big business and the national government is an established fact of life in Japan, and, despite criticism from countries with a more

laissez-faire approach to business, it will undoubtedly continue into the future, because it has served Japan well.

The prolonged recession in Japan uncovered inconsistencies in the structure of Japanese society. For instance, we have seen that the Japanese are an inquisitive, change-oriented people, yet the administrative structure of the government and of the business community seems impervious to change. Decisions take an agonizingly long time to make, and outsiders as well as the Japanese public find the government's mode of operation to be far from open and transparent. For instance, in late 1999, one of Japan's 51 nuclear reactors leaked radiation and endangered the lives of many people. At first the government decided not to announce the leak at all. Then, when it finally did, it claimed that only a few people had been affected. In fact, more than 400 people had been affected, and the government seemed incapable of telling the whole truth about the matter. Many blamed bureaucratic rigidity. Similarly, in 1999–2000, the government injected more than US$62 billion into the economy (the ninth such stimulus package since the recession started in 1989), yet the economy remained sluggish. Again, many blamed the snail-paced decision-making process as a factor that made it difficult for companies to quickly take advantage of new opportunities in the marketplace. Whether Japan can make the structural changes necessary to remain competitive in the new global economy remains to be seen.

There are, of course, many other factors besides Japan's decision-making process that slow or disrupt economic growth. For instance, like other Pacific Rim countries, Japan is regularly afflicted by serious natural disasters. Japan is located in such a seismically active area that the country experiences some 300 small earthquakes (many of them unfelt by the average person) per day. Tsunami warnings were posted in early 2005 when an offshore earthquake raised wave action along Japan's lengthy coastline. Gale-force winds occasionally hit eastern Japan, causing blackouts and stranding ships. Fourteen people were injured in a 7.3-magnitude earthquake in western Japan in 2004, and the strongest typhoon in several decades hit the island of Okinawa. In the same year, tropical storm Meari caused the deaths of 22 people and triggered landslides that drove 10,000 from their homes. Another storm killed 77 people. Also in 2004, one of Japan's most active volcanoes, Mount Asama, erupted, spewing molten rocks and smoke into the air and covering Tokyo—the world's largest city with over 35 million inhabitants—for several days with volcanic ash. Several powerful earthquakes also killed 39 people in northern Japan, derailing a high-speed train and destroying many homes and roads. Like the tsunami that hit South Asia in late 2004, a large tsunami hit southern Japan in 1993 killing hundreds and obliterating entire towns, and in 1995, a large earthquake hit the heavily populated Kobe area killing more than 6,000 people. Of course, no natural disaster has been more devastating than the earthquake and tsunami of 2011, which, as we have seen, took the lives of 23,000 people and obliterated scores of towns along the northeastern coast. The only positive benefit anyone can see from this disaster is the slight uptick in the economy that may come from the government's efforts to rebuild the area.

Sometimes, Japan's disasters are man-made. Mercury poisoning in the 1980s, for example, killed some 1,700 people and caused mothers to give birth to severely deformed babies. The pollutant, dumped by an irresponsible Japanese company, affected drinking water for an entire town. It took two decades to resolve the matter in the Japanese courts. Accidents at some of Japan's many nuclear power plants have also killed workers and forced many residents from their homes. So angry are the citizens over incidents like this that it is now virtually impossible for new power plants to be approved by the government (some 11 new plants have been proposed), yet the old plants are aging and will likely produce even more serious problems. Finally, despite Japan's strong advocacy of the Kyoto Protocol on global warming, experts warn that Japan, itself, is not likely to meet greenhouse emission targets, as it continues to send tons of pollutants into the air each day. Japan is the fifth largest greenhouse gas polluter (the United States and China emit 40 percent of the pollutants) in the world, but the Democratic Party of Japan aims to change that by setting strict emission standards.

■ THE FUTURE

In the postwar years of political stability, the Japanese have accomplished more than anyone, including themselves, thought possible. Japan's literacy rate is 99 percent, almost 90 percent of all households own their own automobiles, and in many other ways Japan remains firmly in the top category of advanced nations. Nationalized health care covers every Japanese citizen, and the Japanese have nearly the longest life expectancy in the world. In other ways too, Japan has dramatically improved life for its people. For instance, the average size of Japanese homes, once criticized by Westerners as too small to live in, has been increasing every year until it is now just about the same as the average size of European homes. Home ownership also compares favorably or even exceeds other industrialized nations: 60 percent of homes are owned in Japan compared to 66 percent in the United States, 53 percent in France and 38 percent in Germany. Japan's success has been noted and emulated by people all over the world. For instance, as of 2010, over 140,000 foreign students (60 percent of them Chinese) were studying in Japan—the highest number in 30 years. In 2010, 8.6 million tourists from almost 200 countries visited Japan—also a record number. While interest in China has certainly grown in recent years, it would be inaccurate to say that people around the world are no longer interested in Japan.

With only half the population of the United States, a land area about the size of Great Britain, and extremely limited natural resources (it has to import 99.6 percent of its oil, 99.8 percent of its iron, 86.7 percent of its coal, and about 60 percent of its food), Japan has nevertheless created one of the largest economies in the world. But there are problems on the horizon. There is great concern over the amount of public debt, some 225 percent of GDP in 2009—the highest in the world.

A less immediate but potentially more serious problem is the decline in population. Unlike some countries in the Pacific Rim, Japan's population is actually going down each year. Around 127 million today, it will drop to 109 million by 2050, and by 2025, nearly 30 percent of all Japanese will be over the age of 65. While many people think that fewer people in Japan's crowded cities would be a relief, the reality is that Japan is going to have a very difficult time keeping its economy going without new entrants into the labor force. Will the nation reverse its long-standing aversion to in-migration? Will there be a sudden change of heart by couples to return Japan to "replacement level" population stability? This is an important issue for Japan to cope with in the near future.

Japan also has to re-think its raison d'être. When the Spanish were establishing hegemony over large parts of the globe, they were driven in part by the desire to bring Christianity to the "heathen." The British, for their part, believed that they were taking "civilization" to the "savages" of the world. China and the former Soviet Union were once strongly committed to the ideals of communism, while the United States has felt that its mission is that of expanding democracy and capitalism.

What about Japan? For what reason do Japanese businesses buy up hotels in New Zealand and skyscrapers in New York? What role does Japan have to play in the world in addition to spawning economic development? What values will guide and perhaps temper Japan's political and military decision in the future?

These are questions that the Japanese people themselves are attempting to answer; but, finding no ready answers, they are beginning to encounter more and more difficulties with the world around them and within their own society. Animosity over the persistent trade imbalance in Japan's favor continues to simmer in Europe and North America as well as in some countries of the Pacific Rim. To deflect these criticisms, Japan has substantially increased its gift-giving to foreign governments, including allocating money for the stabilization or growth of democracy in Central/Eastern Europe and for easing the foreign debt burden of Mexico and other countries.

? DID YOU KNOW?

A residual unease with foreigners remains deep in the DNA of some Japanese, despite 250 years of Westernization. When Japanese in the tsunami-hit zone of Tohoku were asked if there had been any looting, they answered, "no,"—except for foreigners, they said, who were probably responsible for any such problems. In some places, such as the city of Otaru in Hokkaido, "Japanese only" signs have been posted for years in public bathhouses, noodle shops, discos, and restaurants.

(© Dean W. Collinwood)

A busy street in Tokyo.

(© Dean W. Collinwood)

Modern skyscrapers in Shinjuku area of Tokyo.

What Japan has been loathe to do, however, is remove the "structural impediments" that make it difficult for foreign companies to do business in Japan. For example, 50 percent of the automobiles sold in Iceland are Japanese, which means less profit for the American and European manufacturers who used to dominate car sales there. Yet because of high tariffs and other regulations, very few American and European cars have been sold in Japan. Beginning in the mid-1980s, Japan reluctantly began to dismantle many of these trade barriers, and the process has been so successful that Japan now has a lower overall average tariff on nonagricultural products than the United States—its severest critic in this arena.

But Japanese people worry that further opening of their markets may destroy some fundamentals of Japanese life. Rice, for instance, costs much more in Japan than it should, because the Japanese government protects rice farmers with subsidies and limits most rice imports from abroad. The Japanese would prefer to pay less for rice at the supermarket, but they also argue that foreign competition would prove the undoing of many small rice farmers, whose land would then be sold to housing developers. This, in turn, would destroy more of Japan's scarce arable land and weaken the already shaky traditions of

the Japanese countryside—the heart of traditional Japanese culture and values.

Stagnation in the Japanese economy has slowed foreign direct investment (the United States and China are number one and number two in inflow of foreign investment money), but Japan is determined to recapture its role as Asia's number one business hub. Many foreign firms have generated (and continue to generate) massive profits from their Japanese operations. In the case of the United States, the profit made by American firms doing business in Japan (the list of companies reads like

DID YOU KNOW?

After the Fukushima nuclear power plant meltdown in 2011, the Japanese government, which had once planned to increase its electrical dependency on nuclear power from 30 percent to 50 percent, officially announced that it will abandon that plan and promote energy conservation and green energy instead. The plan now is to phase out nuclear power plants altogether.

Snapshot: JAPAN

Summarized below is a quick look at the country with regard to its development, freedom, health/welfare, and achievements.

Development

Despite the devastation of World War II, Japan's economy eclipsed some European economies by the mid-1960s. For several decades it has been the second-largest economy in the world, having just recently been bypassed by China. The country's infrastructure is among the most modern in the world. Offshoring of smokestack industries has become commonplace. Public works have produced high government debt, and a tenacious recession has weakened the economy, but Japan will continue to be a major player in Pacific Rim affairs.

Freedom

Japanese citizens enjoy full civil liberties, and opposition parties and ideologies are seen as natural and useful components of democracy. A revision of the U.S.-drafted 1947 constitution is underway. Certain minority groups such as Ainus and those of Korean ancestry have been subject to social and official discrimination, but improvement in the treatment of such groups is now evident.

Health/Welfare

The Japanese live longer on average than most people on Earth. Every citizen is provided with inexpensive medical care under a national health-care system, but many people still prefer to save substantial portions of their income for health emergencies and old age.

Achievements

Japan has achieved virtually complete literacy. Although there are poor areas and some "tent cities" created by homeless people, there are no slums inhabited by a permanent underclass. The gaps between the social classes appear to be less pronounced than in many other societies. The country seems to be entering an era of remarkable educational accomplishment, and Japanese scientists are at the forefront of inventions that are improving life for people everywhere. Forty Japanese have been awarded Nobel Prizes in such fields as physics, chemistry, and literature.

---

a Who's Who of American business: IBM, Coca-Cola, Microsoft, Apple Computer, and hundreds of others) in a single year is just about equal to the amount of the trade imbalance between Japan and the United States. Japanese supermarkets are filled with foreign foodstuffs, and the radio and television airwaves are filled with the sounds and sights of Western music and dress. Japanese youths are as likely to eat at McDonald's or Kentucky Fried Chicken outlets as at traditional Japanese restaurants, and many Japanese have never worn a kimono nor learned to play a Japanese musical instrument. It is clear to many observers that, culturally, Japan already imports much more from the West than the West does from Japan.

Given this overwhelming Westernization of Japan as well as Japan's current capacity to continue imbibing Western culture, even the change-oriented Japanese are beginning to ask where they, as a nation, are going. Will national wealth, as it slowly trickles down to individuals, produce a generation of hedonistic youths who do not appreciate the sacrifices of those before them? Will there ever be a time when, strapped for resources, the Japanese will once again seek hegemony over other nations? What future role should Japan assume in the international arena, apart from economic development? If these questions remain to be answered, circumstances of international trade have at least provided an answer to the question of Japan's role in the Pacific Rim countries: It is clear that, for the next several decades, Japan, albeit increasingly looking over its shoulder at the new China, will continue to shape the pace and nature of economic development, and thus the political environment, of the entire Pacific Rim.

Statistics

Geography

Area in Square Miles (Kilometers): 145,914 (377,915) (about the size of California)
Capital (Population): Tokyo (Greater Tokyo: 36.5 million)
Environmental Concerns: air and water pollution; acidification; depletion of global resources due to Japanese demand; nuclear radiation effects
Geographical Features: mostly rugged and mountainous
Climate: tropical in the south to cool temperate in the north

People

Population

Total: 126,475,664
Annual Growth Rate: −0.3%
Rural/Urban Population Ratio: 33/67
Ethnic Makeup: 98.5% Japanese; 0.5% Korean; 0.4% Chinese; 0.6% other
Major Language: Japanese
Religions: primarily Shinto and Buddhist; 2% Christian; 8% other

Health

Life Expectancy at Birth: 79 years (male); 86 years (female)
Infant Mortality: 2.8/1,000 live births
Physicians Available: 1/485 (2006)
HIV/AIDS Rate in Adults: less than 0.1%

Education

Adult Literacy Rate: 99%
Compulsory (Ages): 6–15; free

Communication

Telephones: 44.364 million main lines
Mobile Phones: 115 million
Internet Users: 99.2 million (2011)
Internet Penetration (% of pop): 78%

Transportation

Roadways in Miles (Kilometers): 747,992 (1,203,777)
Railroads in Miles (Kilometers): 16,426 (26,435)
Usable Airfields: 176

Government

Type: constitutional monarchy
Independence Date: traditional founding 660 B.C.; constitutional monarchy established May 3, 1947
Head of State/Government: Emperor Akihito; Prime Minister Yoshihiko Noda
Political Parties: Liberal Democratic Party; Social Democratic Party of Japan; New Komeito; Democratic Party of Japan; Japan Communist Party; People's New Party; Your Party
Suffrage: universal at 20

Military

Military Expenditures (% of GDP): 1%
Current Disputes: various territorial disputes with Russia, South Korea, China, and Taiwan

Economy

Currency ($ U.S. Equivalent): 87.78 yen = $1
Per Capita Income/GDP: $34,000/$4.31 trillion
GDP Growth Rate: 3.9%
Inflation Rate: −0.7%
Unemployment Rate: 5.1%
Labor Force by Occupation: 70% services; 26% industry; 4% agriculture
Population Below Poverty Line: 15.7% (2007)
Natural Resources: fish
Agriculture: rice; sugar beets; vegetables; fruit; pork; poultry; dairy and eggs; fish
Industry: metallurgy; engineering; electrical and electronics; textiles; chemicals; automobiles; food processing
Exports: $765 billion (primary partners United States, China, South Korea)
Imports: $637 billion (primary partners China, United States, Australia, Saudi Arabia)

Suggested Websites

www.japantimes.co.jp/
www.jnto.go.jp/

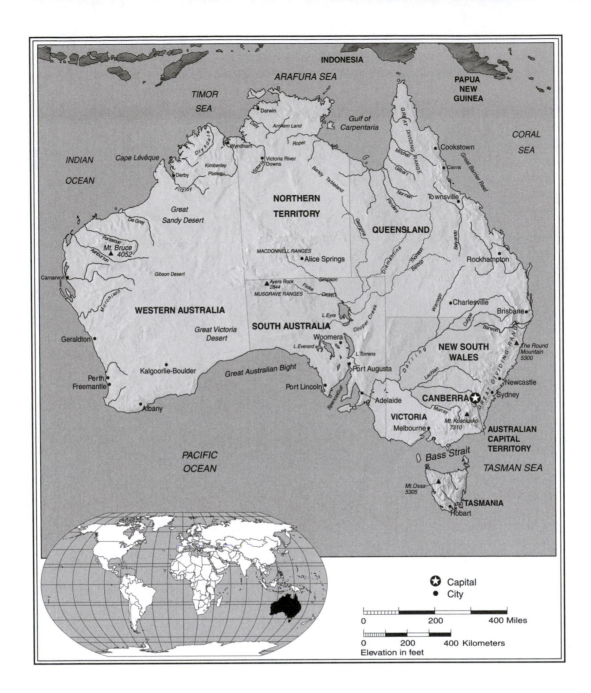

INDONESIA

ARAFURA SEA

TIMOR
SEA

PAPUA
NEW
GUINEA

CORAL
SEA

INDIAN
OCEAN

Darwin

Cape Lévêque

Arnhem Land

Gulf of
Carpentaria

Drysdale

Roper

Cookstown

Wyndham

Kimberley
Plateau

Victoria River
Downs

Cairns

Derby

Fitzroy

NORTHERN
TERRITORY

Townsville

Great
Sandy Desert

QUEENSLAND

De Grey

Fortescue

Mt. Bruce
4052

MACDONNELL RANGES

Rockhampton

Carnarvon

Gibson Desert

Alice Springs

WESTERN AUSTRALIA

Ayers Rock
2844

MUSGRAVE RANGES

Simpson
Desert

Finke

Charlesville

Brisbane

Murchison

Great Victoria
Desert

SOUTH AUSTRALIA

L.Eyre

Cooper Creek

Geraldton

Woomera

NEW SOUTH
WALES

The Round
Mountain
5300

Kalgoorlie-Boulder

Great Australian Bight

L.Everard

L.Torrens

Port Augusta

Darling

Newcastle

Perth
Freemantle

Port Lincoln

Adelaide

CANBERRA

Sydney

Albany

Murray

Mt. Kosciusko
7310

AUSTRALIAN
CAPITAL
TERRITORY

PACIFIC
OCEAN

VICTORIA

Melbourne

Bass Strait

TASMAN SEA

Mt.Ossa
5305

TASMANIA

Hobart

★ Capital
● City

| 0 | 200 | 400 Miles |

| 0 | 200 | 400 Kilometers |

Elevation in feet

Australia

(Commonwealth of Australia)

Despite its out-of-the-way location far south of the main trading routes between Europe and Asia, seafarers from England, Spain, and the Netherlands began exploring parts of the continent of Australia in the seventeenth century. The French later made some forays along the coast, but it was the British who first found something to do with a land that others had disparaged as useless: they decided to send their prisoners there. The British had long believed that the easiest solution to prison overcrowding was expulsion from Britain. Convicts had been sent to the American colonies for many years, but after American independence was declared in 1776, Britain needed another place to send convicts, and Australia was it.

Australia seemed like the ideal spot for a penal colony: It was isolated from the centers of civilization; it had some good harbors; and, although much of the continent was a flat, dry, riverless desert with only sparse vegetation, the coastal fringes were well suited to human habitation. Indeed, although the British did not know it in the 1700s, they had come across a huge continent endowed with abundant natural resources. Along the northern coast (just south of present-day Indonesia and Papua New Guinea) was a tropical zone with heavy rainfall and tropical forests. The eastern coast was wooded with valuable pine trees, while the western coast was dotted with eucalyptus and acacia trees. Minerals, especially coal, gold, nickel, petroleum, iron, uranium, and bauxite, were plentiful, as were the many species of unique animals: kangaroos, platypus, and koalas, to name a few. (As it happens, these unique animals are lucky to be alive, for it was just off the northern coast of Australia that a meteorite six miles across smashed into the earth 250 million years ago, leaving a 125-mile wide crater and wiping out some 70 percent of the world's land species).

Today, grazing and agricultural activities generally take place in the central basin of the country, which consists of thousands of miles of rolling plains. In 2006, the area was afflicted with the worst drought in over 100 years, forcing the government to provide federal relief to some 70,000 farmers. The dry spell was followed in 2010 and 2011 by the wettest year and the worst flooding on record. Thousands of acres of farmland and urban areas were inundated.

The bulk of the population resides along the coast, where the relatively few good harbors are found. Many of Australia's lakes are saltwater lakes left over from an ancient inland sea, much like the Great Salt Lake in North America. Much of the interior is watered by deep artesian wells that are drilled by local ranchers, since Australia's mountain ranges are generally not high enough to provide substantial water supplies.

The British chose to build their first penal colony alongside a good harbor that they called Sydney. By the 1850s, when the practice of transporting convicts stopped, more than 150,000 prisoners, including hundreds of women, had been sent there and to other colonies. Most of them were illiterate English and Irish from the lower socioeconomic classes. Once they completed their sentences, they were set free to settle on the continent. These released prisoners, their guards, gold prospectors, and of course, the Aborigines, constituted the foundation of today's Australian population. The population of 21.8 million people is still quite small compared to the size of the continent.

TIMELINE

1600s
European exploration of the Australian coastline begins

1688
British explorers first land in Australia

1788
The first shipment of English convicts arrives

1851
A gold rush lures thousands of immigrants

1901
Australia becomes independent within the British Commonwealth

1940s
Australia is threatened by Japan during World War II

1947
Australia proposes the South Pacific Commission

1951
Australia joins New Zealand and the United States in the ANZUS military security agreement

1954
Australia joins the South East Asian Treaty Organization

Relations with the United States are strained over the Vietnam War **1960s**

1972
The Australian Labour Party wins for the first time in 23 years; Gough Whitlam is prime minister

1975
After a constitutional crisis, Whitlam is replaced by opposition leader J. M. Fraser

1980s
Australia begins to strengthen its economic ties with Asian countries

1990s
After 13 years in power, the Labour Party is defeated by Liberal Party leader John Howard
Australia condemns nuclear testing in the Pacific; dam project on sacred aboriginal land blocked by the government

1996
World's first voluntary-euthanasia law passed

2000s
Aboriginal rights are increasingly part of the public debate in Australia

(continued)

TIMELINE

■ RACE RELATIONS

But convicts certainly did not constitute the beginning of human habitation on the continent. Tens of thousands of Aborigines (literally, "first inhabitants") inhabited Australia and the nearby island of Tasmania when Europeans first made contact. Living in scattered tribes and speaking hundreds of entirely unrelated languages, the Aborigines, whose origin is unknown (various scholars see connections to Africa, the Indian subcontinent, or the Melanesian Islands), survived by fishing and nomadic hunting. Succumbing to European diseases, violence, forced removal from their lands, and, finally, neglect, thousands of Aborigines died during the first centuries of contact. Indeed, the entire Tasmanian grouping of people (originally numbering 5,000) is now extinct. Today's Aborigines continue to suffer discrimination. A 1997 bill that would have liberalized land rights for Aborigines failed in Parliament. In 2002, Australia's highest court dismissed a land claim by the Yorta Yorta tribe that would have guaranteed their access to 800 square miles of ancestral lands. Even after 10 years of effort, the Council on Aboriginal Reconciliation was not able to persuade the government to issue a formal apology for past abuses. With 62 percent of the population remaining opposed to a formal government apology, the

FURTHER INVESTIGATION

Over the next five years, Australia will remove gender barriers in its military. Positions formerly reserved for men only, including front-line combat roles, will be available to both men and women. The move follows similar steps taken in New Zealand and Canada. To learn more, see: www.guardian.co.uk/world/2011/sep/27/australian-military-women-frontline-roles

It Used to be "Attack Dogs," Now It's "Attack Kangaroos"

94-year-old Phyllis Johnson of Queensland was outside hanging up her laundry when she was attacked by a kangaroo. She tried to fight it off with her broom, but it wasn't until the police arrived with pepper spray that the animal was subdued. Janet Karson of southwestern Australia had a similar thing happen when she was walking her dogs. A large red kangaroo attacked her and one of her dogs. The dogs fought it off but not before she had received enough cuts to require 20 stitches. In New South Wales, a female eastern grey kangaroo with razor-like claws attacked a 2-year-old boy. The animal later had to be euthanized. Animal experts say that kangaroo attacks are rare, but an English tourist observing kangaroos near Melbourne probably would not agree. An unprovoked kangaroo attack left him with 30 stitches. No one knows why attacks are on the rise, but one likely cause is the construction of homes in areas formerly available to kangaroos.

conservative coalition government of John Howard—despite threats of a violent backlash from Aborigines—abolished the 14-year Council in 2005.

With only one Aboriginal member in Parliament to represent the views of the 400,000 Aboriginals (2 percent of the total population), it has always been difficult for the Aboriginal voice to be heard in the halls of government. Occasionally, with Aborigines living in high crime areas and having to deal with alcoholism, drug abuse and high unemployment, frustrations boil over. In 2004, rioting continued for 9 hours in a poor Aboriginal neighborhood in Sydney after a teenager was killed in a chase with police. Rioters threw gas bombs and set fire to a train station, injuring 40 police officers.

After Europeanization of the continent, most Aborigines adopted European ways, including converting to Christianity. Today, many live in the cities or work for cattle and sheep ranchers. Others reside on reserves (tribal reservations) in the central and northern parts of Australia. Yet modernization has affected even the reservation Aborigines—some have telephones, and some dispersed tribes in the Northern Territories communicate with one another by satellite-linked video conferencing. In the main, however, Aborigines continue to live as they have always done, organizing their religion around plant or animal sacred symbols, or totems, and initiating youth into adulthood through lengthy and sometimes painful rituals.

Whereas the United States began with 13 founding colonies, Australia started with 6, none of which felt a compelling need to unite into a single British nation until the 1880s, when other European powers began taking an interest in settling the continent. It was not until 1901 that Australians formally separated from Britain (while remaining within the British Commonwealth,

(© Jan Smith)

Most Aborigines eventually adapted to the Europeans' customs, but some continue to live in their traditional ways.

with the Queen of England as head of state). Populated almost entirely by whites from Britain or Europe (people of European descent still constitute about 92 percent of Australia's population despite a recent influx of Asians and Middle Easterners), Australia has traditionally maintained close cultural and diplomatic links with Britain and the West, at the expense of ties with the geographically closer nations of Asia. That stance is now changing, as Australia slowly recognizes the advantages of trade with its Asian neighbors.

Reaction against Polynesians, Chinese, and other Asian immigrants in the late 1800s produced an official "White Australia" policy, which remained intact until the 1960s and effectively excluded nonwhites from settling in Australia. During the 1960s, however, the government made an effort to relax these restrictions and to restore some of the confiscated land and some measure of self-determination to Aborigines. In the 1990s, Aborigines

DID YOU KNOW?

New South Wales passed a law in 2011 that requires anyone, including Muslim women wearing veils, to remove their face coverings when asked to do so by a police officer. The controversial new law was the result of a vehicle accident caused in part by a woman wearing a niqab. Penalties for those who refuse to remove the veil could include a year in prison and a fine of US$6,000. There are about 400,000 Muslims in Australia, and it is thought that about 2,000 of them wear veils.

successfully persuaded the federal government to block a dam project on Aboriginal land that would have destroyed sacred sites. In this rare case, the federal government sided with the Aborigines against white developers and local government officials. In 1993, despite some public resistance, the government passed laws protecting the land claims of Aborigines and set up a fund to assist Aborigines with land purchases.

These small steps toward racial equality have not altered white Australians' dislike of "outsiders." Evidence of continued racism can be found, for example, in graffiti painted on walls of high-rise buildings (all Asians, regardless of nationality, are referred to as "Japs" or "wogs"), and in the starkly inferior quality of life for Aborigines. The unemployment rate of Aborigines is four times that of the nation as a whole, and there are substantially higher rates of chronic health problems and death by infectious diseases among this population. The health and living conditions of Aborigines in many locations became so precarious in 2006 that the government had to send in administrators to run local Aboriginal councils, a move only begrudgingly accepted by the Aborigines. In 2009, students from India (there are 70,000 of them in Australia) suffered a series of attacks against them by local Australians.

■ ECONOMIC PRESSURES

Despite lingering discriminatory attitudes against nonwhites, events since World War II have forced Australians to reconsider their position, at least economically, vis-à-vis Asia and Southeast Asia. Despite its historic links to the developed economies of Europe, and despite never having directly suffered the destruction of war as have most of its Asian neighbors, the Australians are worried that their economy may not be able to withstand the competitive pressures coming from Japan, China, and other Asian nations. Hong Kong's per capita GDP, for example, is already substantially ahead of Australia's, and Japan's, despite years of a sluggish economy, is competitive with Australia's. Like New Zealand, Australia historically relied on the export of primary goods (minerals, wheat, beef, and wool) rather than more lucrative manufactured goods to support its economy. In more recent years, the country has developed a strong manufacturing sector. But if the Asian economies continue to grow at rates much faster than Australia's, the day may come when Australia will find itself lagging behind its neighbors, and that will be a new and difficult experience for Australians, whose standard of living has been the highest in the Pacific Rim for decades. Building on a foundation of sheep (imported in the 1830s and now supplying more than a quarter of the world's supply of wool), mining (gold was discovered in 1851), and agriculture (Australia is nearly self-sufficient in food), the country has now developed its manufacturing sector such that Australians are able to enjoy a standard of living equal in most respects to that of North Americans.

But Australians are wary of the growing global tendency to create mammoth regional trading blocs, such as the North American Free Trade Association, consisting of the United States, Canada, Mexico, and others; the European Union (formerly the European Community), now including many countries of Eastern Europe; the ASEAN nations of Southeast Asia; and an informal "yen bloc" in Asia, headed by Japan. These blocs might exclude Australian products from preferential trade treatment or eliminate them from certain markets altogether. Beginning in 1983, the Labour government of then–prime minister Robert Hawke began to establish collaborative trade agreements with Asian countries, a plan that seemed to have the support of the electorate, even though it meant reorienting Australia's foreign policy away from its traditional posture Westward.

In the early 1990s, under Labour prime minister Paul Keating, the Asianization plan intensified. The Japanese prime minister and the governor of Hong Kong visited Australia, while Australian leaders made calls on the leaders of South Korea, China, Thailand, Vietnam, Malaysia, and Laos. Trade and security agreements were signed with Singapore and Indonesia, and a national curriculum plan was implemented whereby 60 percent of Australian school children would be studying Japanese and other Asian languages by 2010. So far, the plan has failed, and the project may be cancelled. The "Asia-literate" effort attracted so few students that the Japanese and Indonesian language programs may be eliminated from the school curriculum. Despite comments from the Prime Minster that now is "the Asian century," Australian education remains solidly Eurocentric.

Under Liberal Party prime minister, John Howard, who had remained in office for 11 years, Australia moderated its former restrictive immigration policy, but Howard's support of the Iraq War and his refusal to push for climate legislation (he refused to sign the Kyoto Protocol, for example) cost him the election in 2007. He was replaced by Labor's Kevin Rudd who has promoted environmental issues and whose stimulus package in 2009 successfully buoyed the economy despite the global economic downturn.

Under the Howard administration, strict new laws were passed against illegal immigrants from Afghanistan, Iraq, Iran, and other countries who found their way to Australia from nearby Indonesia. The government's "Pacific Solution" allows the navy to capture refugee ships at sea and send the occupants to other Pacific islands. Then the Australian government pays the smaller islands for handling the refugees. Nauru, for example, has received more than $15 million for taking in 1,200 refugees. The government resorted to these measures after a hunger strike by 500 of the 2,000 refugees in a detention camp drew negative international publicity. Some detainees sewed their mouths shut, while others tried suicide in order to pressure the government to grant them asylum. A bill to crack down on asylum seekers was withdrawn in 2006 after protests from parliamentarians of both parties who argued that Australia should not have been resorting to inhumane measures to solve immigration problems. These lawmakers also believed it was wrong to send refugees to South Pacific islands for processing. Handling these matters is made more difficult by the anti-immigrant, ultraright One Nation Party, which advocates shutting the door to immigrants and returning to Australia's former "whites only" policy. The country's first female Prime Minister, Julia Gillard, elected in 2010, is bothered by the increasing number of asylum-seekers and wants to expand off-site processing of people wanting entrance to Australia.

Despite government efforts to bring Australia within the orbit of Asian trade, the economic threat to Australia remains. Even in the islands of the Pacific, an area that Australia and New Zealand generally have considered their own domain for economic investment and foreign aid, investments by Asian countries are beginning to winnow Australia's sphere of influence. Former U.S. president Bill Clinton, in a 1996 visit, promised Australian leaders that they would not be left out of the emerging economic structures of the region. Indeed, economic conditions have improved; recent budgets have shown surpluses and even allowed for tax cuts. The recession of 2009 and 2010 reduced revenues, but projections are the government will return to budget surpluses as early as 2013. A free trade agreement with the United States that eliminated tariffs on 99 percent of U.S. goods imported in Australia went into effect in 2005, as did a similar agreement with Thailand. These measures have helped maintain the country's economy even in the midst of the world economic crisis of 2008 and 2009. Unemployment has not fallen to the levels of the 1990s when some 2 million people were living in poverty.

In the 1990s, labor tension erupted when dock-workers found themselves locked out of work by employers who claimed they were inefficient workers. Eventually the courts found in favor of the workers, but not until police and workers clashed and national attention was drawn to the protracted sluggish economy. Calls for new protectionist measures against multibillion-dollar foreign acquisitions of Australian companies (the iconic Myer department store chain was purchased in 2006 by a Texas company, for example, and the government chose to intervene to prevent the take-over of its largest public oil and gas company by Royal Dutch Shell, a British/Dutch conglomerate) show that Australians will continue to keep a wary and protective eye on their economy.

■ THE AMERICAN CONNECTION

By any standard, Australia is a democracy solidly embedded in the traditions of the West. Political power is shared back and forth between the Labour Party and the Liberal–National Country Party coalition, and the Constitution is based on both British parliamentary tradition and the U.S. model. Thus, it has followed that Australia and the United States have built a warm friendship as both political and military allies. A military mutual-assistance agreement, ANZUS (for Australia, New Zealand, and the United States), was concluded after World War II (New Zealand withdrew in 1986). And just as it had sent troops to fight Germany during World Wars I and II, Australia sent troops to fight in the Korean War in 1950, the Vietnam War in the 1960s, the anti-Taliban war in Afghanistan in 2002, and the U.S.-led war in Iraq in 2003. Former Prime Minister Howard strongly supported then U.S. President George Bush and defiantly refused to even consider withdrawing Australian troops, despite some Australians having been taken hostage and executed by militants in Iraq. Australia also joined the United States and other countries in 1954 in establishing the Southeast Asia Treaty Organization, an Asian counterpart to the North Atlantic Treaty Organization designed to contain the spread of communism.

In 1991, when the Philippines refused to renew leases on U.S. military bases there, there was much discussion about transferring U.S. operations to the Cockburn Sound Naval Base in Australia. Singapore was eventually chosen for some of the operations, but the two nations continue extensive military cooperation. U.S. military aircraft already land in Australia, and submarines and other naval craft call at Australian ports. The Americans also use Australian territory for surveillance facilities. In 2011, President Barack Obama, in a move widely interpreted as an effort to curb China's military influence in the region, announced that Australia would host a full U.S. Marine Task force by 2016. Some 2,500 troops will be stationed in Australia along with military aircraft.

There is historical precedence for this level of close cooperation: Before the U.S. invasion of the Japanese-controlled Philippines in the 1940s, the United States based its Pacific-theater military headquarters in Australia; moreover, Britain's inability to lead the fight against Japan forced Australia to look to the United States.

A few Australians resent the violation of sovereignty represented by the U.S. bases, but most regard the United States as a solid ally. Indeed, many Australians regard their country as the Southern Hemisphere's version of the United States: Both countries have immense space and vast resources, both were founded as disparate colonies that eventually united and obtained independence from Britain, and both share a common language and a Western cultural heritage.

There is yet another way that some Australians would like to be similar to the United States: They want to be a republic. Polls in advance of a 1999 referendum to decide whether or not Australia should remain a constitutional monarchy, headed by the king or queen of England, showed that more than 60 percent of the population favored severing ties with England. However, the actual vote found 55 percent in favor of the status quo—just enough to keep Queen Elizabeth II as head of state. The queen visited Australia right after the vote to show her thanks to the residents of working-class and rural areas, where support for the monarchy was strongest. Queen Elizabeth has visited Australia 16 times as monarch, most recently in 2011 when she spent 10 days in the country and was cheered by thousands of well-wishers.

Unlike New Zealand, which has distanced itself from the United States by refusing to allow nuclear-armed ships to enter its ports and has withdrawn from the ANZUS security treaty with the United States, Australia has sided with the U.S. in attempting to dissuade South Pacific states from declaring the region a nuclear-free zone. Yet it has also maintained good ties with the small and vulnerable societies of the Pacific through its leadership in such regional associations as the South Pacific Commission, the South Pacific Forum, and the ever-more-influential Asia-Pacific Economic Cooperation group (APEC). It has also condemned nuclear-bomb testing programs in French-controlled territories.

■ AUSTRALIA AND THE PACIFIC

Australia was not always possessed of good intentions toward the islands around it. For one thing, white Australians thought of themselves as superior to the

Snapshot: AUSTRALIA

Summarized below is a quick look at the country with regard to its development, freedom, health/welfare, and achievements.

Development

Mining of nickel, iron ore, uranium, and other metals continues to supply a substantial part of Australia's gross domestic product. The United States, Canada, South Korea, Taiwan, and China buy uranium, and Japan continues to buy beef and many other products. Seven out of 10 of Australia's largest export markets are now Asian countries. With 27 million head of cattle, Australia is the world's largest exporter of beef.

Freedom

Australia is a parliamentary democracy adhering to the ideals incorporated in English common law. Constitutional guarantees of human rights apply to all of Australia's citizens. However, social discrimination continues, and, despite improvements since the 1960s, the Aborigines remain a neglected part of society.

Health/Welfare

Australia has developed a complex and comprehensive system of social welfare. Education is the province of the several states. Public education is compulsory. Australia boasts several world-renowned universities. The world's first voluntary-euthanasia law passed in Northern Territory in 1996, but legal challenges have thus far prevented its use. The government has proposed a nationwide filter to prevent pornographic images from being accessed on the Internet.

Achievements

The vastness and challenge of Australia's interior lands, called the "outback," have inspired a number of Australian writers to create outstanding poetry and fictional novels. In 1973, Patrick White became the first Australian to win a Nobel Prize in Literature. Jill Ker Conway, Thomas Keneally, and Colleen McCullough are other well-known Australian authors. Nicole Kidman and Hugh Jackman are two of many famous Hollywood actors with Australian roots.

---

brown-skinned islanders; and for another, Australia preferred to use the islands' resources for its own economic gain, with little regard for the islanders themselves. At the end of World War I, for example, the phosphate-rich island of Nauru, formerly under German control, was assigned to Australia as a trust territory. Until phosphate mining was turned over to the islanders in 1967, Australian farmers consumed large quantities of the island's phosphates but paid just half the market price. Worse, only a tiny fraction of the proceeds went to the people of Nauru. Similarly, in Papua New Guinea, Australia controlled the island without taking significant steps toward its domestic development until the 1960s, when, under the guidance of the United Nations, it did an about-face and facilitated changes that advanced the successful achievement of independence in 1975.

In addition to forgoing access to cheap resources, Australia was reluctant to relinquish control of these islands because it saw them as a shield against possible military attack. It learned this lesson well in World War II. In 1941, Japan, taking advantage of the Western powers' preoccupation with Adolf Hitler, moved quickly to expand its imperial designs in Asia and the Pacific. The Japanese first disabled the U.S. Navy by attacking its warships docked in Pearl Harbor, Hawaii. They then moved on to oust the British in Hong Kong and the Gilbert Islands, and the Americans in Guam and Wake Island. Within a few months, the Japanese had taken control of Burma, Malaya, Borneo, the Philippines, Singapore, and hundreds of tiny Pacific islands, which they used to create an immense defensive perimeter around the home islands of Japan. They also had captured part of New Guinea and were keeping a large force there, which greatly

Solar Power From Down Under

Under pressure because of its poor environmental record and refusal to sign the Kyoto Protocol on greenhouse gas emissions, Australia has budgeted AU$125 million to build the world's largest solar photovoltaic power plant. The country's heavy reliance on cheap (but heavily polluting) coal makes it reluctant to pass strict environmental regulations. The planned solar power plant will use mirrors positioned over many acres to concentrate the rays of the sun on a single point. It will reach full operating capacity by 2013. Critics think the project is a rather insignificant step compared to the magnitude of the pollution problem in the country, but they at least prefer solar power generation over nuclear power plants.

---

concerned the Australians. Yet fighting was kept away from Australia proper when the Japanese were successfully engaged by Australian and American troops in New Guinea. Other Pacific islands were regained from the Japanese at a tremendous cost in lives and military hardware. Japan's defeat came only when islands close enough to Japan to be attacked by U.S. bomber aircraft were finally captured. Japan surrendered in 1945, but the colonial powers had learned that possession of small islands could have strategic importance. This experience is part of the reason for colonial powers' reluctance to grant independence to the vast array of islands over which they have exercised control. Australia is now faced with the question of whether or not to grant independence to the 4,000 inhabitants of Christmas Island who recently voted to become a self-ruling territory within Australia.

There is no doubt that stressful historical periods and events such as World War II drew the English-speaking countries of the South Pacific closer together and closer to the United States. But recent realignments in the global economic system are creating strains. For example, in the early 1990s, when the United States took steps to reduce its trade imbalance with Japan, Australia came out the loser. It happened like this: Both Australia and the United States are producers of coal. Given the nearly equal distance between those two countries and Japan, it would be logical to assume that Japan would be able to buy coal at about the same price from both countries. In fact, however, Japan paid substantially less to Australia per ton of coal than it did to the United States, a direct result of U.S. pressure on Japan over the then-mammoth trade imbalance with the United States.

Statistics

Geography

Area in Square Miles (Kilometers): 2,988,902 (7,741,220) (slightly smaller than the United States)

Capital (Population): Canberra (384,000)

Environmental Concerns: depletion of the ozone layer; pollution; soil erosion and excessive salinity; desertification; wildlife habitat loss; degradation of Great Barrier Reef; limited freshwater resources; drought

Geographical Features: mostly low plateau with deserts; fertile plain in southeast

Climate: generally arid to semiarid; temperate to tropical

People

Population

Total: 21,766,711

Annual Growth Rate: 1.2%

Rural/Urban Population (Ratio): 11/89

Major Languages: English; indigenous languages

Ethnic Makeup: 92% Caucasian; 7% Asian; 1% Aboriginal and others

Religions: 26% Roman Catholic; 19% Anglican; 19% other Christian; 2% Buddhist; 2% Muslim; 32% unaffiliated or other

Health

Life Expectancy at Birth: 79 years (male); 84 years (female)

Infant Mortality: 5/1,000 live births

Physicians Available: 1/333 people

HIV/AIDS Rate in Adults: .01%

Education

Adult Literacy Rate: 99%

Compulsory (Ages): 6–15; free

Communication

Telephones: 9.02 million main lines

Mobile Phones: 24.22 million (2009)

Internet Users: 17 million (2011)

Internet Penetration (% of Pop.): 78%

Transportation

Roadways in Miles (Kilometers): 505,157 (812,972)

Railroads in Miles (Kilometers): 23,889 (38,445)

Usable Airfields: 462

Government

Type: federal parliamentary democracy

Independence Date: January 1, 1901 (federation of U.K. colonies)

Head of State/Government: Queen Elizabeth II; Prime Minister Julia E. Gillard

Political Parties: Australian Greens; Australian Labor Party; Family First Party; Liberal Party; The Nationals

Suffrage: universal and compulsory at 18

Military

Military Expenditures (% of GDP): 3%

Current Disputes: territorial claim in Antarctica; boundary issues with East Timor and Indonesia

Economy

Currency ($ U.S. Equivalent): 1.09 Australian dollars = $1

Per Capita Income/GDP: 41,000/$882.4 billion

GDP Growth Rate: 2.7%

Inflation Rate: 2.9%

Unemployment Rate: 5.1%

Labor Force by Occupation: 75% services; 21% industry; 3.6% agriculture

Population Below Poverty Line: na

Natural Resources: bauxite; diamonds; coal; copper; iron ore; oil; natural gas; other minerals

Agriculture: wheat; barley; sugarcane; fruit; livestock

Industry: mining; industrial and transportation equipment; food processing; chemicals; steel; tourism

Exports: $210.7 billion (primary partners developing countries, China, Japan, South Korea, India)

Imports: $200.4 billion (primary partners China, United States, Japan)

Suggested Websites

www.australia.gov.au
www.australia.com

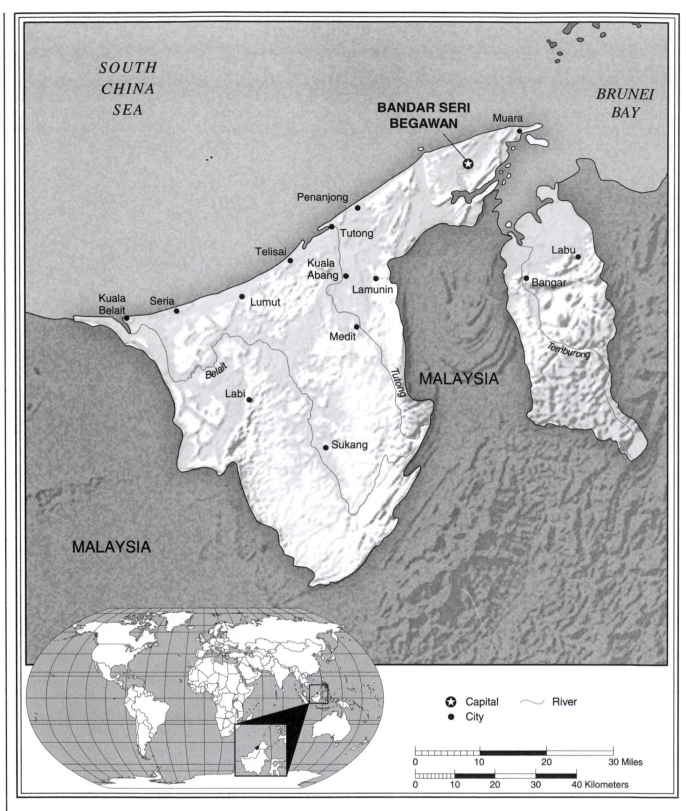

SOUTH
CHINA
SEA

BRUNEI
BAY

**BANDAR SERI
BEGAWAN**

Muara

Penanjong

Tutong

Telisai

Kuala
Abang

Lamunin

Labu

Bangar

Kuala
Belait

Seria

Lumut

Medit

MALAYSIA

Belait

Temburong

Tutong

Labi

Sukang

MALAYSIA

★ Capital   ∿ River
● City

| 0 | 10 | 20 | 30 Miles |
| 0 | 10 | 20 | 30 | 40 Kilometers |

Brunei

(Negara Brunei Darussalam)

Home to only 402,000 people, Brunei rarely captures the headlines. But perhaps it should, for despite its tiny size (about the size of Delaware or Prince Edward Island), the country boasts one of the highest living standards in the world. Moreover, the sultan of Brunei, with assets of over $20 billion, is considered one of the richest persons on earth. The secret? Oil. First exploited in Brunei in the late 1920s, today petroleum and natural gas almost entirely support the sultanate's economy. The government's annual income is nearly twice its expenditures, and there is money left over despite the provision of free education through university schooling, free medical care, subsidized food and housing, and no personal or corporate income taxes. Brunei is currently in the middle of a five-year plan designed to diversify its economy, but 90 percent of the nation's revenues continue to derive from the sale of oil and natural gas.

Muslim sultans ruled over the entire island of Borneo and other nearby islands during the sixteenth century. Advantageously located on the northwest coast of the island of Borneo, along the sea lanes of the South China Sea, Brunei was a popular resting spot for traders. During the 1700s, it became known as a haven for pirates. Tropical rain forests and swamps occupy much of the country—conditions that are maintained by heavy monsoon rains for about five months each year. Oil and natural-gas deposits are found both on- and offshore.

In the 1800s, the sultan then in power agreed to the kingdom becoming a protectorate of Britain, in order to safeguard his domain from being further whittled away by aggressors bent on empire-building. The Japanese easily overtook Brunei in 1941, when they launched their Southeast Asian offensive in search of oil and gas for their war machine. Today, the Japanese Mitsubishi Corporation has a one-third interest in the Brunei gas company, and Shell Oil is also a major player in the economy.

In the 1960s, it was expected that Brunei, which is cut in two and surrounded on three sides by Malaysia, would join the newly proposed Federation of Malaysia; but it refused to do so, preferring to remain under British control. The decision to remain a colony was made by Sultan Sir Omar Ali Saifuddin. Educated in British Malaya, the sultan retained a strong affection for British culture and frequently visited the British Isles. (Brunei's 1959 Constitution, promulgated during Sir Omar's reign, reflected this attachment: It declared Brunei a self-governing state, with its foreign affairs and defense remaining the responsibility of Great Britain.)

In 1967, Sir Omar voluntarily abdicated in favor of his son, who became the 29th ruler in succession. Sultan (and Prime Minister) Sir Hassanal Bolkiah Mu'izzaddin Waddaulah (a shortened version of his name!) oversaw Brunei's gaining of independence, in 1984. When the present Sultan leaves the throne, his son, Crown Prince Al-Muhtadee Billah Bolkiah, will assume the office. The present sultan is the 29th in one of the world's longest reigning royal lines, now over 500 years old. The Crown Prince made news in 2004 when he married a 17-year-old half-Swiss commoner. The traditional Malay Muslim ceremony was held at the enormous main palace and was attended by 2,000 royals and

TIMELINE

A.D. 1521
Brunei is first visited by Europeans
1700s
Brunei is known as haven for pirates
1800s
Briton James Brooke is given Sarawak as reward for help in a civil war
1847
The island of Labuan is ceded to Britain
1849
Britain attacks and ends pirate activities in Brunei
1888
The remainder of Brunei becomes a British protectorate
1963
Brunei rejects confederation with Malaysia
1984
Brunei gains its independence
1990s
Foreign workers are "imported" to ease the labor shortage; Brunei joins the International Monetary Fund.
2000s
The sultan's brother agrees to return billions of dollars in stolen state assets
Brunei declares itself a nuclear-free zone
The sultan signs a new Constitution allowing for direct elections of 15 members of a redesigned 45-person Legislative Council
The Crown Prince weds a half-Swiss commoner in a lavish ceremony
Brunei aims for 60% self-sufficiency in rice by 2015
In 2010 the Sultan appointed the first female Deputy Minister

other dignitaries from all over the world. Those present had an opportunity to view the Sultan's massive automobile collection (estimated at between 1,000 to 5,000 cars), which includes some 500 Rolls Royces. They also had an opportunity to see the largest residential palace in the world, the Istana Nural Iman. Located on the banks of the Brunei River, the istana has 1,788 rooms and a banquet hall that can seat 4,000 people. A prayer hall for the sultan's family and state religious ceremonies accommodates 1,500 worshipers. The Islamic architecture and sweeping Southeast Asian roofs of the palace are capped with glistening domes tiled with 22-carat gold leaf.

Not all Bruneians have been pleased with the present sultan's control over the political process. Despite a constitutional provision calling for regular elections, the last general elections were held over four decades ago, in 1962. Voters who were dissatisfied with the outcome at that time launched an armed revolt, which was put down with British assistance. The sultan declared a state of emergency, which has remained in effect ever since. The 21-member appointed Legislative Council did not meet for the next 20 years, and there were, in effect, no operative political parties. At last, in 2004, the sultan convened the legislature to consider several constitutional amendments. These were subsequently signed into law and included a new provision for direct election of up to 15 members of a newly redesigned Legislative Council of 45 members.

Modern Brunei is officially a constitutional monarchy, headed by the sultan, a chief minister, and a Council; in reality, however, the sultan and his family control all aspects of state decision making. The extent of the sultan's control of the government is revealed by his multiple titles: in addition to sultan, he is Brunei's prime minister, minister of defense, and minister of finance. The Constitution provides the sultan with supreme executive authority in the state. The concentration of near-absolute power in the hands of one family has produced some unsavory episodes. In 2000, the sultan's brother agreed to return billions of dollars of assets that he had stolen from the state while heading Brunei's overseas investment company. Court documents in the $15 billion lawsuit against him claimed that he had personally consumed $2.7 billion (!) to support his lavish lifestyle. The case was being tried in England, when, unexpectedly, the sultan dropped all charges against his brother.

Brunei's largest ethnic group is Malay, accounting for 66 percent of the population. Indians and Chinese constitute sizable minorities, as do indigenous peoples such as Ibans and Dyaks. About one-fifth of the population has non-Bruneian roots: British, Australians, New Zealanders, Americans, Filipinos and others, giving Brunei one of the largest Christian populations (10 percent) in Asia.

Brunei is an Islamic nation with Hindu roots. Islam is the official state religion, and in recent years, the sultan has proposed bringing national laws more closely in line with Islamic ideology and has declared that the monarchy is the protector of the Islamic faith. In 2009, Brunei launched a new brand of halal foods (foods prepared according to Islamic law) which it hopes to sell to other countries with large Muslim populations. Brunei prohibited the consumption of alcohol in the 1990s, although foreigners are allowed small quantities.

Brunei, along with China and several other Southeast Asian nations, claims the Spratly Islands. Recently, the various claimants agreed to work amicably on a long-term solution to the problem. Similarly, Brunei has stopped offshore and deep-sea oil and gas exploration because control of the area is disputed by Malaysia. The matter may eventually end up in an international court. In the meantime, Brunei is proceeding with the development of six petrochemical plants in the Sungai Liang Industrial Park. The venture is headed by Japanese giant Mitsui and will cost US$2.8 billion. The government has offered investors in some of its development projects the opportunity to enjoy tax holidays up to 20 years. In other international matters, Brunei joined with other ASEAN nations in 1995 to declare its country a nuclear-free zone. And in 2001, at the ASEAN meetings held in Brunei, the country agreed with others (including China) to establish a free-trade zone in the region within 10 years.

In recent years, Brunei has been plagued by a chronic labor shortage. The government and Brunei Shell (a consortium owned jointly by the Brunei government and Shell Oil) are the largest employers in the country. They provide generous fringe benefits and high pay. Non-oil private-sector companies with fewer resources find it difficult to recruit within the country and have, therefore, employed many foreign workers. Indeed, one third of all workers today in Brunei are foreigners. This situation is of considerable concern to the government, which is worried that social tensions between foreigners and residents may flare up at any time.

FURTHER INVESTIGATION

To learn more about Brunei's economy, which is strong enough to have no personal income taxes while still providing free education (including university) and medical care to all its citizens, see: www.byebyeblighty.com/tax-havens/living-in-brunei.html

Snapshot: BRUNEI

Summarized below is a quick look at the country with regard to its development, freedom, health/welfare, and achievements.

Development

Brunei's economy is a mixture of the modern and the ancient: foreign and domestic entrepreneurship, government regulation and welfare statism, and village tradition. Chronic labor shortages are managed by the importation of thousands of foreign workers.

Freedom

Although Islam is the official state religion, the government practices religious tolerance. The Constitution was amended in 2006 to give the sultan complete infallibility, which, along with his supreme executive authority, provides him with the ability to continue to suppress the growth of opposition groups and political parties—but now with immunity from public or legal complaint. In the 1990s in nearby Malaysia, where sultans enjoy similar royal immunity, the abuse of such a policy gained global attention when one sultan, while golfing, became so angry at his caddy that he allegedly beat him to death with his club—yet he could not be charged with murder.

Health/Welfare

The country's massive oil and natural-gas revenues support wide-ranging benefits to the population, such as subsidized food, fuel, and housing, as well as free medical care and education. This distribution of wealth is reflected in Brunei's generally favorable quality-of-life indicators.

Achievements

An important project has been the construction of a modern university accommodating 1,500 to 2,000 students. Since independence, the government has tried to strengthen and improve the economic, social, and cultural life of its people.

Statistics

Geography

Area in Square Miles (Kilometers): 2,228 (5,770) (about the size of Delaware)
Capital (Population): Bandar Seri Begawan (22,000)
Environmental Concerns: water pollution; seasonal smoke/haze resulting from forest fires in Indonesia
Geographical Features: flat coastal plain rises to mountains in east; hilly lowlands in west
Climate: tropical; hot, humid, rainy

People

Population

Total: 401,890
Annual Growth Rate: 1.7%
Rural/Urban Population (Ratio): 24/76
Major Languages: Malay; English; Chinese; Iban; native dialects
Ethnic Makeup: 66% Malay; 11% Chinese; 3% indigenous; 20% others
Religions: 67% Muslim; 13% Buddhist; 10% Christian; 10% indigenous and others

Health

Life Expectancy at Birth: 74 years (male); 78.5 years (female)
Infant Mortality: 12/1,000 live births
Physicians Available: 1/706 people
HIV/AIDS Rate in Adults: <0.1%

Education

Adult Literacy Rate: 93%
Compulsory (Ages): 5–17; free

Communication

Telephones: 80,500 main lines (2009)
Internet Users: 319,000 (2010)
Internet Penetration (% of Pop.): 81%

Transportation

Roadways in Miles (Kilometers): 1,846 (2,971)
Railroads in Miles (Kilometers): na
Usable Airfields: 2

Government

Type: constitutional sultanate (monarchy)
Independence Date: January 1, 1984 (from the United Kingdom)
Head of State/Government: Sultan and Prime Minister Sir Hassanal Bolkiah is both head of state and head of government
Political Parties: National Development Party is the only legal party. Brunei National Solidarity Party and People's Awareness Party were deregistered.
Suffrage: universal at 18 for village elections

Military

Military Expenditures (% of GDP): 4.5%
Current Disputes: dispute over the Spratly Islands and border disputes with Malaysia

Economy

Currency ($ U.S. Equivalent): 1.36 Brunei dollars = $1
Per Capita Income/GDP: $51,600/$20.38 billion
GDP Growth Rate: 4.1%
Inflation Rate: 2.7%
Unemployment Rate: 3.7%
Labor Force by Occupation: 40% government; 63% industry; 33% services; 4% agriculture
Natural Resources: petroleum; natural gas; timber
Agriculture: rice; cassava (tapioca); bananas; water buffalo; chickens; cattle; goats; eggs
Industry: petroleum; natural gas; construction
Exports: $10.67 billion (primary partners Japan, Indonesia, South Korea)
Imports: $2.61 billion (primary partners Singapore, Malaysia, Japan)

Suggested Websites

www.brunei.gov.bn
www.tourismbrunei.com

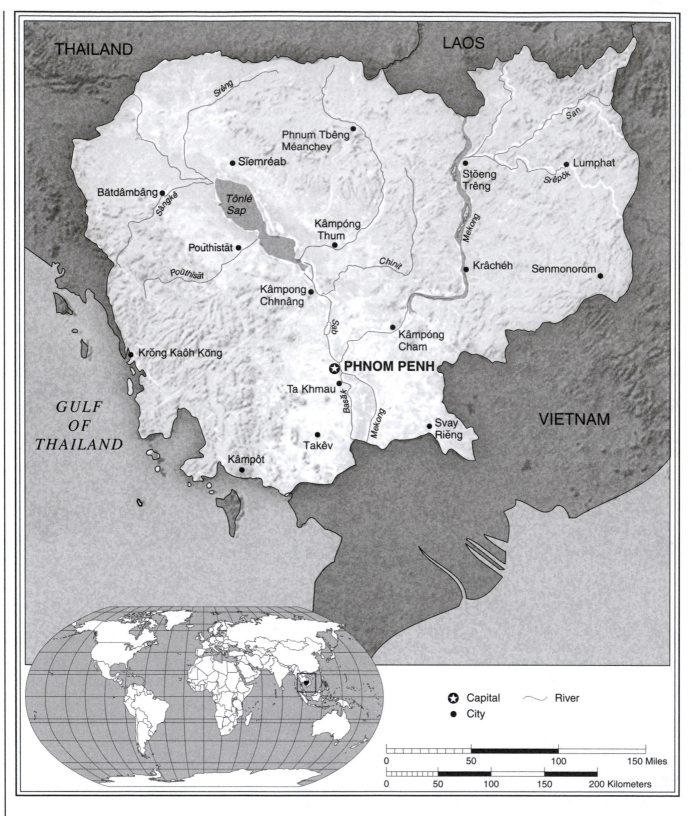

THAILAND

LAOS

Srềng

Phnum Tbêng
Méanchey

Sĩemréab

Lumphat

San

Bătdâmbâng

Stŏeng
Trêng

Srêpôk

Sângkê

Tônlé
Sap

Kâmpóng
Thum

Mekong

Poŭthĭstăt

Krâchéh

Senmonorom

Poŭthĭsăt

Chinĭt

Kâmpong
Chhnâng

Krŏng Kaôh Kŏng

Sab

Kâmpóng
Cham

★ PHNOM PENH

Ta Khmau

Basăk

VIETNAM

GULF
OF
THAILAND

Svay
Riĕng

Mekong

Takêv

Kâmpôt

★ Capital   ∿ River
● City

| 0 | 50 | 100 | 150 Miles |
| 0 | 50 | 100 | 150 | 200 Kilometers |

Cambodia

(Kingdom of Cambodia)

Two hundred miles of steamy jungle from the Cambodian capital city of Phnom Penh is one of Southeast Asia's most impressive architectural complexes: the Hindu temples of Angkor. Built nearly a thousand years ago, when Cambodia ruled an empire stretching from the borders of China to the Bay of Bengal, the 145-acre Angkor complex became the seat of government of a long line of Khmer or Cambodian kings, as well as a pilgrimage destination for thousands of Hindu faithful. Today, the government is attempting to parlay the fame of Angkor into the basis for the country's fledgling tourist industry. Until the world economic crisis of 2009 reduced tourists from South Korea and other Asian countries, the promotional plan worked very well, with hundreds of thousands of tourists visiting the site and the nearby provincial capital of Siem Reap each year. It is estimated that in 2008, the number of tourists was 2 million, and the revenue they generated was US$1.6 billion—a massive infusion by recent Cambodian standards. By 2010, tourism numbers had increased to 2.15 million.

In Khmer (Cambodian), the word *Kampuchea,* which for a time during the 1980s was the official name of Cambodia, means "country where gold lies at the foothill." But Cambodia is certainly not a land of gold, nor of food, freedom, or stability. The average Cambodian is malnourished, illiterate, and will die before age 62, having experienced neither freedom nor peace. Despite a new Constitution, massive United Nations aid, and a formal cease-fire, the horrific effects of Cambodia's bloody civil war continue.

Cambodia was not always a place to be pitied. In fact, at times it was the dominant power of Southeast Asia. Around the fourth century A.D., India, with its pacifist Hindu ideology, began to influence in earnest the original Chinese base of Cambodian civilization. The Indian script came to be used, the name of its capital city came to be an Indian word, its kings acquired Indian titles, and many of its Khmer people came to believe in the Hindu religion. The mile-square Hindu temple Angkor Wat, built in the twelfth century, still stands as a symbolic reminder of Indian influence, Khmer ingenuity, and the Khmer Empire's glory.

But the Khmer Empire, which at its height included parts of present-day Myanmar (Burma), Thailand, Laos, and Vietnam, was gradually reduced both in size and power until, in the 1800s, it was paying tribute to both Thailand and Vietnam. Continuing threats from these two countries as well as wars and domestic unrest at home led the king of Cambodia to appeal to France for help. France, eager to offset British power in the region, was all too willing to help. A protectorate was established in 1863, and French power grew apace until, in 1887, Cambodia became a part of French Indochina, a conglomerate consisting of the countries of Laos, Vietnam, and Cambodia.

The Japanese temporarily evicted the French in 1945 and, while under Japanese control, Cambodia declared its "independence" from France. Heading the country was the young King Norodom Sihanouk. Controlling rival ideological factions, some of which were pro-West while others were pro-Communist, was difficult for Sihanouk, but he built unity

TIMELINE

A.D. 1863
France gains control of Cambodia

1940s
Japanese invasion; King Norodom Sihanouk is installed

1953
Sihanouk wins Cambodia's Independence of France

1970
General Lon Nol takes power in a U.S.-supported coup

1975
The Khmer Rouge, under Pol Pot, overthrows the government and begins a reign of terror

1978
Vietnam invades Cambodia and installs a puppet government

1990s
Vietnam withdraws troops from Cambodia; a Paris cease-fire agreement is violated by the Khmer Rouge; Pol Pot dies and other Khmer Rouge leaders surrender

2000s
Hun Sen is reelected
Prince Ranariddh becomes president of the National Assembly
Local elections are held for the first time in more than 20 years
Hun Sen loses parliamentary majority in the 2003 elections and agrees to share power with opposition parties
King Sihanouk abdicates the throne. His son, Norodom Sihamoni, age 51, is selected to replace him
Hun Sen's deal with Vietnam gives land to solve a long-standing border dispute, but causes violent demonstrations in the capital city
Cambodian troops kill three Thai soldiers who entered a disputed border area
The four surviving leaders of the Khmer Rouge were put on trial in a U.N.-supported tribunal

A row of Buddhist carvings line the bridge of the entrance to Angkor Wat.

around the idea of permanently expelling the French, who finally left in 1955. King Sihanouk then abdicated his throne in favor of his father so that he could, as premier (prime minister), personally enmesh himself in political governance. He used the title Prince Sihanouk until 1993, when he declared himself, once again, King Sihanouk.

From the beginning, Sihanouk's government was bedeviled by border disputes with Thailand and Vietnam and by the incursion of Communist Vietnamese soldiers into Cambodia. Sihanouk's ideological allegiances were confusing at best; but, to his credit, he was able to keep Cambodia officially out of the Vietnam War, which raged for years (1950–1975) on its border. In 1962, Sihanouk announced that his country would remain neutral in the Cold War struggle.

Neutrality, however, was not seen as a virtue by the United States, whose people were becoming more and more eager either to win or to quit the war with North Vietnam. A particularly galling point for the U.S. military was the existence of the so-called Ho Chi Minh Trail, a supply route through the tropical mountain forests of Cambodia. For years, North Vietnam had been using the route to supply its military operations in South Vietnam, and Cambodia's neutrality prevented the United States, at least legally, from taking military action against the supply line.

All this changed in 1970, when Sihanouk, out of the country at the time, was evicted from office by his prime minister, General Lon Nol, who was supported by the United States and South Vietnam. Shortly thereafter, the United States, then at its peak of involvement in the Vietnam War, began extensive military action in Cambodia. The years of official neutrality came to a bloody end.

■ THE KILLING FIELDS

Most of these international political intrigues were lost on the bulk of the Cambodian population, only half of whom could read and write, and almost all of whom survived, as their forebears had before them, by cultivating rice along the lush Mekong River Valley. Always an agricultural economy, Cambodia's annual monsoon rains have contributed to the numerous large rivers that sustain farming on the country's large central plain and support the growth of tropical forests in the mountains. Thus, Cambodia is abundantly blessed with such useful products as coconuts, palms, bananas, and rubber. Despite these resources, the country has almost always been poor, so rural villagers have long since learned that, even in the face of war, they could survive by hard work and reliance on extended-family networks. During the Vietnam War, most farmers probably thought that the war next door would not seriously alter their lives. But they were profoundly wrong, for, just as the United States had an interest in having a pro-U.S. government in Cambodia, the North Vietnamese desperately wanted Cambodia to be pro-Communist.

North Vietnam wanted the Cambodian government to be controlled by the Khmer Rouge, a Communist guerrilla army led by Pol Pot, one of a group of former students influenced by the left-wing ideology taught in French universities during the 1950s. After five years of bloody battles with government troops, the rebel Khmer Rouge took control of the country and launched, in 1975, over 3 years of hellish extermination of the people, resulting in the deaths of between 1 million and 3 million fellow Cambodians—that is, between one-fifth and one-third of the entire Cambodian population. The official goal was to eliminate anyone who had been "polluted" by prerevolutionary thinking, but what actually happened was random violence, torture, and murder.

It is impossible to fully convey the mayhem and despair that engulfed Cambodia during those years. Cities were emptied of people. Teachers and doctors were killed or sent as slaves to work in the rice paddies. Despite the centrality of Buddhism to Cambodian culture (Hinduism having long since been displaced by Buddhist thought), thousands of Buddhist monks were killed or died of starvation as the Khmer Rouge carried out its program of eliminating religion. Some people were killed for no other reason than to terrorize others into submission. Explained Leo Kuper in *International Action Against Genocide:*

> Those who were dissatisfied with the new regime
> were . . . "eradicated," along with their families, by
> disembowelment, by beating to death with hoes, by
> hammering nails into the backs of their heads and by
> other cruel means of economizing on bullets. Persons
> associated with the previous regime were special
> targets for liquidation. In many cases, the executions
> included wives and children. There were summary
> executions too of intellectuals, such as doctors,
> engineers, professors, teachers and students, leaving
> the country denuded of professional skills.

FURTHER INVESTIGATION

During four years in the 1970s, the Khmer Rouge caused the deaths of some 1.7 million people, nearly a fourth of the population. To learn about the cause of this massive violence, see:http://www.edwebproject.org/sideshow/khmeryears/index.html

DID YOU KNOW?

About 95 percent of Cambodians are Theravada Buddhists, but most of the famous religious sites, such as the Preah Vihear temple, are of Hindu origin. Preah Vihear was designated a UNESCO World Heritage Site in 2008. But if the site is supposed to be sacred, one would hardly know it. Displeased with Cambodia's claim to the site, Thailand, on whose border the temple sits, has engaged in repeated firefights with Cambodia for the past several years. Soldiers have been killed, and the temple badly damaged.

The Khmer Rouge wanted to alter the society completely. Children were removed from their families, and private ownership of property was eliminated. Money was outlawed. Even the calendar was started over, at year 0. Vietnamese military leader Bui Tin explained just how totalitarian the rulers were:

> [In 1979] there was no small piece of soap or handkerchief anywhere. Any person who had tried to use a toothbrush was considered bourgeois and punished. Any person wearing glasses was considered an intellectual who must be punished.

It is estimated that before the Khmer Rouge came to power in 1975, Cambodia had 1,200 engineers, 21,000 teachers, and 500 doctors. After the purges, the country was left with only 20 engineers, 3,000 teachers, and 54 doctors.

A kind of bitter relief came in late 1978, when Vietnamese troops (traditionally Cambodia's enemy) invaded Cambodia, drove the Khmer Rouge to the borders of Thailand, and installed a puppet government headed by Hun Sen, a former Khmer Rouge soldier who had defected and fled to Vietnam in the 1970s. Although almost everyone was relieved to see the Khmer Rouge pushed out of power, the Vietnamese intervention was almost universally condemned by other nations. This was because the Vietnamese were taking advantage of the chaos in Cambodia to further their aim of creating a federated state of Vietnam, Laos, and Cambodia. Its virtual annexation of Cambodia eliminated Cambodia as a buffer state between Vietnam and Thailand, destabilizing the relations of the region even more.

■ COALITION GOVERNANCE

The United States and others refused to recognize the Vietnam-installed regime, instead granting recognition to the Coalition Government of Democratic Kampuchea. This entity consisted of three groups: the Communist Khmer Rouge, led by Khieu Samphan and Pol Pot and backed by China; the anti-Communist Khmer People's National Liberation Front, led by former prime minister Son Sann; and the Armee Nationale Sihanoukiste, led by Sihanouk. Although it was doubtful that these former enemies could constitute a workable government for Cambodia, the United Nations granted its Cambodia seat to the coalition and withheld its support from the Hun Sen government.

Vietnam had hoped that its capture of Cambodia would be easy and painless. Instead, the Khmer Rouge and others resisted so much that Vietnam had to send in 200,000 troops, of whom 25,000 died. Moreover, other countries, including the United States and Japan, strengthened their resolve to isolate Vietnam in terms of international trade and development financing. After 10 years, the costs to Vietnam of remaining in Cambodia were so great that Vietnam announced it would pull out its troops.

A 1992 diplomatic breakthrough allowed the United Nations to establish a peacekeeping force in the country of some 22,000 troops, including Japanese soldiers—the first Japanese military presence outside Japan since World War II. These troops were to keep the tenacious Khmer Rouge faction under control. The agreement, signed in Paris by 17 nations, called for the release of political prisoners; inspections of prisons; and voter registration for national elections, to be held in 1993. Most important, the warring factions, consisting of some 200,000 troops, including 25,000 Khmer Rouge troops, agreed to disarm under UN supervision.

Unfortunately, the Khmer Rouge, although a signatory to the agreement, refused to abide by its provisions. With revenues gained from illegal trading in lumber and gems with Thailand, it launched new attacks on villages, trains, and even the UN peacekeepers themselves, and it refused to participate in the elections of 1993, although it had been offered a role in the new government if it would cooperate.

Despite a violent campaign, 90 percent of those eligible voted in elections that, after some confusion, resulted in a new Constitution; the reenthronement of Sihanouk as king; and the appointment of Sihanouk's son, Prince Norodom Ranariddh of the Royalist Party, as first prime minister and Hun Sen of the Cambodian People's Party as second prime minister.

The new Parliament outlawed the Khmer Rouge, but relations between the two premiers was rocky at best. Both began negotiating separately with the Khmer Rouge to entice them to lay down arms in return for amnesty. Thousands accepted the offer, fatally weakening the rebel army. But Hun Sen soon claimed that Norodom Ranariddh was recruiting soldiers for a future coup attempt. Despite Norodom Ranariddh's standing as the son of King Sihanouk and one of the two premiers, Hun

Snapshot: CAMBODIA

Summarized below is a quick look at the country with regard to its development, freedom, health/welfare, and achievements.

Development

In the past, China, the United States, France, and others built roads and industries in Cambodia, but as a result of two decades of internecine warfare, the country remains an impoverished state whose economy rests on fishing, farming, and massive infusions of foreign aid. Beginning in 2001, the government launched an effort to attract tourists to some of the famous temples of ancient Cambodia. The plan seems to be working, although the infrastructure is not in place to support the burgeoning tourist population.

Freedom

Few Cambodians can remember political stability, much less political freedom. Every form of human-rights violation has been practiced in Cambodia since even before the arrival of the barbaric Khmer Rouge. Suppression of dissent continues: Journalists have been killed, and opponents of the government have been expelled from the country.

Health/Welfare

Almost all of Cambodia's doctors were killed or died during the Khmer Rouge regime, and warfare disrupted normal agriculture. Thus, disease was rampant, as was malnutrition. Today, there is 1 doctor for every 4,405 people. International relief workers are hard-pressed to make a dent in the country's enormous problems, especially the increasing problem of drug-resistant malaria which is growing in villages along the Thai-Cambodian border.

Achievements

Despite violence and intimidation, 90 percent of the Cambodian people voted in the 1993 elections. Successful elections were also held in 1998 and 2003. The first indirect elections for the Senate took place in 2006. Parliamentary elections were once again successfully held in 2008. These elections restored an elected government, a limited monarchy, and a new Constitution. The government has finally eliminated the Khmer Rouge, and some measure of stability is returning to the country.

Sen deposed him in a bloody coup, and Ranariddh fled the country. Eventually, the king persuaded Hun Sen to allow his son to return to Cambodia, where he became head of the National Assembly.

International attempts to bring the aging leadership of the Khmer Rouge to justice have been frustrated by Premier Hun Sen, who apparently has persuaded many of the leaders to lay down arms in return for amnesty. Notorious leaders such as Khieu Samphan and Nuon Chea had surrendered by the late 1990s, and Ta Mok had been captured. At age 80, Ta Mok died in government custody while awaiting trial on charges of genocide. The United Nations and the United States wanted the top 12 leaders to be tried for genocide by an international court, but Hun Sen refused to cooperate on the grounds that having outsiders as judges would violate Cambodia's sovereignty.

Hun Sen opposed all but a few prosecutions, but in February 2009, the first trial of a member of the Khmer Rouge took place. The man, Kang Kek Ieu (called "Duch") had headed a prison in Phnom Penh at which 14,000 deaths occurred. Only 12 people survived Security Prison 21. But Hun Sen put severe limitations on the trial to protect former Khmer Rouge members who had joined the government, and he pressured the tribunal to rule that the length of "Duch's" detention so far was illegal, thus eliminating the possibility of his being given a life sentence. Finally, in 2010, thirty-five years after the genocide took place, "Duch" was convicted of crimes against humanity and sentenced to 35 years in prison. Four more senior leaders were brought to trial in 2011.

Cambodia's third national elections were held in 2003. Hun Sen's Cambodian People's Party (CPP) was pitted against two opposition parties, the Sam Rainsy Party (SRP) and Ranariddh's royalist party or FUNCINPEC. The opposition campaigned for human rights and the elimination of corruption in government. As usual, the election was filled with violence; several members of the SRP were killed, as was the head of the 30,000-member Cambodian Free Trade Union of Workers. However, compared to earlier local elections in which some 20 candidates were killed, the 2003 election was relatively rational. Yet, the results were unexpected: Hun Sen's party lost its parliamentary majority and was required to negotiate with the opposition. After nearly a year of bitterness, an agreement was reached whereby Hun Sen's CPP would hold 60 percent of all cabinet seats, and Ranariddh's party would control the remaining 40 percent—some of which it would share with the SRP.

Barely had the dust settled on the election when King Norodom Sihanouk, 81, announced that, due to his failing health, he was abdicating the throne. A nine-member Throne Council of politicians and Buddhist monks was created to pick the royal successor. Although, under the constitution, the kingship is not hereditary, the Council nevertheless selected one of the king's sons, Norodom Sihamoni, age 51, who had served as Cambodian ambassador to UNESCO. His career had taken him to North Korea and Europe where he became a ballet dancer and professor of classical dance in Paris. He was fluent in French, Czech, English, and Russian, and had spent most of his life away from Cambodia, living in a modest Paris apartment as a bachelor with no children.

In the 2008 election, Hun Sen's ruling Cambodian People's Party captured 60 percent of the vote; most opposition parties were unable to make any headway.

The CPP's access to television broadcasting and other resources seemed to bias the election toward the incumbents from the beginning. Despite claims of an unfair election process, campaigning appeared to be vigorous and relatively violence free.

Statistics

Geography

Area in Square Miles (Kilometers): 68,898 (181,035) (about the size of Oklahoma)

Capital (Population): Phnom Penh (1.519 million)

Environmental Concerns: habitat and biodiversity loss from illegal logging; soil erosion; lack of access to potable water; declining fish stocks from overfishing

Geographical Features: mostly low, flat plains; mountains in the southwest and north

Climate: tropical; rainy and dry seasons

People

Population

Total: 14,701,717

Annual Growth Rate: 1.7%

Rural/Urban Population (Ratio): 80/20

Major Languages: Khmer; French; English

Ethnic Makeup: 90% Khmer (Cambodian); 5% Vietnamese; 1% Chinese; 4% others

Religions: 96.5% Buddhist; 2% Muslim; 1.5% other

Health

Life Expectancy at Birth: 60 years (male); 65 years (female)

Infant Mortality: 55/1,000 live births

Physicians Available: 1/4,405 people

HIV/AIDS Rate in Adults: 0.5% (2009)

Education

Adult Literacy Rate: 74%

Compulsory Ages: 6–12

Communication

Telephones: 358,800 main lines

Mobile Phones: 8.151 million (2010)

Internet Users: 300,00 (2011)

Internet Penetration (% of Pop.): 2%

Transportation

Roadways in Miles (Kilometers): 23,670 (38,093)

Railroads in Miles (Kilometers): 423 (680)

Usable Airfields: 17

Government

Type: multiparty democracy under a constitutional monarchy

Independence Date: November 9, 1953 (from France)

Head of State/Government: King Norodom Sihamoni; Prime Minister (Premier) Hun Sen

Political Parties: Cambodian People's Party; Human Rights Party; National United Front (FUNCINPEC); Nationalist Party; Sam Rangsi Party

Suffrage: universal at 18

Military

Military Expenditures (% of GDP): 3%

Current Disputes: border disputes with Thailand and Vietnam; concern over Laos' upstream dam

Economy

Currency ($U.S. Equivalent): 4,145 riels = $1

Per Capita Income/GDP: $2,100/$30 billion

GDP Growth Rate: 6%

Inflation Rate: 4.1%

Unemployment Rate: 3.5%

Labor Force by Occupation: 58% agriculture; 16% industry; 26% services

Population Below Poverty Line: 31% (2007)

Natural Resources: timber; gemstones; iron ore; manganese; phosphates; hydropower potential; oil and gas

Agriculture: rice; rubber; corn; vegetables; cashews; tapioca; silk

Industry: rice processing; fishing; wood and wood products; rubber; cement; gem mining; textiles and garments; tourism; construction

Exports: $4.7 billion (primary partners Hong Kong, United States, Singapore)

Imports: $6 billion (primary partners China, Vietnam, Hong Kong, Thailand)

Suggested Websites

www.cambodia.org

www.tourismcambodia.com/

China

(People's Republic of China)

In early 2003, the world's first commercial magnetic levitation train or "maglev" floated silently out of Shanghai China's glitzy downtown and glided at 260 miles per hour to the ultramodern Pudong International Airport 19 miles away. Constructed by a German firm, the US$1.2 billion train, which floats above a single elevated monorail track, is not only the world's most modern form of land transportation, but it is also the face of the new China. Together with the US$1.24 billion Shanghai Formula 1 racetrack, and the launching of satellites and other unmanned spacecraft in preparation for China's go-it-alone space station, the train symbolizes a China that most people never thought would exist.

But to arrive at this point, the Chinese have had to endure many years of suffering throughout their long history. One important characteristic of China is that it is a very old culture. Human civilization appeared in China as early as 20,000 years ago, and the first documented Chinese dynasty, the Shang, began about 1523 B.C. Unproven legends suggest the existence of an even earlier Chinese dynasty (about 2000 B.C.), making China one of the oldest societies with a continuing cultural identity. Over the centuries of documented history, the Chinese people have been ruled by a dozen imperial dynasties; have enjoyed hundreds of years of stability and amazing cultural progress; and have endured more centuries of chaos, military mayhem, and hunger. Yet China and the Chinese people remain intact—a strong testament to the tenacity of human culture.

A second major characteristic is that the People's Republic of China (P.R.C.) is very big. It is the fourth-largest country in the world, accounting for 6.5 percent of the world's landmass. Much of China—about 40 percent—is mountainous; but large, fertile plains have been created by the country's numerous rivers, most of which flow toward the Pacific Ocean. China is blessed with substantial reserves of oil, minerals, and many other natural resources. Its large size and geopolitical location—it is bordered by Russia, Kazakhstan, Pakistan, India, Nepal, Bhutan, Myanmar, Laos, Vietnam, North Korea, and Mongolia—have caused the Chinese people over the centuries to think of their land as the "Middle Kingdom": that is, the center of world civilization.

However, its unwieldy size has been the undoing of numerous emperors who found it impossible to maintain its borders in the face of outside "barbarians" determined to possess the riches of Chinese civilization. During the Ch'in Dynasty (221–207 B.C.), a 1,500-mile-long, 25-foot-high wall—the so-called Great Wall—was erected along the northern border of China, in the futile hope that invasions from the north could be stopped. Although most of China's national boundaries are now recognized by international law, recent Chinese governments have found it necessary to "pacify" border areas by settling as many Han Chinese there as possible (for example, in Tibet), to prevent secession by China's numerous ethnic minorities.

Another important characteristic of modern China is its huge population. With 1.3 billion people, China is home to about one-fifth of all human beings alive today. About 92 percent of China's people are Han Chinese; the remaining 8 percent are divided into more than 50 separate

TIMELINE

1523–1027 B.C.
The Shang Dynasty is the first documented Chinese dynasty

1027–256 B.C.
The Chou Dynasty and the era of Confucius, Laotze, and Mencius

211–207 B.C.
The Ch'in Dynasty, from which the word China is derived

202 B.C.–A.D. 220
The Han Dynasty

A.D. 220–618
The Three Kingdoms period; the Tsin and Sui Dynasties

618–906
The T'ang Dynasty, during which Confucianism flourished

906–1279
The Five Dynasties and Sung Dynasty periods

1260–1368
The Yuan Dynasty is founded by Kublai Khan

1368–1644
The Ming Dynasty

1644–1912
The Manchu or Ch'ing Dynasty

1834
Trading rights and Hong Kong Island are granted to Britain

1894–1895
The Sino-Japanese War

1912
Sun Yat-sen's republican revolution ends centuries of imperial rule; the Republic of China is established

1921
The Chinese Communist Party is organized

1926
Chiang Kai-shek begins a long civil war with the Communists

1949
Mao Zedong's Communist Army defeats Chiang Kai-shek

1958
A disastrous economic reform, the "Great Leap Forward," is launched by Mao

1966–1976
The Cultural Revolution; Mao dies

(continued)

1980s

Economic and political liberalization begins under Deng Xiaoping; the P.R.C. and Britain agree to return Hong Kong to the Chinese

China expands its relationship with Taiwan; the Tiananmen Square massacre provokes international outrage

1990s

Crackdowns on dissidents and criminals result in hundreds of arrests and executions

Deng Xiaoping dies; Jiang Zemin becomes president

2000s

China bans the Falun Gong religion

The pace of China's modernization and political influence accelerates; military spending increases

Hu Jintao becomes president, in another peaceful transfer of power

The 2,900-member Legislature amends the Constitution, adding phrases about the preservation of human rights and the inviolability of private property rights

China joins Southeast Asian nations in a free trade accord that will take effect in 2010 and will become the world's largest free trade area with nearly 2 billion consumers

China is the second largest economy in the world

Beijing hosts the 2008 Olympic Games

China is the world's largest greenhouse gas polluter

China hosts World Expo 2010 in Shanghai with 70 million attendees, most of them Chinese middle class

minority groups. Many of these ethnic groups speak mutually unintelligible languages, and although they often appear to be Chinese, they derive from entirely different cultural roots; some are Muslims, some are Buddhists, some are animists. As one moves away from the center of Chinese civilization and toward the western provinces, the influence of the minorities increases. The Chinese government has accepted the reality of ethnic influence and has granted a degree of limited autonomy to some provinces with heavy populations of minorities.

Often the autonomy is in name only. For example, the far western province of Xinjiang is home to 13 ethnic groups, the largest of which is the Uighurs, a Muslim community which has briefly been independent from China in the past. The Uighurs consider themselves to have been forcefully taken over by China and frequently engage in resistance against the government. China has responded by brutally suppressing dissent and by sending large numbers of Han Chinese to live in the region to dilute the Uighur influence.

Another example of autonomy in name only is Tibet. A land of rugged beauty northeast of Nepal, India, and Bhutan, Tibet was forcefully annexed by China in 1959. Many Tibetans regard the spiritual leader, the Dalai Lama, not the Chinese government, as their true leader, but the Dalai Lama and thousands of other Tibetans live in exile in India, reducing the percentage of Tibetans in Tibet to only 44 percent. China suppresses any dissent against its rule in Tibet (even photos of the Dalai Lama, for example, cannot be displayed in public) and has diluted Tibetan culture by resettling thousands of Han Chinese in the region. Whereas the world community has

acted to protect Taiwan from an aggressive Chinese takeover, it has done very little to protest China's takeover of Tibet. In 2010, U.S. President Barack Obama officially greeted the Dalai Lama at the White House, an act which received the condemnation of the Chinese.

In the 1950s, Chinese Communist Party (CCP) chairman Mao Zedong encouraged couples to have many children, but this policy was reversed in the 1970s, when a formal birth-control program was inaugurated in China. Urban couples today are permitted to have only one child (the rule does not apply to Hong Kong or Macau). If they have more, penalties include expulsion from the Communist Party, dismissal from work, or a 10 percent reduction in pay for up to 14 years after the birth of the second child. The policy is strictly enforced in the cities, but it has had less demographic impact in the countryside where three quarters of China's people live and where families are allowed more children to help with the farmwork. In the city of Shanghai, with a population of about 17 million people, authorities have recently removed second-child privileges for farmers living near the city and for such former exceptional cases as children of revolutionary martyrs and workers in the oil industry. Despite these and other restrictions, it is estimated that 15 million to 17 million people are now born each year in China—as many people as constitute the entire population of the Netherlands or Chile.

Over the centuries, millions of people have found it necessary or prudent to leave China in search of food, political stability, or economic opportunity. Those who emigrated a thousand or more years ago are now fully assimilated into the cultures of Southeast Asia and elsewhere, and identify themselves accordingly. More recent émigrés (in the past 200 years or so), however, constitute visible, often wealthy, minorities in their new host countries, where they have become the backbone of the business community. Ethnic Chinese constitute the majority of the population in Singapore and a sizable minority in Malaysia. Important world figures such as Corazon Aquino, the former president of the Philippines, and Goh Chok Tong, the former prime minister of Singapore, are ethnically part or full Chinese. The Chinese constituted the first big wave of the 13 million Asians or part-Asians who call the United States home. Large numbers of Hong Kong Chinese immigrated to Canada in the mid-1990s, raising the number of Asians in Canada to over 3.5 million (11 percent of the population) by 2006. Of those, 1.4 million are Chinese. Thus, the influence of China continues to spread far beyond its borders, due to the influence of what are called "overseas Chinese."

Another crucial characteristic of China is its history of imperial and totalitarian rule. Except for a few years in the early 1900s, China has been controlled by imperial decree, military order, and patriarchal privilege. Confucius taught that a person must be as loyal to the government as a son should be to his father. Following Confucius by a generation or two was Shang Yang, of a school of governmental philosophy called Legalism,

The Teachings of Confucius

Confucius (550–478 B.C.) was a Chinese intellectual and minor political figure. He was not a religious leader, nor did he ever claim divinity for himself or divine inspiration for his ideas. As the feudalism of his era began to collapse, he proposed that society could best be governed by paternalistic kings who set good examples. Especially important to a stable society, he taught, were respect and reverence for one's elders. Within the five key relationships of society (ruler and subject, husband and wife, father and son, elder brother and younger brother, and friend and friend), people should always behave with integrity, propriety, and goodness.

The writings of Confucius—or, rather, the works written about him by his followers and entitled the *Analects*—eventually became required knowledge for anyone in China claiming to be an educated person. However, rival ideas such as Legalism were at times more popular with the elite; at one point, 460 scholars were buried alive for teaching Confucianism. Nevertheless, much of the hierarchical nature of Asian culture today can be traced to Confucian ideas.

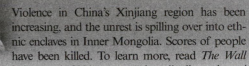

FURTHER INVESTIGATION

Violence in China's Xinjiang region has been increasing, and the unrest is spilling over into ethnic enclaves in Inner Mongolia. Scores of people have been killed. To learn more, read *The Wall Street Journal*, August 1, 2011 or go to: http://online.wsj.com/article/SB10001424053111904800304576479212126041134.html

which advocated unbending force and punishment against wayward subjects. Compassion and pity were not considered qualities of good government.

Mao Zedong, building on this heritage as well as that of the Soviet Union's Joseph Stalin and Vladimir Lenin, exercised strict control over both the public and private lives of the Chinese people. Dissidents were summarily executed (generally people were considered guilty once they were arrested), the press was strictly controlled, and recalcitrants were forced to undergo "reeducation" to correct their thinking. Religion of any kind was suppressed, and churches were turned into warehouses. It is estimated that, during the first three years of CCP rule, more than 1 million opponents of Mao's regime were executed. During the so-called Cultural Revolution (1966–1976), Mao, who apparently thought that a new mini-revolution in China might restore his eroding stature in the Chinese Communist Party, encouraged young people to report to the authorities anyone suspected of owning books from the West or having contact with Westerners. Even party functionaries were purged from the ranks if it were believed that their thinking had been corrupted by Western influences.

◼ ORIGINS OF THE MODERN STATE

Historically, authoritarian rule in China has been occasioned, in part, by China's mammoth size; by its unwieldy population; and by the ideology of some of its intellectuals. The modern Chinese state has arisen from those same pressures as well as some new ones. It is to these that we now turn.

The Chinese had traded with such non-Asian peoples as the Arabs and Persians for hundreds of years

before European contact. For example, about the year 1400, almost a hundred years before Columbus reached the New World, Chinese explorer Admiral Zheng He made multiple sailings to both India and Africa. Some claim that Chinese treasure fleets also sailed to North and South America some 70 years before Columbus did. In the 1700s and 1800s, the British and others extracted something new from China in exchange for merchandise from the West: the permission for foreign citizens to live in parts of China without being subject to Chinese authority. Through this process of granting extraterritoriality to foreign powers, China slowly began to lose control of its sovereignty. The age of European expansion was not, of course, the first time in China's long history that its ability to rule itself was challenged; the armies of Kublai Khan successfully captured the Chinese throne in the 1200s, as did the Manchurians in the 1600s. But these outsiders, especially the Manchurians, were willing to rule China on-site and to imbibe as much Chinese culture as they could. Eventually they became indistinguishable from the Chinese.

The European powers, on the other hand, preferred to rule China (or, rather, parts of it) from afar, as a vassal state, with the proceeds of conquest being drained away from China to enrich the coffers of the European monarchs. Aggression against Chinese sovereignty increased in 1843, when the British forced China to cede Hong Kong Island. Britain, France, and the United States all extracted unequal treaties from the Chinese that gave them privileged access to trade and ports along the eastern coast. By the late 1800s, Russia was in control of much of Manchuria, Germany and France had wrested special economic privileges from the ever-weakening Chinese government, and Portugal had long since controlled Macau. Further affecting the Chinese economy was the loss of many of its former tributary states in Southeast Asia. China lost Vietnam to France, Burma (today called Myanmar) to Britain, and Korea to Japan. During the violent Boxer Rebellion of 1900, the Chinese people showed how frustrated they were with the declining fortunes of their country.

Thus weakened internally and embarrassed internationally, the Manchu rulers of China began to initiate reforms that would strengthen their ability to compete with the Western and Japanese powers. A constitutional monarchy was proposed by the Manchu authorities but was preempted by the republican revolutionary

During Mao Zedong's "Great Leap Forward," huge agricultural communes were established, and farmers were denied the right to grow crops privately. The government's strict control of these communes met with chaotic results; there were dramatic drops in agricultural output.

movement of Western-trained (he received most of his education in Hawaii) Sun Yat-sen. Sun and his armies wanted an end to imperial rule; their dreams were realized in 1912, when Sun's Kuomintang (Nationalist Party, or KMT) took control of the new Republic of China, and the last emperor was forced to relinquish the throne.

Sun's Western-style approach to government was received with skepticism by many Chinese who distrusted the Western European model and preferred the thinking of Karl Marx and the philosophy of the Soviet Union. In 1921, Mao Zedong and others organized the Soviet-style Chinese Communist Party (CCP), which grew quickly and began to be seen as an alternative to the Kuomintang. After Sun's death, in 1925, Chiang Kai-shek assumed control of the Kuomintang and waged a campaign to rid the country of Communist influence. Although Mao and Chiang cooperated when necessary—for example, to resist Japanese incursions into Manchuria—they eventually came to be such bitter enemies that they brought a ruinous civil war to all of China.

Mao derived his support from the rural areas of China, while Chiang depended on the cities. In 1949, facing defeat, Chiang Kai-shek's Nationalists retreated to the island of Taiwan, where, under the name Republic of China (R.O.C.), they continued to insist on their right to rule all of China. The Communists, however, controlled the mainland and insisted that Taiwan was just a renegade province of the People's Republic of China. These two antagonists are officially (but not in actuality) still at war. Sometimes tensions between Taiwan and China reach dangerous levels. In the 1940s, the United States had to intervene to prevent an attack from the mainland. In 1996, U.S. warships once again patrolled the 150 miles of ocean named the Taiwan Strait to warn China not to turn its military exercises, including the firing of missiles in the direction of Taiwan, into an actual invasion. China used the blatantly aggressive actions as a warning to the newly elected Taiwanese president not to take any steps toward declaring Taiwan an independent nation. Each Chinese ruler since 1949 has re-stated the demand that Taiwan return to the fold of China; for their part, most of the people of Taiwan seem to prefer to live with de facto independence but without officially declaring it.

For many years after World War II, world opinion sided with Taiwan's claim to be the legitimate government of China. Taiwan was granted diplomatic recognition by many nations and given the China seat in the United Nations. In the 1970s, however, world leaders came to believe that it was dysfunctional to withhold recognition and standing from such a large and potentially powerful nation as the P.R.C. Because both sides insisted that there could not be two Chinas, nor one China and one Taiwan, the United Nations proceeded to give the China seat to mainland China. Dozens of countries subsequently broke off formal diplomatic relations with Taiwan in order to establish a relationship with China. Some countries continue to recognize only Taiwan, and the island relates and does business with other nations as if it were legally recognized by all. It has maintained de facto independence and is protected in that condition by the United States.

■ PROBLEMS OF GOVERNANCE

The China that Mao came to control was a nation with serious economic and social problems. Decades of civil war had disrupted families and wreaked havoc on the economy. Mao believed that the solution to China's ills was to wholeheartedly embrace socialism. Businesses were nationalized, and state planning replaced private initiative. Slowly, the economy improved. In 1958, however, Mao decided to enforce the tenets of socialism more vigorously so that China would be able to take an economic "Great Leap Forward." Workers were assigned to huge agricultural communes and were denied the right to grow their soybeans, cabbage, corn, rice, onions, and other crops privately. All enterprises came under the strict control of the central government. The result was economic chaos and a dramatic drop in both industrial and agricultural output.

Exacerbating these problems was the growing rift between the P.R.C. and the Soviet Union. China insisted that its brand of communism was truer to the principles of Marx and Lenin, and criticized the Soviets for selling out to the West. As relations with (and financial support from) the Soviet Union withered, China found itself increasingly isolated from the world community, a circumstance worsened by serious conflicts with India, Indonesia, and other nations. To gain friends, the P.R.C. provided substantial aid to Communist insurgencies in Vietnam and Laos, thus contributing to the eventual severity of the Vietnam War.

In 1966, Mao found that his power was waning in the face of Communist Party leaders who favored a more moderate approach to internal problems and external relations. To regain lost ground, Mao urged young students called Red Guards to fight against anyone who might have liberal, capitalist, or intellectual leanings. He called it the Great Proletarian Cultural Revolution, but it was an *anti*cultural purge: Books were burned, and educated people were arrested and persecuted. In fact, the entire country remained in a state of domestic chaos for more than a decade.

Soon after Mao died, in 1976, Deng Xiaoping, who had been in and out of Communist Party power several times before, came to occupy the senior position in the CCP. A pragmatist, he was willing to modify or forgo strict socialist ideology if he believed that some other approach would work better. Despite pressure by hardliners to tighten governmental control, he nevertheless was successful in liberalizing the economy and permitting exchanges of scholars with the West. In 1979, he accepted formalization of relations with the United States—an act interpreted as a signal of China's opening up to the world.

China's opening has been dramatic, not only in terms of its international relations but also internally. During the 1980s, the P.R.C. joined the World Bank, the International Monetary Fund, the Asian Development Bank, and other multilateral organizations. It also began to welcome foreign investment of almost any kind and permitted foreign companies to sell their products within China itself (although many companies have pulled out of China, in frustration over unpredictable business policies or because they were unexpectedly shut down by the authorities). Trade between Taiwan and China—mostly carried on via Hong Kong, but now also permitted through several small Taiwanese islands adjacent to the mainland—was nearly $6 billion by the early 1990s, and has increased exponentially since then. And while Hong Kong was investing some $25 billion in China, China was investing $11 billion in Hong Kong. More Chinese firms were permitted to export directly and to keep more of the profits. Special Economic Zones (SEZs)—capitalist enclaves adjacent to Hong Kong and along the coast into which were sent the most educated of the Chinese population—were established to catalyze the internal economy. In coastal cities, especially in south China, construction of apartment complexes, new manufacturing plants, and roads and highways began in earnest. Indeed, China, along with Hong Kong and Taiwan, seemed to be emerging as a mammoth trading bloc—"Greater China"—which economists predict will one day eclipse the economy of the United States and has already forged ahead of the economy of Japan.

? DID YOU KNOW?

According to many sources, China has an abysmal human rights record. When journalist Qi Chonghuai wrote an expose of corruption among local party officials in Tengzhou, he was arrested and sent to prison for eight years. Amnesty International says that despite China's statements that it is improving human rights, the country is actually increasing its attacks on activist lawyers, journalists, and others who speak out against abuses. Lawyers are often banned from assisting pro-democracy dissidents or members of banned religious groups. In 2011, China barred writers from joining a literary celebration in Hong Kong. One writer, Dai Qing, was banned from participating in the literary conference because she had written a story about the environmental impact of the Three Gorges Dam.

When capitalism was introduced in China, stock exchanges were opened in Shanghai and Shenzhen, and other dramatic changes were implemented, even in the western rural areas. The collectivized farm system imposed by Mao was replaced by a household contract system with hereditary contracts (that is, one step away from actual private land ownership), and free markets replaced most of the system of mandatory agricultural sales to the government. New industries were established in rural villages, and incomes improved such that many families were able to add new rooms onto their homes or to purchase two-story and even three-story homes. Glowing assessments of China's economic future must be tempered by recognition of its internal market conditions and demography. Once a country of poor, rural peasants who were willing to work for low wages, 50 percent of Chinese today live in cities where they expect higher pay. People remaining in the countryside prefer work in factories or in the service sector rather than the dirty and difficult work of farm life. The 2010 census also revealed that China's population is aging quickly. Once a country of hungry, desperate, young people, the population's share of people under 14 has dropped from 23 percent in 2000 to 16 percent in 2010, while those over 60 years old now make up over 16 percent of the population. By 2035, 20 percent of China's population will be 65 and older. Thus, China's era of abundant, cheap labor is slowly coming to an end. But the country's economy remains a force to be reckoned with. China's entry into the World Trade Organization (WTO) in 2001, though causing some difficult internal adjustments, will, in the long run, strengthen China's position in the world economy.

■ TIANANMEN SQUARE

A strong spirit of entrepreneurship took hold throughout the country in the 1980s. Many people, especially the growing body of educated youth, interpreted economic liberalization as the overture to political democratization. College students, some of whom had studied abroad, pressed the government to crack down on corruption in the Communist Party and to permit greater freedom of speech and other civil liberties.

In 1989, tens of thousands of college students staged a prodemocracy demonstration in Beijing's Tiananmen Square. The call for democratization received wide international media coverage and soon became an embarrassment to the Chinese leadership, especially when, after several days of continual protest, the students constructed a large statue in the square similar in appearance to the Statue of Liberty in New York Harbor. Some party leaders seemed inclined at least to talk with the students, but hard-liners apparently insisted that the prodemocracy movement be crushed in order that the CCP remain in control of the government. The official policy seemed to be that it would be the Communist Party, and not some prodemocracy movement, that would, paradoxically, lead China to capitalism.

The CCP leadership had much to fear; it was, of course, aware of the quickening pace of Communist party power dissolution in the Soviet Union and Central/Eastern Europe, but it was even more concerned about corruption and the breakdown of CCP authority in the rapidly capitalizing rural regions of China, the very areas that had spawned the Communist Party under Mao. Moreover, economic liberalization had spawned inflation, higher prices, and spot shortages, and the general public was disgruntled. Therefore, after several weeks of pained restraint, the authorities moved against the students in what has become known as the Tiananmen Square massacre. Soldiers injured thousands and killed hundreds, perhaps thousands, of students; hundreds more were systematically hunted down and brought to trial for sedition and for spreading counterrevolutionary propaganda.

In the wake of the brutal crackdown, many nations reassessed their relationships with the People's Republic of China. The United States, Japan, and other nations halted or canceled foreign assistance, exchange programs, and special tariff privileges. The people of Hong Kong, anticipating the return of their British colony to P.R.C. control in 1997, staged massive demonstrations against the Chinese government's brutality, and they continue to do so annually, much to the frustration of the Chinese leadership, who, according to the agreement with Britain, must allow such demonstrations in Hong Kong, but brutally suppresses them within China proper. Foreign tourism all but ceased, and foreign investment declined abruptly.

The withdrawal of financial support and investment was particularly troublesome to the Chinese leadership, as it realized that China's economy was far behind other nations. Even Taiwan, with a similar heritage and a common history until the 1950s, but having far fewer resources and much less land, had long since eclipsed the mainland in terms of economic prosperity. The Chinese understood that they needed to modernize (although not, they hoped, to Westernize), and they knew that large capital investments from such countries as Japan, Hong Kong, and the United States were crucial to their economic reform program. Moreover, they knew that they could not tolerate a cessation of trade with their new economic partners. By the end of the 1980s, about 13 percent of China's imports came from the United States, 18 percent from Japan, and 25 percent from Hong Kong. Similarly, Japan received 16 percent of China's exports, and Hong Kong received 43 percent.

Once the worldwide furor over the 1989 Tiananmen massacre died down, and fortunately for the Chinese economy, the investment and loan-assistance programs from other countries were reinstated in most cases. China was even able to close a US$1.2 billion contract with McDonnell Douglas Corporation to build 40 jetliners; and as a result of decisions to separate China's human-rights issues from trade issues, the United States repeatedly renewed China's "most favored nation" trade status. U.S. President Bill Clinton and China's leader, Jiang Zemin, engaged in an unprecedented public debate on Chinese television in June 1998; and Clinton was allowed to engage students and others in direct dialogue in which he urged religious freedom, free speech, and the protection of other human rights.

These and other events suggest that China is trying to address some of the major concerns voiced against it by the industrialized world—one of which is copyright violations by Chinese companies. Some have estimated that as much as 88 percent of China's exports of CDs consists of illegal copies. A 1995 copyright agreement is having some effect, but still, much of China's trade deficit with the United States comes from illegal products.

One such illegal activity was the fake Apple stores discovered in 2010. In a situation people are calling "iFraud," numerous retail stores looking almost like authorized Apple stores have been selling illegally imported Apple products but without Apple's knowledge or permission. Problems like this are massive in China, and continue despite official statements to the contrary. These things did not stop the United States, Japan, Australia, and other countries from pushing for China to join the World Trade Organization. In fact, many business leaders believe that WTO membership will force China to adhere more faithfully to international rules of fair production and trade. Still, China's reputation as an honest trade partner continues to suffer. For instance, in 2009, it was revealed that China (and perhaps the government itself) was behind a mammoth scheme to cyber-spy on computers in Vietnam, Taiwan, India, and some 100 other countries. Before that, in 2008, additional damage was inflicted on China's reputation when it was learned that milk tainted with the chemical melamine was being sold. Altogether some 300,000 children were sickened and 10 died.

Improved trade notwithstanding, the Tiananmen Square massacre and continuing instances of brutality against citizens have convinced many people, both inside and outside China, that the Communist Party has lost, not necessarily its legal, but certainly its moral, authority to govern. Amnesty International's annual reports consistently report hundreds of people who are arrested for believing in a religion of which the government does not approve, or for advocating greater freedoms of expression in Tibet (where photos of the Dalai Lama are outlawed), or in Uighur areas of west China. The organization has reported violations of human rights "on a massive scale," and noted that torture is used on political prisoners held in *laogai*, Chinese gulags similar to those in the former Soviet Union.

FURTHER INVESTIGATION

In 2011, two 18-year-old Tibetan monks set themselves on fire in China while declaring, "Long live His Holiness the Dalai Lama; we need religious freedom immediately." Evidence of China's restrictions on religious freedom may be found at: www.christianpersecution.info/china.php

Although some religions are permitted in China, others are suppressed, the most notable being the Falun Gong, a meditative religion that China banned in 2000 and whose adherents are systematically arrested and jailed. In 2010 troops in armored personnel carriers shut down a five-story mega-church attended by 50,000 Christians, and pressure against Christians in so-called "house religions" has also increased.

The cases of China's human-rights violations would fill volumes, from Tibetan leaders imprisoned for "revolutionary incitement," to U.S. prodemocracy activists deported for talking to striking workers, to the jamming of Radio Free Asia and the Voice of America, to the blocking of Internet search engines and websites. One Chinese dissident, Xu Wenli, was imprisoned for 16 years just for organizing an independent political party. He was released in 2002 after the United States pressured the Chinese government. In just one recent year, numerous instances of suppression of free speech and other civil rights were reported: a Chinese Buddhist leader was dragged away for attempting to hold a religious ceremony not approved by the government; a Catholic bishop was arrested twice; a military surgeon was jailed for writing a letter in which he urged the government to take responsibility for the Tiananmen massacre; three Protestant activists were sent to prison for three years for "leaking state secrets"; some 54 people were jailed when they wrote their opinions on the Internet about the SARS epidemic and other topics; and, after some of the 400 million mobile phone users started sending text messages that exposed the national cover-up of the SARS epidemic, the Chinese government shut down 20 Internet service providers and began censoring text messages.

There are so many mobile phone users in China (an average of 7,000 messages every second, or more than the rest of the world combined), that the government regards the technology as a direct threat to its ability to control the flow of information and thus to govern. When villagers in Wangying protested illegal taxes on their crops, the government countered by ransacking their homes, dousing people with boiling water, and beating and whipping them. In 2009, Tibetan monks were arrested for possessing "reactionary music." Many countries have condemned these actions (the U.S. House of Representatives approved an almost unanimous declaration to demand that China stop its harsh policies toward Tibet), but few press the point vigorously for fear of negative impacts on trade with China.

It is not only outsiders who protest the government's policies. In 2008, for example, an earthquake in Sichuan province killed some 87,000 people, 5,000 of whom were school children who died when some 7,000 poorly constructed schools collapsed on them. When the parents tried to protest the shoddy construction of the so-called "tofu-dregs schools," some of them were arrested or forcibly returned to their homes.

Accurate figures are hard to obtain, but in one year alone, 2005, it is estimated that there were over 87,000 "public order" disturbances—that is, riots and other violent attacks against government authorities. Many of those were spawned by the government's taking of land from farmers. A 2002 law gives farmers the right to a 30-year contract for their land, but fewer than half the farmers have received a contract, and nearly 30 percent of villages report cases of land being forcibly taken by the government.

The government has little patience for public protest. When residents protested the construction of a power plant, for example, the local Communist Party head hired a gang to attack and beat the protestors. In an unusual case of justice, he was later convicted to life in prison and four others were executed (China leads the world in the number of annual executions; many are of political dissenters). Generally, however, maltreatment of citizens goes unpunished. Police beat rural protestors at one sit-in and killed many others in an incident in which farmers had built dikes to irrigate land without government permission. Protestors burned police cars and threw bricks at police. Thousands of such protests occur each year, but the general public does not hear about them because the government has increasingly restricted the reporting of public unrest in the newspapers. Thus, while the number of newspapers has increased, the amount of press freedom has not. Foreign companies operating in China have also succumbed to pressure from the government to restrict the free flow of information. Yahoo, Google, Microsoft, and Cicso Systems have all submitted to Chinese government pressure on this. Google, in a protracted struggle with Chinese censorship, finally had all its users re-directed to its servers in Hong Kong because of cyber attacks on it, apparently from the Chinese government.

Then, of course, there is the issue of Tibet, where repression of religion has kept the Dalai Lama and many others in exile for more than 40 years, and where police in a western province shot and killed the head of a Buddhist monastery in 2004 after he complained about being beaten in custody. One political prisoner, a Tibetan nun, was released in 2002 after 11 years in prison. She said that since the age of 13 she was tortured with beatings, solitary confinement, and electric shocks while in custody in the Drapchi prison. Despite all the economic liberalization, it is very clear to all observant China-watchers that China remains a dictatorship where basic human freedoms are regularly denied, and where most challenges to the government, however mild, will be met with suppression.

Despite China's controlled press, reports of other forms of social unrest are occasionally heard. For example, in 1999, some 3,000 farmers in the southern Hunan Province demonstrated against excessive taxation. One protester was killed. Labor riots break out from time to time as workers protest the growing disparity between rich and poor. The government, despite its socialist rhetoric, admits that it is trying to create a middle-income group, but corruption at every level of society causes seething resentments. In western regions with large Islamic populations, anti-Beijing sentiment sometimes erupts in the form of bombings of government buildings, and through underground antigovernment meetings. In Langchenggang, a town in central China, deadly street fights broke out between Muslims and Han Chinese in 2004, and many other villages have erupted in violence, although news about the events is usually suppressed.

■ THE SOCIAL ATMOSPHERE

In 1997, the aged Deng Xiaoping died. He was replaced by a decidedly more forward-looking leader, Jiang Zemin. Under his charge, the country was able to avoid many of the financial problems that affected other Asian nations in the late 1990s (although many Chinese banks are dangerously overextended, and real-estate speculation in Shanghai and other major cities has left many high-rise office buildings severely underoccupied). Despite many problems yet to solve, including serious human-rights abuses, it is clear that the Chinese leadership has actively embraced capitalism and has effected a major change in Chinese society. Historically, the loyalty of the masses of the people was placed in their extended families and in feudal warlords, who, at times of weakened imperial rule, were nearly sovereign in their own provinces. Communist policy has been to encourage the masses to give their loyalty instead to the centrally controlled Communist Party. The size of families has been reduced to the extent that "family" as such has come to play a less important role in the lives of ordinary Chinese.

Historical China was a place of great social and economic inequality between the classes. The wealthy feudal lords and their families and those connected with the imperial court and bureaucracy had access to the finest in educational and cultural opportunities. While around them lived illiterate peasants who often could not feed themselves, let alone pay the often heavy taxes imposed on them by feudal and imperial elites. The masses often found life to be bitter, but they found solace in the teachings of the three main religions of China (often adhered to simultaneously): Confucianism, Taoism, and Buddhism. Islam, animism, and Christianity have also been significant to many people in China.

The Chinese Communist Party under Mao, by legal decree and by indoctrination, attempted to suppress people's reliance on religious values and to reverse the ranking of the classes; the values of hard, manual work and rural simplicity were elevated, while the refinement and education of the urban elites were denigrated. Homes of formerly wealthy capitalists were taken over by the government and turned into museums, and the opulent life of the capitalists was disparaged. During the Cultural Revolution, high school students who wanted to attend college had first to spend two years doing manual labor in factories and on farms to help them learn to relate to the peasants and the working class. So much did revolutionary ideology and national fervor take precedence over education that schools and colleges were shut down for several years during the 1960s and 1970s, and the length of compulsory education was reduced.

One would imagine that after six decades of communism, the Chinese people would have discarded the values of old China. However, the reverse seems to be true.

Woman washing clothes by hand in the Yangtze River. At least 1 million villagers living along the Yangtze have been displaced by the rising waters of the mammoth Three Gorges Dam.

When the liberalization of the economy began in the late 1970s, many of the former values also came to the fore: the Confucian value of scholarly learning and cultural refinement, the desirability of money, and even Taoist, Buddhist, and Christian religious values. Also returning is extreme inequality. In Shanghai, some of China's 535,000 millionaires (China is fourth in the world in the number of millionaires) ride in chauffeured cars and enjoy dinner at the Hard Rock Café, while their children attend concerts of the Rolling Stones. In the countryside, farmers live with no electricity and have minimal health care and schooling. Government leaders are aware of, and comment publicly about, the growing inequality, yet the desire to make money causes many urban dwellers to ignore the problems of the farmers.

Thousands of Chinese are studying abroad with the goal of returning to China to establish or manage profitable businesses. Indeed, some Chinese, especially those with legitimate access to power, such as ranking Communist Party members, have become extremely wealthy. Along with the privatization of state enterprises has come the unemployment of hundreds of thousands of "redundant" workers (2 million workers lost their jobs in one province in a single year in the early 1990s). Many others have had to settle for lower pay or unsafe work conditions as businesses strive to enter the world of competitive production. Each year, numerous accidents at various workplaces remind everyone that China's physical and managerial infrastructure

is woefully inadequate: a gas explosion in a coal mine in Shaanxi province traps hundreds of miners; a chlorine gas leak and explosion at a chemical plant in Chongqing kills 233 and forces the evacuation of 150,000; an accident in a tin mine kills 81 people, and the Communist Party secretary pays bribes to conceal the tragedy; and 2,000 fireworks factories are closed after a series of fatal explosions. Some 4,000 to 5,000 miners die each year in mine accidents. Hundreds of demonstrations and strikes each year highlight the difficulty of life for laborers; even those with good jobs find it difficult to keep up with inflation, which in 2010 was 15 percent and has been as high as 22 percent. Many of the most modern features of the new China are simply unattainable by the average Chinese person. For instance, the cost of riding the maglev train mentioned at the beginning of this chapter was so high that few were buying tickets on it; the government recently had to slash prices. Likewise, many modern high-rise office buildings in Shanghai and Beijing remain largely empty because business owners cannot afford the high rent. Nevertheless, those with an entrepreneurial spirit are finding ways to make more money than they had ever dreamed possible in an officially Communist country.

Some former values may help revitalize Chinese life, while others, once suppressed by the Communists, may not be so desirable. For instance, despite being an unabashed womanizer himself, Mao attempted to eradicate prostitution, eliminate the sale of women as brides, and prevent child marriages. Some of those customs are returning, and gender-based divisions of labor are making their way into the workplace.

Particularly hard hit by the results of recent population policies are females. So many couples are having female fetuses aborted in favor of males that there are about 117 males for every 100 females. Some women would rather sell their female babies than have abortions, and some of these children end up being smuggled and sold by baby-trafficking gangs. Many other girls and women are exploited on hundreds of Internet pornography sites (which the government is trying hard to eliminate by shutting down Internet cafes and making nudity on the Internet illegal), while prostitution has made a comeback all over China.

Predicting China's future is difficult, but in recent years, the Chinese government has accelerated the pace of China's modernization and its role as a major power. The results are evident at every turn, from the US$8 billion upgrading of the railway system, to the purchase by Chinese companies of major gas and oil fields in Indonesia and Nigeria, to the hosting of the 2008 Summer Olympics in Beijing and the World Expo 2010 in Shanghai. China is now the world's largest recipient of foreign direct investment, overtaking the United States. With more than US$50 billion annually flowing into China from such companies as Goldman Sachs, General Motors, Volkswagen, Ford Motor, and Toyota Motor, to name a few, China's annual growth exceeds that of most countries of the world (but when Coca-Cola tried to buy a Chinese brand, a Chinese court stopped the purchase because the public did not like a foreign firm taking control). Many in the West claim that the rapid growth is due to an unfair evaluation of the Chinese yuan compared to other currencies; this produces inexpensive products that compete successfully against similar products made in other countries where labor costs are higher. China has twice changed the value of the yuan, but only by small increments.

The rapid modernization of China is staggering: The Three Gorges Dam, completed in 2009, cost US$25 billion and employed 20,000 people. It will generate so much electricity that it will pay for itself by 2014. In addition, the digital age has come to China, with more Internet users in China than anywhere else except the United States. Also, areas that were just swamps a few years ago, for example, Pudong in Shanghai, are now filled with modern glass-and-steel skyscrapers or with industrial parks, or with modern, beautifully landscaped highways and freeways. Of interest to Westerners is the changing role of Mao in Chinese history. Once revered around the country, Mao's role in China's development, as chronicled in new history textbooks in Shanghai, has been reduced to less than a page, and Mao's famous little red book of sayings is now purchased in second-hand shops only by tourists.

Along with other sectors, China, with the second-largest military budget in the world, is moving quickly to modernize its military, a fact that is causing consternation among all other Asian states, which worry about China's intentions with its new missiles and submarines. China has also been quietly flexing its muscles in the South China Sea, and it bristles at any suggestion of a breach of its sovereignty, one example being the emergency landing (after a forced collision with a Chinese fighter jet) of a U.S. surveillance plane on Hainan Island. The Chinese government exploited the situation fully, detaining the crew for two weeks and refusing to allow the repaired plane to take off; rather, it had to be dismantled and removed piece by piece. Another example was the accidental bombing of the Chinese Embassy in Yugoslavia by NATO planes engaged in the Kosovo conflict. Upon news of the bombing, the Chinese government organized anti-U.S. and anti-NATO demonstrations all over China. The U.S. Embassy in Beijing and the consulate in Shanghai were damaged, and the United States, in particular, was vilified in the press.

Responding to China's growing tendency to flex its muscles in the region, U.S. President Barack Obama, while admitting that China cooperates with the United States on such tense matters as the North Korean nuclear issue, nevertheless declared that America will never stop being a Pacific power and never give China a free hand to do as it wishes in the region. With China spending US$160 billion in one year (2010) on military growth, including acquiring its first aircraft carrier and new stealth fighter jet, Obama indicated that the U.S. presence in the Asia Pacific was a top priority. He announced an agreement with Australia to move 2,500 Marines onto a base there, and indicated that the United States would join a new regional free trade group that does not include China.

Regardless of how China develops in the future, every country in the world now recognizes that it will have to find new ways of dealing with Asia's colossus.

SNAPSHOT

Snapshot: CHINA

Summarized below is a quick look at the country with regard to its development, freedom, health/welfare, and achievements.

Development

In the early years of Communist control, authorities stressed the value of establishing heavy industry and collectivizing agriculture. More recently, China has become the recipient of millions of dollars of investment money from around the world—money that China is using to convert itself into a modern, partly capitalist economy. Labor being cheap in China, many companies outsource their manufacturing to China. The world's largest dam, the Three Gorges Dam, was completed in 2009. China is now the second-largest economy in the world, after the United States.

Freedom

Until the late 1970s, the Chinese people were controlled by Communist Party cadres who monitored both public and private behavior. Some economic and social liberalization occurred in the 1980s, when some villages were allowed to directly elect their leaders (although campaigning was not allowed). However, the 1989 Tiananmen Square massacre reminded Chinese and the world that despite some reforms, China is still very much a dictatorship. worship is severely restricted, and those speaking against the government are often arrested and given lengthy prison terms.

Health/Welfare

The Communist government has overseen dramatic improvements in the provision of social services for the masses. Life expectancy has increased from 45 years in 1949 to 75 years today. Diverse forms of health care are available at low cost to the patient. The government has attempted to eradicate such diseases as malaria and tuberculosis. SARS remains a worrisome problem for China, as does the avian flu. In 2009, China announced it had developed a vaccine for "swine flu."

Achievements

Chinese culture has, for thousands of years, provided the world with classics in literature, art, pottery, ballet, and other arts. Under communism the arts have been marshaled in the service of ideology and have lost some of their dynamism. Since 1949, literacy has increased dramatically and now stands at 92 percent—the highest in Chinese history. Beijing successfully hosted the Olympic Games in 2008 and the World Expo in Shanghai in 2010. China became the world's largest exporter in 2010 and has successfully moved from a closed economy to an open market system for large parts of its economy.

Statistics

Geography

Area in Square Miles (Kilometers): 3,705,386 (9,596,960) (about the same size as the United States)

Capital (Population): Beijing 12.2 million

Environmental Concerns: air and water pollution; water shortages; desertification; trade in endangered species; acid rain; loss of agricultural land; deforestation

Geographical Features: mostly mountains, high plateaus, and deserts in the west; plains, deltas, and hills in the east

Climate: extremely diverse, from tropical to subarctic

People

Population

Total: 1,336,718,015

Annual Growth Rate: 0.5%

Rural/Urban Population Ratio: 53/47

Major Languages: Standard Chinese or Mandarin; Yue; Wu; Minbei; Minnan; Xiang; Gan; Hakka dialects; other minority languages

Ethnic Makeup: 91.5% Han Chinese; 8.5% minority groups including Zhuang, Hui, Uighur, Yi, Miao, Mongol, Manchu, Tibetan, Tujia, Dong, Yao, Buyi, Korean, and others

Religions: officially atheist; but Taoism, Buddhism, Islam, Christianity, ancestor worship, and animism exist

Health

Life Expectancy at Birth: 73 years (male); 77 years (female)

Infant Mortality: 16/1,000 live births

Physicians Available: 1/707 people

HIV/AIDS Rate in Adults: 0.1%

Education

Adult Literacy Rate: 92%

Compulsory (Ages): 7–17

Communication

Telephones: 314 million main lines

Cell Phones: 747 million (2009)

Internet Users: 485 million (2011)

Internet Penetration (% of Pop.): 36%

Transportation

Roadways in Miles (Kilometers): 2,398,990 (3,860,800)

Railroads in Miles (Kilometers): 53,437 (86,000)

Usable Airfields: 502

Government

Type: one-party Communist state

Independence Date: unification in 221 B.C.; People's Republic established October 1, 1949

Head of State/Government: President Hu Jintao; Premier Wen Jiabao

Political Parties: Chinese Communist Party; eight registered small parties controlled by the CCP

Suffrage: universal at 18 in village and urban district elections

Military

Military Expenditures (% of GDP): 4.3% (2006)

Current Disputes: minor border disputes with a few countries, and potentially serious disputes over Spratly and Paracel Islands with several countries

Economy

Currency ($ U.S. Equivalent): 6.78 yuan = $1

Per Capita Income/GDP: $7,600/$10 trillion

GDP Growth Rate: 10.3%

Inflation Rate: 5%

Unemployment Rate: about 4% in urban areas and 25% or more in rural areas

Labor Force by Occupation: 38% agriculture; 34% services; 28% industry

Population Below Poverty Line: 2.8% (mainly in rural areas)

Natural Resources: coal; petroleum; iron; ore; tin; tungsten; antimony; lead; zinc; vanadium; magnetite; uranium; hydropower; natural gas; mercury; aluminum; manganese

Agriculture: food grains; cotton; oilseed; pork; fish; tea; potatoes; peanuts; apples

Industry: iron and steel; coal; machinery; textiles and apparel; food processing; consumer durables and electronics; telecommunications; armaments; transportation equipment; space launch vehicles and satellites; petroleum; cement; chemicals; fertilizers

Exports: $1.5 trillion (primary partners United States, Hong Kong, Japan)

Imports: $1.3 trillion (primary partners Japan, South Korea, United States)

Suggested Websites

www.china.org.cn

www.chinatoday.com/

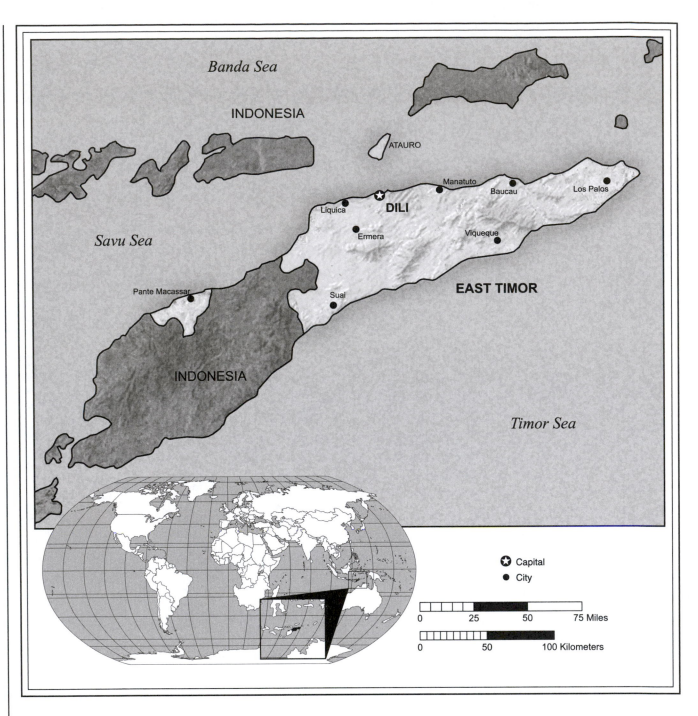

Banda Sea

INDONESIA

ATAURO

Manatuto

Baucau

Los Palos

Liquica

DILI

Savu Sea

Ermera

Viqueque

Pante Macassar

Suai

EAST TIMOR

INDONESIA

Timor Sea

⭐ Capital
● City

| 0 | | | 25 | | 50 | | 75 Miles |

| 0 | | 50 | | 100 Kilometers |

East Timor

(Democratic Republic of Timor-Leste)

Among the athletes competing in Sydney in the 2002 Summer Olympics were four from East Timor, the Pacific Rim's newest nation. With the sovereignty of their country still in the hands of a transitional United Nations administration, the athletes had to march behind the flag of the International Olympic Committee. But they marched to thunderous applause from the spectators, who were obviously pleased that the painful struggle to bring peace and freedom to the troubled island had finally ended in success.

■ THE PORTUGUESE SETTLE IN

It was the fragrant sandalwood forests that first attracted the Portuguese to the tiny island of Timor. While other European explorers were busy trying to find their way to the fabled riches and spices of China by sailing west, the Portuguese and the Dutch tried sailing east around Africa. Along the way, they came across Timor. Located about 400 miles northwest of Australia, the mountainous Timor (280 miles long and 65 miles across, with some mountains nearly 10,000 feet high) is one of the 13,000 islands that comprise the Malay (Indonesian) Archipelago. The sandalwood trees, useful for woodcarvings, as well as teak, rosewood, bamboo, and eucalyptus, convinced the Portuguese to stay.

By 1520 the Portuguese had settled a colony on the island; they would, no doubt, have controlled the whole island had it not been for the Dutch, who claimed the island as well. Eventually, control of Timor was divided between the Netherlands in the west, and Portugal in the east. Thus was born East Timor. When, after World War II, the rest of the Dutch East Indies gained independence as the new nation of Indonesia, East Timor remained a colony of Portugal.

The Portuguese never invested heavily in the development of East Timor. Portuguese-owned companies harvested the sandalwood trees, extracted marble from quarries, and grew coffee on the mountainsides, but Portugal regarded the distant island as somewhat peripheral to its interests. Yet when some of the Timorese people (many of whom were of mixed Portuguese-Timorese descent and most of whom had become Christians) declared their desire for independence, Portugal initially responded with force and then, in 1974, attempted to establish a provisional local government. Fighting broke out between those favoring independence and those preferring integration with Indonesia. Rather than try to solve the problem, Portugal withdrew. In 1975, the Revolutionary Front of Independent East Timor (Fretilin) declared victory and proclaimed the formation of the Democratic Republic of East Timor. After 450 years of outside control, East Timor was free. But the freedom lasted only nine days.

■ INDONESIA INVADES

In December 1975, General Suharto of Indonesia invaded the island with some 35,000 air, land, and naval troops. The attackers were brutal: They dropped napalm bombs on isolated mountain villages; they shot indiscriminately into crowds of unarmed civilians; they tortured, raped, and mutilated their victims. By the time the violence had stopped four years

TIMELINE

A.D. 1000
Traders from Java and China arrive to obtain sandalwood

1520
The Portuguese settle a colony on the island

1613
Dutch traders land on Timor

1859
The island is divided between the Netherlands in the west and Portugal in the east

1894–1912
The Portuguese put down an independence movement

1950s–1970s
East Timorese rebel against Portuguese rule

1975
Portugal withdraws; East Timor declares independence from Portugal; Indonesia invades

1975–1989
East Timor is closed to the outside world

1976
Indonesia proclaims East Timor its 27th province

1978
East Timor's president is killed by Indonesian soldiers

1990s
Indonesia agrees to independence referendum; 79% favor independence; Indonesian militias ravage the country
The UN takes control of the country and restores peace

2000s
"Xanana" Gusmao is elected president, in UN-sponsored elections
East Timor becomes a sovereign nation
A committee to settle maritime boundaries with Indonesia and others continues its work of mapping the actual border of the world's newest country.
Domestic violence caused by unemployment taxes the skills of the country's leadership

2008
Ramos-Horta survives an assassination attempt

FURTHER INVESTIGATION

The Indonesian's military attack on East Timor was horrendous. To learn more about the cause of this carnage, see: http://news.bbc.co.uk/2/hi/asia-pacific/443456.stm

later, some 200,000 people (almost a third of the population) had been killed, including Nicolau Lobato, the East Timorese president. Many resisters had been placed in concentration camps. Furthermore, nearly 80 percent of the farmers (most islanders are farmers) had been displaced from their lands, wreaking havoc on the economy. The world community condemned the invasion, and the United Nations refused to accept Indonesia's claim that East Timor had now become a province of Indonesia. But Indonesia held onto the island for 25 years, until 1999, when a new group of Indonesian leaders reluctantly agreed to a UN-brokered referendum on independence.

When the votes were counted, 79 percent of the people (almost the exact percentage of the non-Indonesian part of the population) voted for independence. The result produced outrage among the mostly Muslim Indonesian troops living in East Timor. Refusing to accept the result, they launched a hellish blitz of random violence and destruction—much of it with anti-Christian overtones. Bands of marauding militias killed anyone in their way, raped women, and destroyed everything they could. They attacked United Nations officials as well as the Australian ambassador, and as they launched their scorched-earth campaign, they declared that "a free East Timor will eat stones." Despite a global outcry against the violence, Indonesian leaders seemed to do nothing to stop it. An international force led by Australia eventually drove out the Indonesians and brought an end to the brutality, but by that time hundreds of civilians had been killed, dozens of villages had been burned, and 70 percent of the buildings of the capital city of Dili had been intentionally destroyed. The militias did so much damage that 70 percent of the economy was also destroyed. From these painful beginnings was born the Pacific Rim's newest nation.

■ INDEPENDENCE

On May 20, 2002, East Timor became a sovereign nation. Former U.S. president Bill Clinton attended the independence ceremony, as did the president of Indonesia, who,

DID YOU KNOW?

In 2006, East Timorese soldiers went on strike for higher pay. When the prime minister fired them (about half of the entire force), they began rioting in the streets. Forces loyal to the prime minister fought back, and soon the violence spread to the police force and the population as a whole. Frightened, 130,000 people fled their homes. The rioting finally stopped when Australian troops arrived and when the prime minister resigned his office.

in a gesture of reconciliation, shared the platform with President Gusmao.

In an interesting development, East Timor adopted the U.S. dollar as its currency. One would assume that the people would be relieved that the fighting has stopped, yet, troops from Australia, New Zealand, Japan, Portugal, and Malaysia have been needed to stabilize the still-restive populace. Although war refugees were finally resettled by late 2005, the economy has struggled to recover. As of 2011, unemployment was still 20 percent. A sign of the country's impoverishment is that out of 1 million inhabitants, only 2,100 people use the Internet. In 2006, when some 600 soldiers demanding more pay were fired by the government, they launched a revolt and demanded the resignation of the premier. Gangs armed with bows and arrows, machetes, and slingshots looted businesses and burned cars and homes. Government troops fired on and killed unarmed police who they claimed were helping the rebels, and the former interior minister was arrested for plotting to kill opponents of the premier. In all, some 30 deaths resulted from the violence, and 150,000 people fled their homes.

Eventually, Premier Mari Alkatiri resigned, but only after President Gusmao threatened to resign himself unless Alkatiri quit his office. Gusmao's popularity with the people remains so high that he was finally able to pacify the opponents and have the rebels turn their arms over to Australian troops. But violence flared once again in 2008 when President José Ramos-Horta was severely wounded by rebel soldiers in what some believed was a coup attempt. Ramos-Horta returned to his post after medical attention in Australia.

With plentiful supplies of oil in the oceans surrounding East Timor, the country could become economically viable if the oil extraction is handled properly. However, a number of interests claim rights to the same oil fields as East Timor. For example, Australia claims it controls a large share of the oil, as does Indonesia, ConocoPhillips, and others. President Gusmao accused Australia of an "oil grab" when it started drilling for oil immediately upon the expiration of the former sea boundary agreement with Indonesia in 1999. Gusmao claims that his country's economy will be doomed if Australia continues to drill. It appears it will be many years before the maritime boundaries and oil rights issues are settled by the East Timor-Indonesia Boundary Committee, but if traditional maritime law is followed, most oil deposits will likely be located inside East Timor's waters. In the meantime, the country has signed a 50-year oil development agreement with Australia. If successful, it could transform the economy.

■ THE PEOPLE OF EAST TIMOR

Full and accurate statistics on the population of East Timor are difficult to come by. In 1975, just before Suharto's invasion, the 680,000 East Timorese were divided as follows: 97 percent Timorese (including mestizos); 2 percent Chinese; 1 percent Portuguese. In 2011, the population was estimated at 1.2 million.

Snapshot: EAST TIMOR

Summarized below is a quick look at the country with regard to its development, freedom, health/welfare, and achievements.

Development

Under Indonesian occupation, new roads, bridges, schools, hospitals, and community health centers were constructed. Government offices were built in Dili and in smaller towns, but militias destroyed many of them in 1999. Australia signed an oil-development deal in 2002, but East Timor wants to clarify maritime boundaries before approving further oil development. In 2002, the government took control of the media, which had been under UN supervision. In 2007, East Timor and Australia signed a 50-year development agreement.

Freedom

The Indonesian government released Xanana Gusmao after 20 years as a political prisoner. He became president of East Timor in 2002 and attempted to restore civil liberties on the basis of Portuguese law and tradition. Under UN supervision, refugees from the 1999 violence were resettled as of 2005, but high unemployment continues to yield pockets of unrest and domestic violence. José Ramos-Horta became the second president in 2006 and survived an assassination attempt in 2008.

Health/Welfare

East Timor is one of the poorest nations on Earth. A very weak infrastructure limits access to medical care. Years of fighting have destroyed hospitals and many governmental functions.

Achievements

The Nobel Peace Prize of 1996 was awarded to two East Timorese: José Ramos-Horta and Bishop Carlos Filipe Ximenes Belo.

Like many other Southeast Asian nations, East Timor is a social gumbo of religions, languages, and ethnicities. The first known inhabitants were the Atoni, a people of Melanesian descent. Most of the people today are of mixed Malay, Polynesian, and Papuan descent, with some Chinese and many Portuguese-Timorese mestizos added into the mix. Official statistics show that about 98 percent of the population is Roman Catholic (the harbor city of Dili boasts the largest Catholic cathedral in Southeast Asia), but animism is still widely practiced in many of the hundreds of isolated villages.

Isolation has meant that not everyone in East Timor speaks the same language. Despite the small size of the country, East Timor is comprised of 12 ethnic groups speaking some sixteen Austronesian and Papuan languages. The most frequently used language is Tetum, spoken by about 60 percent of the people, but Portuguese is also an official language.

Although international attention has focused on the capital city of Dili and the ravages that city-dwellers suffered at the hands of the Indonesian militias, the typical East Timorese is a country person, a farmer using slash-and-burn methods and growing only enough corn and vegetables to support a family. Many grow coffee, which was once the country's most profitable export. Handicraft enterprises (household utensils, clothing, and farm tools) also provide a livelihood for some. But years of warfare and a drought in the early 2000s, have destroyed much of the economy. International aid has helped rebuild part of the country, but, for several years, the economy actually declined each year, and East Timor was ranked as the poorest country on Earth. The economy began to improve in 2006, and with oil revenues coming in sooner than expected, the economy has been showing growth above 6 percent annually. Yet massive problems remain with both infrastructure and with the neophyte civil service.

Solving these economic problems and restoring normalcy will tax the skills of East Timor's leaders, prominent among whom are Prime Minister Gusmao who once headed the Timorese resistance movement and was a political prisoner for 20 years; and President Ramos-Horta, also a member of the resistance leadership and co-winner (with Catholic bishop Carlos Filipe Ximenes Belo) of the Nobel Peace Prize. Other claimants to political leadership have had a hard time allowing these men to rule, but Gusmao, in particular, commands the respect of the people.

Statistics

Geography

Area in Square Miles (Kilometers): 5,743 (14,874) (about the size of Connecticut)
Capital (Population): Dili (166,000)
Environmental Concerns: deforestation; soil erosion
Geographical Features: mountainous; part of the Malay Archipelago
Climate: tropical

People

Population

Total: 1,177,834
Annual Growth Rate: 1.98%
Rural/Urban Population Ratio: 72/28
Major Languages: Tetum; Portuguese; Indonesian; English
Ethnic Makeup: Malayo-Polynesian; Papuan; small Chinese minority

Religions: 98% Roman Catholic; 1% Protestant; 1% Muslim; animism

Health

Life Expectancy at Birth: 65 years (male); 70 years (female)
Infant Mortality: 38/1,000 live births
Physicians Available: 1/10,000 people (2004)
HIV/AIDS Rate in Adults: na

Education

Adult Literacy Rate: 59%

Communication

Telephones: 2,400 main lines (2009)
Mobile Phones: 116,000 (2009)
Internet Users: 2,100 (2010)
Internet Penetration (% of Pop.): 0.2%

Transportation

Roadways in Miles (Kilometers): 3,753 (6,040)
Railroads in Miles (Kilometers): none
Usable Airfields: 6

Government

Type: republic
Independence Date: May 20, 2002 (from Indonesia)
Head of State/Government: President José Ramos-Horta; Prime Minister Kay Rala Xanana Gusmao (formerly known as José Alexandre Gusmao)
Political Parties: Democratic Party; National Congress for Timorese Reconstruction; National Democratic Union of Timorese Resistance; National Unity Party; and many others
Suffrage: universal at 17

Military

Military Expenditures (% of GDP): na
Current Disputes: maritime boundary dispute with Indonesia

Economy

Currency ($ U.S. Equivalent): East Timor uses the U.S. dollar
Per Capita Income/GDP: $2,600/$3.051 billion
GDP Growth Rate: 6.1%
Inflation Rate: 6%
Unemployment Rate: 20%
Labor Force by Occupation: 90% agriculture; 10% other
Population Below Poverty Line: 42%
Natural Resources: petroleum; gold; natural gas; timber; marble; manganese
Agriculture: rice; cassava; coffee; corn; soybeans; fruits; vanilla; cabbage; sweet potatoes
Industry: printing; textiles; handicrafts; soap
Exports: $10 million (excluding oil)
Imports: $202 million

Suggested Websites

www.gov.east-timor.org
www.easttimordirectory.net

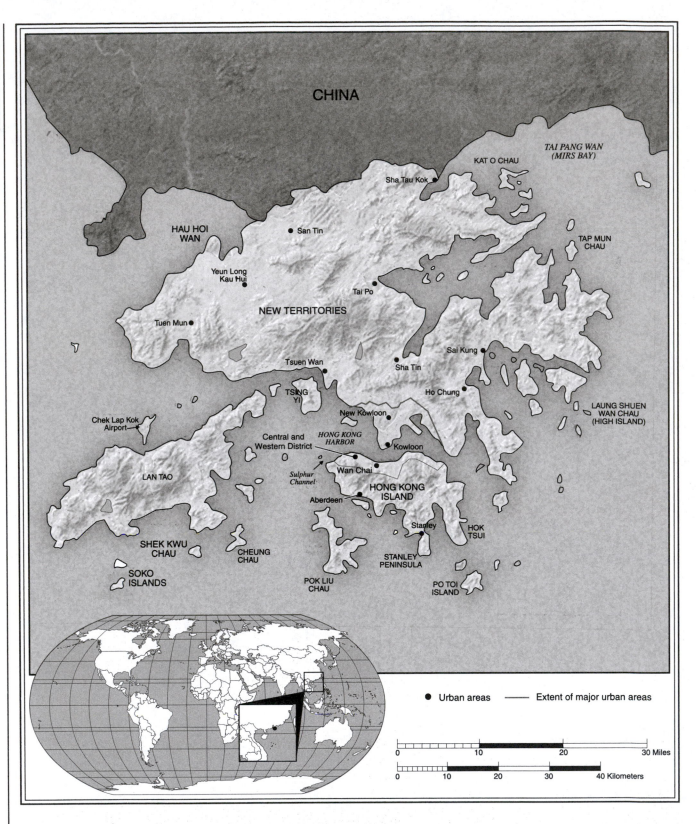

CHINA

TAI PANG WAN
(MIRS BAY)

KAT O CHAU

Sha Tau Kok

HAU HOI
WAN

San Tin

TAP MUN
CHAU

Yeun Long
Kau Hui

Tai Po

NEW TERRITORIES

Tuen Mun

Sai Kung

Tsuen Wan

Sha Tin

TSING
YI

Ho Chung

LAUNG SHUEN
WAN CHAU
(HIGH ISLAND)

Chek Lap Kok
Airport

New Kowloon

Central and
Western District

HONG KONG
HARBOR

Kowloon

LAN TAO

Sulphur
Channel

Wan Chai

HONG KONG
ISLAND

Aberdeen

Stanley

HOK
TSUI

SHEK KWU
CHAU

CHEUNG
CHAU

STANLEY
PENINSULA

SOKO
ISLANDS

POK LIU
CHAU

PO TOI
ISLAND

● Urban areas ——— Extent of major urban areas

0   10   20   30 Miles

0   10   20   30   40 Kilometers

Hong Kong
(Hong Kong Special Administrative Region)

Opium started it all for Hong Kong. The addictive drug from which such narcotics as morphine, heroin, and codeine are made, opium had become a major source of income for British merchants in the early 1800s. Therefore, when the Chinese government declared the opium trade illegal and confiscated more than 20,000 large chests of opium that had been on their way for sale to the increasingly addicted residents of Canton, the merchants persuaded the British military to intervene and restore their trading privileges. The British Navy attacked and occupied part of Canton. Three days later, the British forced the Chinese to agree to their trading demands, including a demand that they be ceded the tiny island of Hong Kong (meaning "Fragrant Harbor"), where they could pursue their trading and military business without the scrutiny of the Chinese authorities.

Initially, the British government was not pleased with the acquisition of Hong Kong; the island, which consisted of nothing more than a small fishing village, had been annexed without the foreknowledge of the authorities in London. Shortly, however, the government found the island's magnificent deepwater harbor a useful place to resupply ships and to anchor military vessels in the event of further hostilities with the Chinese. It turned out to be one of the finest natural harbors along the coast of China. On August 29, 1842, China reluctantly signed the Treaty of Nanking (or Nanjing), which ended the first Opium War and gave Britain ownership of Hong Kong Island "in perpetuity."

Twenty years later, a second Opium War caused China to lose more of its territory; Britain acquired permanent lease rights over Kowloon, a tiny part of the mainland facing Hong Kong Island. By 1898, Britain had realized that its miniscule Hong Kong naval base would be too small to defend itself against sustained attack by French or other European navies seeking privileged access to China's markets. The British were also concerned about the scarcity of agricultural land on Hong Kong Island and nearby Kowloon Peninsula. In 1898, they persuaded the Chinese to lease them more than 350 square miles of land adjacent to Kowloon. Thus, Hong Kong consists today of Hong Kong Island (as well as numerous small, uninhabited islands nearby), the Kowloon Peninsula, and the agricultural lands that came to be called the New Territories.

From its inauspicious beginnings, Hong Kong grew into a dynamic, modern society, wealthier than its promoters would have ever dreamed in their wildest imaginations. Hong Kong is now home to 7 million people. Most of the New Territories are mountainous or are needed for agriculture, so the bulk of the population is packed into about one tenth of the land's space. This gives Hong Kong the dubious honor of being one of the most densely populated human spaces ever created. Millions of people live stacked on top of one another in 30-story-high buildings. Even Hong Kong's excellent harbor has not escaped the population crunch: Approximately 10 square miles of harbor have been filled in and now constitute some of the most expensive real estate on Earth.

Why are there so many people in Hong Kong? One reason is that, after occupation by the British, many Chinese merchants moved their businesses

TIMELINE

A.D. 1839–1842
The British begin to occupy and use Hong Kong Island; the first Opium War

1842
The Treaty of Nanking cedes Hong Kong to Britain

1856
The Chinese cede Kowloon and Stonecutter Island to Britain

1898
England gains a 99-year lease on the New Territories

1898–1900
The Boxer Rebellion

1911
Sun Yat-sen overthrows the emperor of China to establish the Republic of China

1941
The Japanese attack Pearl Harbor and take Hong Kong

1949
The Communist victory in China produces massive immigration into Hong Kong

1980s
Great Britain and China agree to the return of Hong Kong to China

1990s
China resumes control of Hong Kong on July 1, 1997; pro-democracy politicians sweep the 1998 elections

2000s
Hong Kong's economy continues its recovery from the Asian financial crisis

Signs emerge of increasing P.R.C. control

Half a million protestors successfully defeat a proposed anti-subversion law

China refuses to allow democratic election of the Chief Executive in 2007 elections

New electronic surveillance law allows wiretapping

Five pro-democracy Legco members resigned only to be re-elected in the next election in 2010

Land is so expensive in Hong Kong that most residences and businesses today are located in skyscrapers. While the buildings are thoroughly modern, construction crews typically erect bamboo scaffolding to which is attached to protective netting. The skyscrapers that appear darker in color in the forefront of this photo are being built using this technique.

to Hong Kong, under the correct assumption that trade would be given a freer hand there than on the mainland. (Hong Kong continues to boast that it is the world's freest economy). Eventually, Hong Kong became the home of mammoth trading conglomerates. The laborers in these profitable enterprises came to Hong Kong, for the most part, as political refugees from mainland China in the early 1900s. Another wave of immigrants arrived in the 1930s upon the invasion of Manchuria by the Japanese, and yet another influx came after the Communists took over China in 1949. Thus, like Taiwan, Hong Kong became a place of refuge for those in economic or political trouble on the mainland.

Overcrowding plus a favorable climate for doing business have produced extreme social and economic inequalities. Some of the richest people on Earth live in Hong Kong, alongside some of the most wretchedly poor, notable among whom are recent refugees from China and Southeast Asia (more than 300,000 Vietnamese sought refuge in Hong Kong after the Communists took over South Vietnam, although most have now been repatriated—some forcibly). Some of these refugees have joined the traditionally poor boat peoples living in Aberdeen Harbor. Although surrounded by poverty, many of Hong Kong's economic elites have not found it inappropriate to indulge in ostentatious displays of wealth, such as riding in chauffeured, pink Rolls-Royces or wearing full-length mink coats. Some Hong Kong residents are listed among the wealthiest people in the world: Li Ka-Shing worth US$26 billion; the three Kwok brothers, worth US$20 billion; Lee Shau Kee, worth US$19 billion; and Cheng Yu- tung, worth US$9 billion, to name a few.

Workers are on the job six days a week, morning and night, yet the average pay for a worker in industry is only about $5,000 per year. With husband, wife, and older children all working, families can survive; some even make it into the ranks of the fabulously wealthy. Indeed, the desire to make money was the primary reason that Hong Kong was settled in the first place. That fact is not lost on anyone who lives there today. In fact, making money seems to take precedence over other things, such as the state of the environment. Hong Kong has congested and dirty streets, and, in recent years, clouds of soot from mainland manufacturers have worsened Hong Kong's already poor air quality. The city now has particulate matter in the air that is 40 percent higher than in America's most polluted city, Los Angeles. Some companies recently have decided to resettle in Singapore and other places with higher environmental standards, but most people continue to find Hong Kong an excellent place to make money.

Yet materialism has not wholly effaced the cultural arts and social rituals that are essential to a cohesive society. Indeed, with the vast majority of Hong Kong's residents hailing originally from mainland China, the spiritual beliefs and cultural heritage of China's long history abound. Some residents hang small, eight-sided

(Courtesy of Lisa Clyde Nielsen)

A fishing family lives on this houseboat in Aberdeen Harbor. On the roof, strips of fish are hung up to dry.

mirrors outside windows to frighten away malicious spirits, while others burn paper money in the streets each August to pacify the wandering spirits of deceased ancestors. Business owners carefully choose certain Chinese characters for the names of their companies or products, which they hope will bring them luck. Even skyscrapers are designed following ancient Chinese customs of Feng Shui so that their entrances are in balance with the elements of nature.

Buddhist and Taoist beliefs remain central to the lives of many residents. In the back rooms of many shops, for example, small religious shrines are erected; joss sticks burning in front of these shrines are thought to bring good fortune to the proprietors. Elaborate festivals, such as those at New Year's, bring the costumes, art, and dance of thousands of years of Chinese history to the crowded streets of Hong Kong. And the British legacy may be found in the cricket matches, ballet troupes, philharmonic orchestras, English-language radio and television broadcasts, and the legal system under which capitalism flourished.

■ THE END OF AN ERA

Britain was in control of this tiny speck of Asia for nearly 160 years. Except during World War II, when the Japanese occupied Hong Kong for about four years, the territory was governed as a Crown colony of Great Britain, with a governor appointed by the British sovereign. In 1997, China recovered control of Hong Kong from the British. The events happened in this way: In 1984, British prime minister Margaret Thatcher and Chinese leader Deng Xiaoping concluded two years of acrimonious negotiations over the fate of Hong Kong upon the expiration of the New Territories' lease in 1997. Great Britain

claimed the right to control Hong Kong Island and Kowloon forever—a claim disputed by China, which argued that the treaties granting these lands to Britain had been imposed by military force. Hong Kong Island and Kowloon, however, constituted only about 10 percent of the colony; the other 90 percent was to return automatically to China at the expiration of the lease. The various parts of the colony having become fully integrated, it seemed desirable to all parties to keep the colony together as one administrative unit. Moreover, it was felt that Hong Kong Island and Kowloon could not survive alone.

The British government had hoped that the People's Republic of China would agree to the status quo, or that it would at least permit the British to maintain administrative control over the colony should it be returned to China. Many Hong Kong Chinese felt the same way, since they had, after all, fled to Hong Kong to escape the Communist regime in China. For its part, the P.R.C. insisted that the entire colony be returned to its control by 1997. After difficult negotiations, Britain very reluctantly agreed to return the entire colony to China as long as China would grant important concessions. Foremost among these were that the capitalist economy and lifestyle, including private-property ownership and basic human rights, would not be changed for 50 years. The P.R.C. agreed to govern Hong Kong as a "Special Administrative Region" (SAR) within China and to permit British and local Chinese to serve in the administrative apparatus of the territory. The first direct elections for the 60-member Legislative Council were held in September 1991, while the last British governor, Christopher Patten, attempted to expand democratic rule in the colony as much as possible before the 1997 Chinese takeover—reforms that the Chinese dismantled to some extent after 1997.

The Joint Declaration of 1984 was drafted by top governmental leaders, with very little input from the people of Hong Kong. This fact plus fears about what P.R.C. control would mean to the freewheeling lifestyle of Hong Kong's ardent capitalists caused thousands of residents, with billions of dollars in assets in tow, to abandon Hong Kong for Canada, Bermuda, Australia, the United States, and Great Britain. Surveys found that as many as one third of the population of Hong Kong wanted to leave the colony before the Chinese takeover. In the year before the change to Chinese rule, so many residents—16,000 at one point—lined up outside the immigration office to apply for British passports that authorities had to open up a nearby sports stadium to accommodate them. About half of Hong Kong residents already held British citizenship, but many of the rest, particularly recent refugees from China, wanted to secure their futures in case life under Chinese rule became repressive. Immigration officials received more than 100,000 applications for British passports in a single month in 1996!

Emigration and unease over the future have unsettled, but by no means ruined, Hong Kong's economy. According to the World Bank, Hong Kong is home to the world's seventh-largest stock market, the fifth-largest banking center and foreign-exchange market (and the second largest in Asia after Japan), and its per capita GDP at US$45,900 is one of the highest in the world and very similar to that of the United States. In 2005, Hong Kong became the home of Hong Kong Disneyland, Disney's third, but very small, theme park outside the United States. It was hoped that the 766-acre project on Lantau Island would attract 10 million visitors each year, but, so far, the annual visitor count has been less than 5 million, and the project lost over a billion dollars in 2009. Despite the objections of the Chinese government that the British were depleting government coffers before the handover to China, the outgoing British authorities embarked on several ambitious infrastructural projects that would allow Hong Kong to continue to grow economically in the future. Chief among these was the airport on Chek Lap Kok Island. At a cost of $21 billion, the badly needed airport was one of the largest construction projects in the Pacific Rim. In 2010, the Hong Kong government announced its own massive development project, a US$8.6 billion railway that will eventually link Hong Kong with several mainland cities.

Opinion surveys showed that despite fears of angering the incoming Chinese government, most Hong Kong residents supported efforts to improve the economy and to democratize the government by lowering the voting age and allowing direct election rather than appointment of officials by Beijing. In the 1998 elections, more than 50 percent of registered voters cast ballots—more than voted in Hong Kong's last election under British rule. In 1999, the opposition Democratic Party gained a substantial number of seats in district-level elections. These indicate a desire by the people of Hong Kong for more democracy.

Unfortunately, that desire may be under attack. Although in 1997 the Chinese did not impose massive changes to the structure of democracy in Hong Kong, little by little the instances of restrictions are piling up. For example, the Beijing bureau chief for the *South China Morning Post,* a Hong Kong newspaper, was fired after complaining about restrictions on freedom of the press. The Hong Kong government continues to feel tremendous pressure from Beijing to clamp down on the Falun Gong religious movement in Hong Kong, as Beijing did on the mainland. One of the biggest flashpoints since reversion to the mainland was the proposed anti-subversion law, which would have given the mainland-influenced government more power to define ordinary journalists, labor activists, and even academics as subversives. Street demonstrations in 2003 in which 500,000 protestors marched, were successful in preventing the proposal from becoming law. However, in 2006 a new electronic surveillance law was passed allowing wire-tapping and other hidden listening devices. Eighteen opposition lawmakers walked out of Legco (the legislature) in protest, but the bill passed nonetheless.

The 2007 election was another flashpoint. Voters wanted the chief executive to be chosen in direct elections rather than have the position appointed by Beijing. Beijing flatly refused this demand, and the voters took to the streets again. They also wanted all lawmakers to be democratically elected by 2008 rather than have a large percentage of them appointed by Beijing. In 2004, voters also gave pro-democracy parties 3 more seats in Legco, Hong Kong's legislature, although 30 seats are still appointed by a China-controlled group of business leaders. Annual protests on the anniversary of the Tiananmen Square massacre of 1989 draw tens of thousands of people, as they did on the 20th anniversary in 2009. Indeed, Hong Kong is the only place in China where demonstrations can be legally held. Other types of protests happen in Hong Kong that could not happen on the mainland. For instance, in 2009, a critical book on the Tiananmen massacre by a former Secretary-General of the Chinese Communist Party was published.

These protests may suggest that the people of Hong Kong would have preferred to remain a British colony. However, while there were large British and American communities in Hong Kong, and although English has been the medium of business and government for many years (China is now proposing the elimination of English as a language of instruction in most schools), many residents over the years had little or no direct emotional involvement with British culture and no loyalty to the British Crown. They asserted that they were, first and foremost, Chinese. This, of course, does not amount to a popular endorsement of Beijing's rule, but it does imply that some residents of Hong Kong feel that if they have to be governed by others, they would rather it be by the Chinese. Moreover, some believe that the Chinese government may actually help rid Hong Kong of financial corruption and allocate more resources to the poor—although with tourism down since the handover to China and the Hong Kong stock market still suffering the effects of the general Asian financial problems of 1998 and 1999, it is clear that Hong Kong's economy is highly impacted by world events.

Snapshot: HONG KONG

Summarized below is a quick look at the country with regard to its development, freedom, health/welfare, and achievements.

Development

Hong Kong is one of the preeminent financial and trading dynamos of the world. Annually, it exports billions of dollars' worth of products. Hong Kong's political future may be uncertain, but its fine harbor as well as its recently built $21 billion airport, a new Disney theme park, and information technology "cyberport," are sure to continue to fuel its economy.

Freedom

Hong Kong was an appendage to one of the world's foremost democracies for 160 years. Thus, its residents enjoyed the civil liberties guaranteed by British law. Under the new Basic Law of 1997, the Chinese government has agreed to maintain the capitalist way of life and other freedoms for 50 years. Many residents believe that, slowly, democracy is being compromised by China's repressive policies.

Health/Welfare

Schooling is free and compulsory in Hong Kong through junior high school. The government has devoted large sums for low-cost housing, aid for refugees, and social services such as adoption. Housing, however, is cramped and inadequate.

Achievements

Hong Kong has the capacity to hold together a society where the gap between rich and poor is enormous. The so-called boat people have long been subjected to discrimination, but most other groups have found social acceptance and opportunities for economic advancement. Hong Kong's per capita income is one of the highest in the world.

Hong Kong's natural links with China had been expanding steadily for years before the handover. In addition to a shared language and culture, there are in Hong Kong thousands of recent immigrants with strong family ties to China. In 2010, 22 million mainland Chinese visited Hong Kong, and in July 2011, the number of visitors broke an all-time record with 3.8 million visitors in one month. The Chinese tourists, part of the affluent new Chinese middle class, spend more per person than tourists from any other country.

There are also increasingly important commercial ties. Every day, some 30,000 management-level workers travel from Hong Kong to the mainland to run businesses there, and there are thousands of ordinary employees who also cross the border each day. The trains, ferries, and buses carrying workers into China each day are packed. Some 100,000 Hong Kong companies have operations in the mainland, and nearly 45 percent of outside investment in China's main cities (Beijing, Shanghai, and Guangzhou) comes from those companies. These connections have made Hong Kong one of the world's top financial centers and the number one or number two exporter of such hot commodities as cell phones, televisions, and digital cameras. Hong Kong has always served as south China's entrepôt to the rest of the world for both commodity and financial exchanges.

For instance, for years Taiwan circumvented its regulations against direct trade with China by transshipping its exports through Hong Kong. Commercial trucks plying the highways between Hong Kong and the P.R.C. form a bumper-to-bumper wall of commerce between the two regions. Sixty percent of Hong Kong's reexports now originate in China. Despite a tightening of laws regarding human rights, the P.R.C. realizes that Hong Kong needs to remain more or less as it is—therefore, the transition to Chinese rule (despite an unease over growing restrictions on human rights) has been less jarring to residents than was expected. Most people think that Hong Kong will remain a major financial and trading center for Asia, but adjustments are under way as Hong Kong copes with its new status as just one of China's holdings. For instance, the sluggish world economy in 2001 and then in 2008 pushed up unemployment and caused thousands of people—including some former millionaires—to declare bankruptcy. Competition from mainland China's less expensive docking facilities and from the booming, Hong Kong–like metropolis of Shanghai has contributed to some of the downturn. Perhaps the biggest drain on the economy is the devaluation of property values. Still expensive by world standards, Hong Kong property now suffers from "negative equity"; that is, some property is worth less than the mortgage that was used to buy it. There may be more such adjustments in the future, but the people of Hong Kong are hard-working and say that they are determined to make the new system work.

DID YOU KNOW?

The top political leader of Hong Kong is called the Chief Executive. He is elected by a group of 800 citizens appointed with the approval of Beijing, not by the votes of all residents of Hong Kong. It is anticipated that all residents will be able to vote for the Chief Executive by 2017, but those, including students at Hong Kong University, wanting full universal suffrage more quickly have staged various protests, and five elected legislators resigned their seats in 2010 in order to force a sort of referendum (there is no actual right to referenda in Hong Kong) on the subject. A by-election was held, and the five were re-elected, which most people defined as a statement of support by the people for a quicker enactment of the universal suffrage plan.

Statistics

Geography

Area in Square Miles (Kilometers): 686 (1,777) (about six times the size of Washington, D.C.)

Capital (Population): Victoria (na)

Environmental Concerns: air and water pollution

Geographical Features: composed of more than 200 islands; hilly to mountainous, with steep slopes; lowlands in the north

Climate: subtropical monsoon

People

Population

Total: 7,122,508

Annual Growth Rate: 0.45%

Rural/Urban Population Ratio: 100% urban (as of 2010)

Major Languages: Chinese (Cantonese); English; Chinese (Mandarin) and other dialects

Ethnic Makeup: 95% Chinese (mostly Cantonese); 2% Filipino, 1% Indonesian, 2% other

Religions: 90% a combination of Buddhism and Taoism; 10% Christian

Health

Life Expectancy at Birth: 79 years (male); 85 years (female)

Infant Mortality: 3/1,000 live births

Physicians Available: 1/1,000 people

HIV/AIDS Rate in Adults: 0.1%

Education

Adult Literacy Rate: 93.5%

Communication

Telephones: 4.2 million main lines (2009)

Mobile Phones: 12.2 million (2009)

Internet Users: 4.88 million (2009)

Internet Penetration (% of Pop.): 69%

Transportation

Roadways in Miles (Kilometers): 1,284 (2,067)

Railroads in Miles (Kilometers): na

Usable Airfields: 2

Government

Type: limited democracy as Special Administrative Region (SAR) of China

Head of State/Government: Hu Jintao; Chief Executive Donald Tsang

Political Parties: No registered political parties; politically active groups register as societies or companies. These include Democratic Alliance for the Betterment and Progress of Hong Kong; Democratic Party; Association for Democracy and People's Livelihood; Liberal Party; Civic Party; League of Social Democrats; The Frontier (disbanded)

Suffrage: direct elections universal at 18 for residents who have lived in Hong Kong for at least 7 years

Military

Military Expenditures (% of GDP): defense is the responsibility of China

Current Disputes: none

Economy

Currency ($ U.S. Equivalent): 7.79 Hong Kong dollars = $1

Per Capita Income/GDP: $45,900/$325.8 billion

GDP Growth Rate: 6.8%

Inflation Rate: 4.5%

Unemployment Rate: 4.3%

Natural Resources: outstanding deepwater harbor; feldspar

Agriculture: vegetables; poultry; fish; pork

Industry: textiles; clothing; tourism; banking; shipping; electronics; plastics; toys; clocks; watches

Exports: $389 billion (primary partners China, United States, Japan)

Imports: $431 billion (primary partners China, Japan, Taiwan)

Suggested Websites

www.scmp.com/

www.discoverhongkong.com/eng/ index.jsp

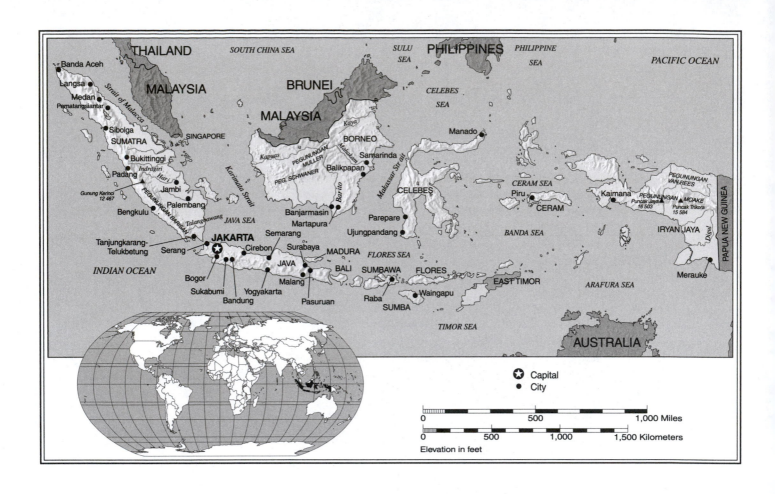

Indonesia

(Republic of Indonesia)

Present-day Indonesia is a kaleidoscope of some 731 language dialects and more than 500 ethnic groups, including such disparate cultures as the Javanese of Java, the Bugis and Toraja people of South Sulawesi, the Dayaks of Kalimantan, the Bataks of Sumatra, the Asmat of Papua, and the Balinese of Bali. Beginning over 5,000 years ago, people of Mongoloid stock settled the islands that today constitute Indonesia, in successive waves of migration from China, Thailand, and Vietnam. Animism—the nature-worship religion of these peoples—was altered substantially (but never completely lost) about A.D. 200, when Hindus from India began to settle in the area and wield the dominant cultural influence. Several hundred years later, Buddhist missionaries and settlers began converting Indonesians in a proselytizing effort that produced strong political and religious antagonisms. In the thirteenth century, Muslim traders began the Islamization of the Indonesian people; today, 86 percent of the population claim the Muslim faith—meaning that there are more Muslims in Indonesia than in any other country of the world, including the states of the Middle East. Commingling with all these influences are cultural inputs from the islands of Polynesia and colonial remnants from the Portuguese and the Dutch.

The real roots of the Indonesian people undoubtedly go back much further than any of these historic cultures. In 1891, the fossilized bones of a hominid who used stone tools, camped around a fire, and probably had a well-developed language were found on the island of Java. Named *Pithecanthropus erectus* ("erect ape-man"), these important early human fossils, popularly called Java Man, have been dated at about 750,000 years of age. Fossils similar to Java Man have been found in Europe, Africa, and Asia.

Modern Indonesia was sculpted by the influence of many outside cultures. Portuguese Catholics, eager for Indonesian spices, made contact with Indonesia in the 1500s and left 20,000 converts to Catholicism, as well as many mixed Portuguese–Indonesian communities and dozens of Portuguese "loan words" in the Indonesian-style Malay language, a language that shares much in common with Malaysian. In the following century, Dutch Protestants established the Dutch East India Company to exploit Indonesia's riches. Eventually and after the deaths of thousands of Dutch and Indonesian soldiers, the Netherlands was able to gain complete political control; it reluctantly gave it up, in the face of insistent Indonesian nationalism, only as recently as 1950. Before that, however, the British briefly controlled one of the islands, and the Japanese ruled the country for three years during the 1940s.

Indonesians, including then-president Sukarno, initially welcomed the Japanese as helpers in their fight for independence from the Dutch. Everyone believed that the Japanese would leave soon. Instead, the Japanese military forced farmers to give food to the Japanese soldiers, made everyone worship the Japanese emperor, neglected local industrial development in favor of military projects, and took 270,000 young men away from Indonesia to work elsewhere as forced laborers (fewer than 70,000 survived to return home). Military leaders who attempted to revolt against Japanese rule were executed. Finally, in August 1945, the Japanese abandoned their control of Indonesia, according to the terms of surrender with the Allied powers.

TIMELINE

750,000 B.C.
Java Man lived here

A.D. 600
Buddhism gains the upper hand

1200
Muslim traders bring Islam to Indonesia

1509
The Portuguese begin to trade and settle in Indonesia

1596
Dutch traders begin to influence Indonesian life

1942
The Japanese defeat the Dutch

1949–1950
Indonesian independence from the Netherlands; President Sukarno retreats from democracy and the West

1965
Failed communist coup; violent purge of communists

1966
General Suharto takes control of the government from Sukarno and establishes his New Order, pro-Western government

1975
Indonesia annexes East Timor

1990s
Suharto steps down after 32 years in power; East Timor votes for independence; violence erupts in Borneo

2000s
The economy remains stalled

East Timor obtains independence

Aceh Province signs a peace agreement ending a 130-year armed rebellion

A peace agreement in separatist Aceh Province mutes, but does not eliminate, some of the violence there

Moderate Susilo Bambang Yudhoyono, age 55, wins a landslide election to become the first Indonesian president directly elected by voters

A series of natural disasters, including a volcanic eruption in northeastern Indonesia, rain-triggered landslides in Sumatra, powerful earthquakes in Papua, and a mammoth tsunami in Aceh kill tens of thousands and add further woes to the struggling economy

Powerful bombs damage Marriott and Ritz-Carlton Hotels and kill 9 people in Jakarta's central business district.

2004
A mammoth tsunami slams into Indonesia killing nearly 200,000 people

95

Consider what all these influences mean for the culture of modern Indonesia. Some of the most powerful ideologies ever espoused by humankind—supernaturalism, Islam, Hinduism, Buddhism, Christianity, mercantilism, colonialism, and nationalism—have had an impact on Indonesia. Take music, for example. Unlike Western music, which most people just listen to, traditional Indonesian music, played on drums and gongs, is intended as a somewhat sacred ritual in which all members of a community are expected to participate. The instruments themselves are considered sacred. Dances are often the main element in a religious service whose goal might be a good rice harvest, spirit possession, or exorcism. A variety of musical styles can be heard here and there around the country. In the eastern part of Indonesia, the Nga'dha peoples, who were converted to Christianity in the early 1900s, sing Christian hymns to the accompaniment of bronze gongs and drums. On the island of Sumatra, Minang Kabau peoples, who were converted to Islam in the 1500s, use local instruments to accompany Islamic poetry singing. Communal feasts in Hindu Bali, circumcision ceremonies in Muslim Java, and Christian baptisms among the Bataks of Sumatra all represent borrowed cultural traditions. Thus, out of many has come one rich culture.

But the faithful of different religions are not always able to work together in harmony. For example, in the 1960s, when average Indonesians were trying to distance themselves from radical Communists, many decided to join Christian faiths. Threatened by this tilt toward the West and by the secular approach of the government, many fundamentalist Muslims resorted to violence. They burned Christian churches, threatened Catholic and Baptist missionaries, and opposed such projects as the construction of a hospital by Baptists. As we have seen, Indonesia has more Muslims than any other country in the world, and the tens of thousands of members of Islamic socioreligious, political, and paramilitary organizations intend to keep it that way. For example, some 9,000 people in the eastern provincial capital of Ambon in the Maluku islands, many of them Christians desiring independence from Indonesia, were killed in Muslim–Christian sectarian violence in 2001, and gangs from both religions fought street battles there again in 2004, leaving more than two dozen dead and scores wounded. In addition to detonating bombs and hacking people to death with swords, the gangs set fire to churches and destroyed a United Nations office. Clearly, some Indonesians have a long way to go in developing mutual respect and tolerance for the diversity of cultures in their midst (although in Jakarta, there are many moderate Muslims whose lifestyle is similar to the moderate Muslims in Turkey and Malaysia).

■ A LARGE LAND, LARGE DEBTS

Unfortunately, Indonesia's economy is not as rich as its culture. About 56 percent of the population live in rural areas; more than half of the people engage in fishing and small-plot rice and vegetable farming. The average income per person is only US$4,200 a year based on gross domestic product. A new law increased the minimum wage in Jakarta to US$130 a month.

Also worrisome is the level of government debt. Indonesia is blessed with large oil reserves (Pertamina is the state-owned oil company) and minerals and timber of every sort (also state-owned), but to extract these natural resources has required massive infusions of capital, most of it borrowed. In fact, Indonesia has borrowed more money than any other country in Asia. The debt burden increased in 2009 when the World Bank gave a US$2 billion loan to the country—one of the largest loans ever given. Indonesia must allocate 40 percent of its national budget just to pay the interest on loans. Low oil prices in the 1980s made it difficult for the country to keep up with its debt burden. Extreme political unrest, an economy that contracted nearly 14 percent (!) during the Asian financial crisis of 1998, and major natural catastrophes (the Aceh tsunami of 2004, the Java earthquakes of 2006, and many others) have seriously exacerbated Indonesia's economic headaches.

An example of the difficulty that Indonesia has in moving forward its economy is the experience of ExxonMobil Corporation. The U.S. company operates a large oil and gas refinery in Aceh Province in north Sumatra Island. But many of the Acehnese wanted independence from Indonesia and, for many years, were locked in bitter battles with the government. Wanting more revenue from the company, Acehnese rebels started a series of pipe bombings, kidnappings for ransom, bus hijackings, and even mortar and grenade attacks. In order to keep business going, the government had to send in 3,000 troops to guard the refinery day and night. Troops lined the company airport for every take-off and landing; and to protect the safety of its staff en route, the company had to purchase 16 armor-plated Land Rovers. The rebellion has now been brought to a peaceful end, but the case illustrates how Indonesia's problems always involve both business and political entanglements.

To cope with these problems, Indonesia has relaxed government control over foreign investment and banking, and it seems to be on a path toward privatization of other parts of the economy. The country is the largest oil producer in Southeast Asia, the world's largest liquefied natural gas exporter, the world's largest plywood exporter, and the second-largest exporter of rubber and palm oil. Many joint ventures are underway with companies from the United States, Germany, and Japan. ExxonMobil, for example, is joining with Pertamina, the state-owned oil company, to develop an oil field on the island of Java—a $2 billion project. The country is also establishing 10 special economic zones to boost the economy. Still, the income and development gap between the 10 modernized cities and the traditional countryside continues to plague government planners.

Indonesia's financial troubles seem puzzling, because in land, natural resources, and population, the country appears quite well-off. Indonesia is the second-largest country in Asia (after China). Were it superimposed on a map of the United States, its 13,677 tropical islands would stretch from California, past New York, and out to Bermuda in the Atlantic Ocean. Oil and hardwoods are plentiful, and the population is large enough to constitute a viable internal consumer market. But transportation and communication are problematic and costly in archipelagic states. Before the Asian financial crisis of the late 1990s hit, Indonesia's national airline, Garuda Indonesia, had hoped to launch a US$3.6 billion development program that would have brought into operation 50 new aircraft stopping at 13 new airports. New seaports were also planned. But the contraction of the economy and the enormous cost of linking together some 6,000 inhabited islands made most of the projects unworkable. Moreover, exploitation of Indonesia's amazing panoply of resources is drawing the ire of more and more people around the world who fear the destruction of one more part of the world's fragile ecosystem. Ten percent of all plant species are thought to be found in Indonesia, and tropical forests cover 75 percent of the land. Thus the stage is set for ongoing conflicts between those desiring development and those attempting to preserve a valuable world ecosystem. In 2006, students demonstrated in Papua province (the country's poorest) to demand the end of operations of the world's largest gold and copper mine. Three were killed in the melee. Another mine shut down in 2005 but was subsequently sued by the government for dumping arsenic and mercury into a bay.

Illiteracy and demographic circumstances also constrain the economy. Indonesia's population of 245.6 million is one of the largest in the world (it's the world's third-largest democracy), but 10 percent of adults (13 percent of females) cannot read or write. Only about 600 people per 100,000 attend college, as compared to 3,580 in nearby Philippines. Moreover, since almost 70 percent of the population reside on or near the island of Java, on which the capital city, Jakarta, is located, educational and development efforts have concentrated there at the expense of the communities on outlying islands. Many children in the out-islands never complete the required six years of elementary school. Some ethnic groups, in the remote provinces of Papua (formerly called Irian Jaya) on the island of New Guinea, and Kalimantan (Borneo), for example, continue to live isolated in small tribes, much as they did thousands of years ago. By contrast, the modern city of Jakarta, with its classical European-style buildings, is home to millions of people, many of whom are well-educated and live lifestyles not that different from upper and middle class people in the West. Over the past 20 years, poverty has been reduced from 60 percent (the current poverty rate is 18 percent), but Indonesia was seriously damaged by the Asian financial crisis, by the global economic downturn in 2009, and by a seemingly never-ending series of natural and man-made disasters including earthquakes (a magnitude 7.9 quake hit Sumatra in 2007), mudslides (a burst dam near Jakarta in 2009 drowned nearly 80 people), and the

mammoth tsunami that devastated Aceh province in 2004). Experts say that the economy will not return to normal for many years.

A big blow to Indonesia's tourism revenues came in October 2002, when terrorists linked to the al-Qaeda network bombed a nightclub in the popular resort community of Bali. Killing dozens of locals and tourists—many of them Australians—the bombings inflicted major damage to the tourism industry and convinced the world that Indonesia was not doing enough to support the international war on terrorism. Muslim Jemaah Islamiyah militants from Malaysia detonated a bomb outside the Australian embassy in Jakarta in 2004, wounding 173 people and killing 9. Other foreign interests, such as the Jakarta Marriott Hotel, have also been targets of deadly bomb attacks, and although the government has arrested several suspects in these and other bombings, militants can often escape to one of the thousands of remote islands that make up the country, where they find refuge among like-minded residents.

With 2.3 million new Indonesians entering the labor force every year, and with half the population under age 20 (27 percent were under age 15 in 2011), serious efforts must be made to increase employment opportunities. The most pressing problem is to finish the many projects for which international bank loans have already been received. With considerable misgivings, the World Bank, the Asian Development Bank, and the government of Japan provided more than US$4 billion in aid to Indonesia to alleviate poverty and help decentralize government authority. As we have seen, US$2 billion more was added by the World Bank in 2009. International confidence in the country is not helped by Indonesia's reputation as one of the most corrupt societies in the world—despite the establishment of a Corruption Eradication Commission in 2003.

■ MODERN POLITICS

Establishing the current political and geographic boundaries of the Republic of Indonesia has been a bloody and protracted task. So fractured is the culture that many people doubt whether there really is a single country that one can call Indonesia. During the first 15 years of independence (1950–1965), there were revolts by Muslims and pro-Dutch groups, indecisive elections, several military coups, battles against U.S.-supported rebels, and serious territorial disputes with Malaysia and the

FURTHER INVESTIGATION

In the Maluku provincial capital of Ambon, Muslims and Christians fought each other with rocks and machetes in 2011 violence that left 5 people dead and 150 injured. To find out why such religion-based violence happens frequently in Maluku, see: http://www.asianews.it/news-en/Clashes-in-Maluku,-45-houses-and-a-church-set-on-fire-13974.html

Netherlands. In 1966, nationalistic president Sukarno, who had been a founder of Indonesian independence, lost power to Army General Suharto. (Many Southeast Asians had no family names until influenced by Westerners; Sukarno and Suharto have each used only one name.) Anti-Communist feeling grew during the 1960s, and thousands of suspected members of the Indonesian Communist Party (PKI) and other Communists were killed before the PKI was banned in 1966.

In 1975, ignoring the disapproval of the United Nations, President Suharto invaded and annexed East Timor, a Portuguese colony. Although the military presence in East Timor was subsequently reduced, separatists were beaten and killed by the Indonesian Army as recently as 1991; and in 1993, a separatist leader was sentenced to 20 years in prison. In late 1995, Amnesty International accused the Indonesian military of raping and executing human rights activists in East Timor. The 20th anniversary of the Indonesian takeover was marked by Timorese storming foreign embassies and demanding asylum and redress for the kidnapping and killing of protesters. In 1996, antigovernment rioting in Jakarta resulted in the arrest of more than 200 opposition leaders and the disappearance of many others. The rioters were supporters of the Indonesian Democracy Party and its leader, Megawati Sukarnoputri, daughter of Sukarno, and the woman who would shortly become president of the country.

Suharto's so-called New Order government ruled with an iron hand, suppressing student and Muslim dissent and controlling the press and the economy. With the economy in serious trouble in 1998, and with the Indonesian people tired of government corruption and angry at the control of Suharto and his six children over much of the economy, rioting broke out all over the country. Some 15,000 people took to the streets, occupied government offices, burned cars, and fought with police. The International Monetary Fund suspended vital aid because it appeared that Suharto would not conform to the belt-tightening required of IMF aid recipients. With unemployed migrant workers streaming back to Indonesia from Malaysia and surrounding countries, with the government unable to control forest fires burning thousands of acres and producing a haze all over Southeast Asia, with even his own lifetime political colleagues calling for him to step down, Suharto at last resigned, ending a 32-year dictatorship. The new leader, President Bacharuddin Jusuf Habibie, pledged to honor IMF commitments and restore dialogue on the East Timor dispute. But protests dogged Habibie, because he was seen as too closely allied with the Suharto leadership. In the first democratic elections in years, a respected Muslim cleric, Abdurrahman Wahid, was elected president, with Megawati Sukarnoputri (daughter of Indonesia's founding father, Sukarno) as vice-president.

In August 1999, the 800,000 residents of East Timor, the majority of whom were Catholic Christians, voted overwhelmingly (78.5 percent) for independence from Indonesia, in a peaceful referendum. Unfortunately,

anti-independence Muslim militias, together with Indonesian troops, launched a hellish drive to prevent East Timor from separating from Indonesia. They drove thousands of residents from their homes, beat and killed them, and then burned their homes and businesses. The militias virtually obliterated the capital city of Dili. The violence became so severe that nearby Australia felt compelled to send in some 8,000 troops to prevent a wholesale bloodbath. Eventually East Timor obtained the independence it desired, but not before Indonesia received worldwide criticism for its ineffectual response to the rampage. Other separatist movements have also challenged the government.

For example, a drive for independence in Papua province continues unabated, and in Aceh province, a Free Aceh movement was launched in the mid-1970s. After proclaiming independence from Indonesia, the rebel leader fled to Sweden, from where he directed violent activities against the government. Up through the end of the 1990s, the government responded with an iron hand, sending troops to attack dissenters. Under President Wahid, a dialogue was launched that involved negotiations in Geneva and employed Japan as a mediator. When the Asian tsunami hit Aceh in 2004, the government sent in troops to help the people instead of repress them, thus softening the tension and resulting in a peace agreement that went into effect in mid-2005, effectively ending the 30-year-old rebellion.

Unfortunately, there have been far more instances in which government troops have exacerbated rather than mollified tense situations. Often, the troops have run "amok" (a Malay word meaning "to erupt in violent rage"). In 1965, for example, an abortive Communist coup precipitated a violent purge of Communists by the Army. Within a few months over 250,000 citizens had been killed. In 1975, as we have seen, General Suharto invaded East Timor; his troops tortured, raped, and mutilated their victims. After four years, some 200,000 people had perished.

■ CRISIS IN BORNEO

In early 1999, violence broke out on the Indonesian island of Borneo. By the time it ended months later, 50,000 people had fled the island, and hundreds of men, women, and children had been killed, their decapitated heads displayed in towns and villages. The attackers even swaggered through towns victoriously holding up the dismembered body parts of their victims.

What caused this violence? It started with overpopulation and poverty. Some 50 years ago, the Indonesian government decided that it had to do something about overpopulation on the soil-poor island of Madura. Located near the island of Java, Madura, with its whitesand beaches and its scores of related islets, was the home of the Madurese, Muslims who had long since found it difficult to survive by farming the rocky ground. The government thus decided to move some of the Madurese to the island of Borneo, where land was more plentiful

Snapshot: INDONESIA

Summarized below is a quick look at the country with regard to its development, freedom, health/welfare, and achievements.

Development

Indonesia continues to be hamstrung by its heavy reliance on foreign loans, a burden inherited from the Sukarno years. Current Indonesian leaders speak of "stabilization" and "economic dynamism," but there are always obstacles—government corruption and such natural disasters as the devastating tsunami of 2004—that hamper smooth economic improvement.

Freedom

Demands for Western-style human rights are frequently heard, but until recently, only the army has had the power to impose order on the numerous and often antagonistic political groups. However, some progress is evident: as of 2009, Indonesians had completed their third successful democratic election. The next election is scheduled for 2014.

Health/Welfare

Indonesia has one of the highest birth rates in the Pacific Rim. Many children will grow up in poverty, never learning even to read or write their national language, Bahasa Indonesian. Many girls are sold into sexual slavery in Southeast Asian countries.

Achievements

Balinese dancers' glittering gold costumes and unique choreography epitomize the "Asianness" of Indonesia as well as the Hindu roots of some of its communities.

---

and fertile. It seemed like a logical solution, but the government failed to consider the ethnic context—that is, the island of Borneo was inhabited by the Dayaks, a people who considered the island their tribal homeland. They had little or no interest in the government in Jakarta, regarding themselves as Dayaks first and Indonesians a distant second, if at all.

Almost immediately the Dayaks began harassing the newcomers, whom they judged to be "hot-headed" and crude. They resented both the loss of their land and, later, the loss of jobs in the villages. Over the years, hundreds of people were killed in sporadic attacks, but the violence in 1999 was worse than ever, in part because the Dayaks were, for the first time, joined by the Malays in "cleansing" the island of the hated Madurese. At first, the Indonesian government seemed to do little to stop the carnage. Eventually, when the Madurese had been chased from their homes, President Wahid promised aid money and assistance with relocation. But he wanted the Madurese to return to Borneo. The Dayaks, with little regard for the Indonesian government, responded that they would kill any who returned. As of this writing, thousands of Madurese remain as refugees on the island of Java.

In mid-2000, in the face of mounting criticism of his seeming inability to restore order and jump-start the economy, Wahid, who was nearly blind due to a serious eye disease, announced that he was turning over day-to-day administration to his vice-president. Wahid was subsequently censured, impeached, and then replaced. The new female president, Megawati Sukarnoputri, made some attempts to clean up the leftover corruption of the Suharto years, and she encouraged foreign investment, particularly from Japan, which buys 23 percent of Indonesia's produce. In the past, companies like Toyota had invested millions in Indonesia's ASTRA automobile

Nature Attacks Indonesia

The world watched in horror in late 2004 when walls of water between 80 and 100 feet high (24–30 meters), and traveling at speeds up to 500 miles per hour, crashed ashore in Indonesia, instantly killing some 170,000 people (plus 60,000 more in other affected countries) and laying waste to miles of coastline. The waves were caused by an undersea earthquake of a 9.3 magnitude—one of the largest in recorded history. But the Asian tsunami was just one of scores of natural disasters that afflicted Indonesia that year. Prior to the tsunami, nearly 200 people had already been killed as a result of 72 storms, 112 floods, 20 earthquakes, 22 forest fires, 5 volcanic eruptions, and 5 tidal waves. It is more or less the same every year, because Indonesia sits atop one of the most seismically active regions in the world. An earthquake in Java killed 6000 people in 2006, and Sumatra Island endured a magnitude 7.9 earthquake in 2007. The country has 500 volcanoes, 128 of which are active. Excessive logging has worsened the situation, with El Niño flooding in 2001 killing 150 people and displacing 150,000. Forest fires in 1997 and 1998 burned thousands of acres. Bird flu, which has killed nearly 50 people, plane and ferry boat accidents (some 500 people perished in early 2007 in two of several such accidents), and other disasters also seem to be on the rise. A volcanic eruption from Mount Lokon in 2011 required the evacuation of 500 people. The National Board of Disaster Management and the Ministry of Health find they are overwhelmed by the magnitude of the problems they face.

---

company, while Japanese banks had supported the expansion of Indonesia's tourist industry. Closer links with the West, particularly the United States, are difficult due to the Iraq War, against which thousands of Indonesians have protested in mass rallies in Jakarta.

In the 2004 elections, in which 24 parties vied for 14,000 seats at all levels of government and fielded 450,000 (!) candidates, Megawati, whom the U.S. accused of having made illegal oil deals with Iraq's Saddam Hussein, was swept from office in a landslide victory by 55-year-old guitar-playing, poetry-reading Susilo Bambang Yudhoyono, the former security minister in Megawati's cabinet. Voters had great hope that Yudhoyono would be more approachable than Megawati, but barely had he taken office when the country was hit by the December 2004 tsunami. Yudhoyono managed to be re-elected in 2009, with the Democrat Party substantially beating the Golkar Party. The 2009 election also saw a decline in the strength of several of the Islamic parties, but on the lips of every member of the electorate is whether or not Yudhoyono has the will to eliminate the widespread political corruption—of which he himself has been accused.

In mid-2011, the president outlined a master plan for development of the economy. He hoped the plan would boost GDP growth to nearly 9 percent a year. He also hoped that foreign investment in excess of US$400 billion would flow in to help the country build infrastructure, particularly roads, ports, and airports.

Statistics

Geography

Area in Square Miles (Kilometers): 735,358 (1,904,569) (nearly three times the size of Texas)
Capital (Population): Jakarta (9.12 million)
Environmental Concerns: air and water pollution; sewage; deforestation; smoke and haze from forest fires
Geographical Features: the world's largest archipelago; coastal lowlands; larger islands have interior mountains
Climate: tropical; cooler in highlands

People

Population

Total: 245,613,043
Annual Growth Rate: 1.07%
Rural/Urban Population Ratio: 56/44
Major Languages: Bahasa Indonesian; English; Dutch; Javanese; many others
Ethnic Makeup: 41% Javenese; 15% Sundanese; 3% Madurese; 3% Minangkabau; 2% Betawi; 2% Bugis; 2% Banten; 2% Banjar; 30% other or unspecified
Religions: 86% Muslim; 6% Protestant; 3% Roman Catholic; 2% Hindu; 3% other

Health

Life Expectancy at Birth: 69 years (male); 74 years (female)
Infant Mortality: 28/1,000 people
Physicians Available: 1/3,472 people
HIV/AIDS in Adults: 0.2%

Education

Adult Literacy Rate: 90%
Compulsory (Ages): 7–16

Communication

Telephones: 34 million main lines (2009)
Mobile Phones: 159 million (2009)
Internet Users: 39.6 million (2010)
Internet Penetration (% of Pop.): 16%

Transportation

Roadways in Miles (Kilometers): 272,011 (437,759)
Railroads in Miles (Kilometers): 3,133 (5,042)
Usable Airfields: 684

Government

Type: republic
Independence Date: August 17, 1945 (from the Netherlands)
Head of State/Government: President Susilo Bambang Yudhoyono is both head of state and head of government
Political Parties: National Mandate Party; Prosperous Justice Party; People's Conscience Party; Democrat Party; Golkar; Great Indonesia Movement Party; Democratic Party-Struggle; National Awakening Party
Suffrage: universal at 17; married persons regardless of age

Military

Military Expenditures (% of GDP): 3%
Current Disputes: territorial disputes with Malaysia, Australia, East Timor, Singapore, others; refugee repatriation issues

Economy

Currency ($ U.S. Equivalent): 9,170 rupiahs = $1
Per Capita Income/GDP: $4,200/$1.03 trillion
GDP Growth Rate: 6.1%
Inflation Rate: 5.1%
Unemployment Rate: 7.1%
Labor Force by Occupation: 38% agriculture; 49% services; 13% industry
Population Below Poverty Line: 13%
Natural Resources: petroleum; tin; natural gas; nickel; timber; bauxite; copper; fertile soils; coal; gold; silver
Agriculture: rice; cassava; peanuts; rubber; cocoa; coffee; copra; poultry; beef; pork; eggs; palm oil
Industry: petroleum and natural gas; textiles; apparel; footwear; mining; cement; chemical fertilizers; plywood; rubber; food; tourism
Exports: $158 billion (primary partners Japan, China, United States, Singapore)
Imports: $127 billion (primary partners China, Singapore, Japan)

Suggested Website

www.thejakartapost.com/
www.indonesia.travel/

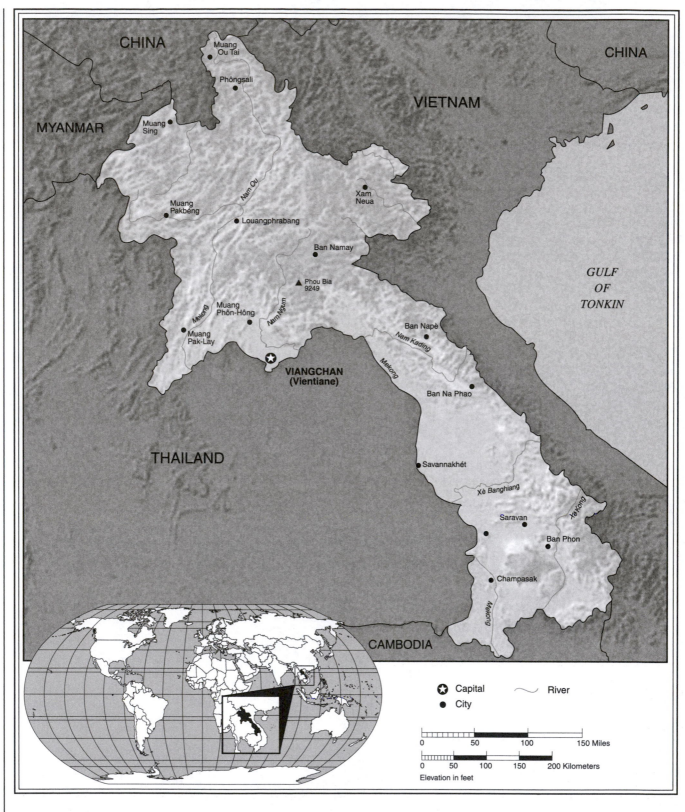

CHINA

Muang
Ou Tai

Phôngsali

MYANMAR

VIETNAM

CHINA

Muang
Sing

Muang
Pakbéng

Xam
Neua

Louangphrabang

Ban Namay

GULF
OF
TONKIN

▲ Phou Bia
9249

Muang
Phôn-Hông

Ban Napè

Nam Kading

Muang
Pak-Lay

Ban Na Phao

★ VIANGCHAN
(Vientiane)

THAILAND

Mekong

Savannakhét

Xè Banghiang

Saravan

Ban Phon

Champasak

Mekong

CAMBODIA

★ Capital   ～ River
● City

0	50	100		150 Miles

| 0 | 50 | 100 | 150 | 200 Kilometers |

Elevation in feet

Laos

(Lao People's Democratic Republic)

Laos seems a sleepy place. Almost everyone lives in small villages where the only distraction might be the Buddhist temple gong announcing the day. Water buffalo plow quietly through centuries-old rice paddies, while young Buddhist monks in saffron-colored robes make their silent rounds for rice donations. Villagers build their houses on stilts for safety from annual river flooding, and top them with thatch or tin. Barefoot children play under the palm trees or wander to the village Buddhist temple for school in the outdoor courtyard. Mothers work at home, weaving brightly colored cloth for the family and preparing meals—on charcoal or wood stoves—of rice, bamboo shoots, pork, duck, and snakes seasoned with hot peppers and ginger. Even the 799,000 person capital city of Viangchan (the name means Sandalwood City) seems laid-back, with chickens wandering the downtown streets.

Below this serene surface, however, Laos is a nation divided. The name Laos is taken from the dominant ethnic group, Lao, which is derived from a word meaning "star," as in people who came from the stars. But there are also large populations of Khmers and Hmongs, along with almost 100 other ethnic groups in the country. Over the centuries, they have battled one another for supremacy, for land, and for tribute money. The constant feuding has weakened the nation and served as an invitation for neighboring countries to annex portions of Laos forcibly, or to align themselves with one or another of the Laotian royal families or generals for material gain. China, Burma (today called Myanmar), Vietnam, and especially Thailand—with which Laotian people share many cultural and ethnic similarities—have all been involved militarily in Laos.

Historically, jealousy among members of the royal family caused most of Laos's bloodshed. More recently, Laos has been seen as a pawn in the battle of the Western powers for access to the rich natural resources of Southeast Asia, or as a "domino" that some did and others did not want to fall to communism. Members of the former royal family, some of whom remain influential, continue to find themselves on the opposite sides of many issues.

The results of these struggles have been devastating. Laos is now one of the poorest countries in the world, with a per capita income of only US$2,500. In Asia only four countries (Papua New Guinea, Cambodia, North Korea, and Myanmar) have lower per capita GDPs. With electricity available in only a few cities, there are few industries in the country, so most people survive by subsistence farming and fishing, raising or catching just what they need to eat rather than growing food to sell. In fact, some "hill peoples" (about half of the Laotian people live in the mountains) in the long mountain range that separates Laos from Vietnam continue to use the most ancient farming technique known, slash-and-burn farming, an unstable method of land use that allows only a few years of good crops before the soil is depleted and the farmers must move to new ground. Today, soil erosion and deforestation pose significant threats to economic growth.

Even if all Laotian farmers used the most modern techniques and geared their production to cash crops, it would still be difficult to export food (or, for that matter, anything else) because of Laos's woefully inadequate transportation network. There are no railroads, and muddy, unpaved

TIMELINE

A.D. 1300s
The first Laotian nation is established

1400s–1700s
Laos is the largest and most powerful country in Southeast Asia

1890s
Laos is under French control

1940s
The Japanese conquer Southeast Asia

1949
France grants independence to Laos

1971
South Vietnamese troops, with U.S. support, invade Laos

1975
Pathet Lao Communists gain control of the government

1977
Laos signs military and economic agreements with Vietnam

1980s
The government begins to liberalize some aspects of the economy

1990s
The Pathet Lao government maintains firm control; efforts to maintain high GDP growth are threatened by deforestation and soil erosion

Funded by Australia, a "Friendship Bridge" connecting Laos and Thailand across the Mekong River opens for commerce.

2000s
Laos works for closer economic ties to Vietnam

The G-7 nations promise debt relief for Laos

Over 1 million tourists visit Laos in 2005

roads make many mountain villages completely inaccessible by car or truck. Only one bridge in Laos, the Thai-Lao Friendship Bridge near the capital city of Viangchan (Vientiane), spans the famous Mekong River. Moreover, Laos is landlocked. In a region of the world where wealth flows toward those countries with the best ports, having no direct access to the sea is a serious impediment to economic growth.

In addition, for years the economy has been strictly controlled by the government. Foreign investment and trade have not been welcomed; tourists were not allowed into the country until 1989. But the economy began to open up in the late 1980s. The government's "New Economic Mechanism" (NEM) in 1986 called for foreign investment in all sectors and anticipated gross domestic product growth of 8 percent per year (it came close to that—7.7 percent—in 2010). The year 1999 was declared "Visit Laos Year," and the government set a goal of attracting 1 million tourists. With its technological infrastructure abysmally underdeveloped, tourism seemed like the easiest way for the country to gain some foreign currency and create jobs. In Luan Prabang province, where the United States is helping locals preserve historical artifacts, tourists can visit a large statue of Buddha and observe a lifestyle that has changed little in centuries. In Viangchan, tourists can visit the mammoth gold Buddhist stupa of Pha That Luang around which saffron-robed monks pray and chant to early-morning drums and then depart for the streets with their begging bowls until the heat becomes too oppressive. Breakfast (and lunch and dinner for that matter) for Laotians often consists of beef soup mixed with rice noodles, fish sauce, bean sprouts, and chili powder. The lingering French influence can be seen in the Fountain Circle area, where foreign restaurants continue to serve French dishes.

Some economic progress has been made in the past decade. Laos is now self-sufficient in its staple crop, rice; and surplus electricity generated from dams along the Mekong River is sold to Thailand to earn foreign exchange. In fact, the potential for generating electricity—as many as 25 hydroelectric generating plants could be built in Laos—is so great that government leaders now talk of Laos becoming "the electric battery" for all of Southeast Asia. Massive infusions of investment monies would be needed to realize this dream.

Some investment money has already started to flow in. China, for example, is cooperating with Laos on a high speed rail system with a price tag of US$7 billion. The project started in 2011. China is now Laos' largest source of foreign investment. Significant too is that, in a first experiment with capitalism, the Vientiane stock market opened in early 2011. Other signs indicate a strong desire on the part of the government to get the economy moving: in 2007, Laos and Thailand signed a joint communiqué covering many matters; in 2010, the government signed a civil aviation agreement with the United States. Perhaps most importantly, Laos now harbors a goal to be removed from the United Nations' list of least-developed countries by 2020.

Laos imports various commodity items from Thailand, Vietnam, Singapore, Japan, and other countries, and it has received foreign aid from the Asian Development Bank and other organizations, including a $40.2 million loan in 2001 from the International Monetary Fund. Exports to Thailand, China, and the United States include teakwood, tin, and various minerals. In 1999, a state-owned bank in Vietnam agreed to set up a joint venture with the Laotian Bank for Foreign Trade. The new bank will mainly deal with imports and exports, particularly between Vietnam and Laos. In 2000, leaders of the world's wealthiest nations, the G-7 group, agreed to help Laos by offering various types of debt relief, but it will be many years before the country can claim that its economy is solid. When Laos joined ASEAN in 1997, it was the group's poorest member, and that remained true until Myanmar dislodged Laos for that dubious honor.

Despite the 1995 "certification" by the United States that Laos is cooperating in the world antidrug effort, Laos continues to supply opium, cannabis, and heroin to users in Europe and North America. The Laotian government is now trying to prevent hill peoples from cutting down valuable forests for opium-poppy cultivation.

■ HISTORY AND POLITICS

The Laotian people, originally migrating from south China, settled Laos in the thirteenth century A.D., when the area was controlled by the Khmer (Cambodian) Empire. Early Laotian leaders expanded the borders of Laos through warfare with Cambodia, Thailand, Burma, and Vietnam. Between 1400 and 1700, Laos was the largest and most powerful kingdom in the region. Buddhist monks from all of Southeast Asia traveled to Laos for higher education. Internal warfare, however, led to a loss of autonomy in 1833, when Thailand forcibly annexed the country (against the wishes of Vietnam, which also had designs on Laos). In the 1890s, France, determined to have a part of the lucrative Asian trade and to hold its own against growing British strength in Southeast Asia, forced Thailand to give up its hold on Laos. Laos, Vietnam, and Cambodia were combined into a new political entity, which the French termed *Indochina*. Between these French possessions and the British possessions of Burma and Malaysia (then called Malaya) lay Thailand; thus, France, Britain, and Thailand effectively controlled mainland Southeast Asia for several decades.

There were several small uprisings against French power, but these were easily suppressed until the Japanese conquest of Indochina in the 1940s. The Japanese, with their "Asia for Asians" philosophy, convinced the Laotians that European domination was not a given. In the Geneva Agreement of 1949, Laos was granted independence, although full French withdrawal did not take place until 1954.

Prior to independence, Prince Souphanouvong (who died in 1995 at age 82) had organized a Communist guerrilla army, with help from the revolutionary Ho Chi Minh of the Vietnamese Communist group Viet Minh. This army called itself *Pathet Lao* (meaning "Lao Country").

Snapshot: LAOS

Summarized below is a quick look at the country with regard to its development, freedom, health/welfare, and achievements.

Development

Communist rule after 1975 isolated Laos from world trade and foreign investment. The planned economy has not been able to gain momentum on its own. In 1986, the government loosened restrictions so that government companies could keep a portion of their profits. A goal is to integrate Laos economically with Vietnam and Cambodia, and the effort appears to be working.

Freedom

Laos is ruled by the political arm of the Pathet Lao Army. Opposition parties and groups as well as opposition newspapers and other media are outlawed. Lack of civil liberties as well as poverty have caused many thousands of people to flee the country.

Health/Welfare

Laos is typical of the least-developed countries in the world. The birth rate is high, but so is infant mortality. Most citizens eat less than an adequate diet. Life expectancy is low, and many Laotians die from illnesses for which medicines are available in other countries. Many doctors fled the country when the Communists came to power.

Achievements

The original inhabitants of Laos, the Kha, have been looked down upon by the Lao, Thai, and other peoples for centuries. But under the Communist regime, the status of the Kha has been upgraded and discrimination formally proscribed.

In 1954, it challenged the authority of the government in the Laotian capital. Civil war ensued, and by 1961, when a cease-fire was arranged, the Pathet Lao had captured about half of Laos. The Soviet Union supported the Pathet Lao, whose strength was in the northern half of Laos, while the United States supported a succession of pro-Western but fragile governments in the south. A coalition government consisting of Pathet Lao, pro-Western, and neutralist leaders was installed in 1962, but it collapsed in 1965, when warfare once again broke out.

During the Vietnam War, U.S. and South Vietnamese forces bombed and invaded Laos in an attempt to disrupt the North Vietnamese supply line known as the Ho Chi Minh Trail. Americans flew nearly 600,000 bombing missions over Laos (many of the small cluster bombs released during those missions remain unexploded in fields and villages and present a continuing danger). Communist battlefield victories in Vietnam encouraged and aided the Pathet Lao Army, which became the dominant voice in a new coalition government established in 1974. The Pathet Lao controlled the government exclusively by 1975. In the same year, the government proclaimed a new "Lao People's Democratic Republic." It abolished the 622-year-old monarchy and sent the king and the royal family to a detention center to learn Marxist ideology.

Vietnamese Army support and flight by many of those opposed to the Communist regime have permitted the Pathet Lao to maintain control of the government. Despite the current leadership's stated desire for the world to consider Laos a republic instead of a Communist state, the ruling dictatorship is determined to prevent the democratization of Laos: In 1993, several cabinet ministers were jailed for 14 years for trying to establish a multiparty democracy. In 2001, the government deported five activists, including a Belgian member of the European Parliament, for handing out prodemocracy leaflets on the streets of Viangchan. A year before, students doing the same thing disappeared; they have never been seen since.

The Pathet Lao government was sustained militarily and economically by the Soviet Union and other East bloc nations for more than 15 years. However, with the end of the cold war and the collapse of the Soviet Union, Laos has had to look elsewhere, including non-Communist countries, for support. In 1992, Laos signed a friendship treaty with Thailand to facilitate trade between the two historic enemy countries. In 1994, the Australian government, continuing its plan to integrate itself more fully into the strong Asian economy, promised to provide Laos with more than $33 million in aid. In 1995, Laos joined with ASEAN nations to declare the region a nuclear-free zone.

Trying to teach communism to a devoutly Buddhist country has not been easy. Popular resistance has caused the government to retract many of the regulations it has tried to impose on the Buddhist Church (technically, the Sangara, or order of the monks—the Buddhist equivalent of a clerical hierarchy). As long as the Buddhist hierarchy limits its activities to helping the poor, it seems to be able to avoid running afoul of the Communist leadership.

Intellectuals, especially those known to have been functionaries of the French administration, have fled Laos, leaving a leadership and skills vacuum. As many as 300,000 people are thought to have left Laos for refugee camps in Thailand and elsewhere. Many have taken up permanent residence in foreign countries, especially France and the United States, where they work as taxi drivers and in other low-income occupations, forever separated from their families. They fear returning to Laos and decry its lack of human rights. Among these refugees are some 15,000 Hmongs, an ethnic group that helped the United States during the Vietnam War. Having been persecuted ever since, the Hmongs are now being re-settled in the United States.

Statistics

Geography

Area in Square Miles (Kilometers): 91,400 (236,800) (about the size of Utah)

Capital (Population): Viangchan (Vientiane) (799,000) (about the size of Utah)

Environmental Concerns: unexploded ordnance; deforestation; soil erosion; lack of access to potable water

Geographical Features: landlocked; mostly rugged mountains; some plains and plateaus

Climate: tropical monsoon

People

Population

Total: 6,477,211

Annual Growth Rate: 1.68%

Rural/Urban Population (Ratio): 71/33

Major Languages: Lao; French; English; ethnic languages

Ethnic Makeup: 55% Lao; 11% Khmou; 8% Hmong; 26% other ethnic groups

Religions: 67% Buddhist; 1.5% Christian; 31.5% indigenous beliefs and others

Health

Life Expectancy at Birth: 60.5 years (male); 64.4 years (female)

Infant Mortality Rate: 59/1,000 live births

Physicians Available: 1/3,333 people

HIV/AIDS Rate in Adults: 0.2%

Education

Adult Literacy Rate: 73%

Compulsory (Ages): 6–15

Communication

Telephones: 132,200 (2009) main lines

Mobile Phones: 3.24 million (2009)

Internet Users: 527,000 (2009)

Internet Penetration (% of Pop.): 8%

Transportation

Roadways in Miles (Kilometers): 22,885 (36,831)

Railroads in Miles (Kilometers): none

Usable Airfields: 41

Motor Vehicles in Use: 18,000

Government

Type: Communist state

Independence Date: July 19, 1949 (from France)

Head of State/Government: President Choummaly Sayasone; Prime Minister Thongsing Thammavong

Political Parties: Lao People's Revolutionary Party; other parties proscribed

Suffrage: universal at 18

Military

Military Expenditures (% of GDP): 0.5%

Current Disputes: indefinite border with Thailand; concerns with China and Cambodia over Mekong dam construction

Economy

Currency ($ U.S. Equivalent): 8,320 kip = $1

Per Capita Income/GDP: $2,500/$15.7 billion

GDP Growth Rate: 7.7%

Inflation Rate: 7.7%

Unemployment Rate: 2.5%

Labor Force by Occupation: 75% agriculture; 5.5% industry; 19.5% services

Population Below Poverty Line: 26%

Natural Resources: timber; hydropower; gypsum; tin; gold; gemstones

Agriculture: rice; sweet potatoes; vegetables; coffee; tobacco; sugarcane; cotton; livestock

Industry: mining; timber; garments; electric power; agricultural processing; construction; cement; tourism

Exports: $1.95 billion (primary partners Thailand, Vietnam, China)

Imports: $1.753 billion (primary partners Thailand, China, Vietnam)

Suggested Websites

www.laoembassy.com/

www.kpl.net.la

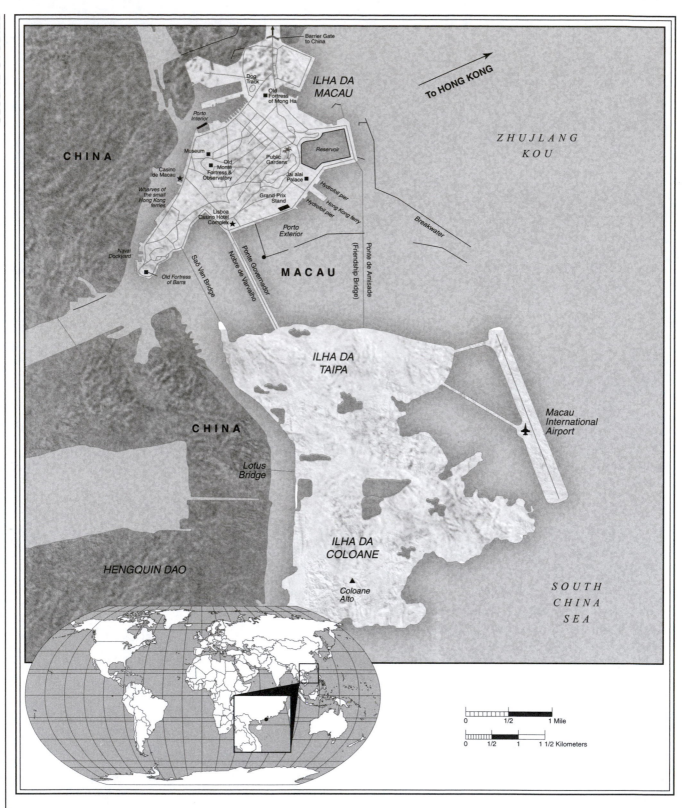

Barrier Gate
to China

Dog
Track

*ILHA DA
MACAU*

Old
Fortress
of Mong Ha

To HONG KONG

*Porto
Interior*

*ZHUJLANG
KOU*

CHINA

Museum

Reservoir

Casino
de Macau

Old
Monte
Fortress &
Observatory

*Public
Gardens*

*Wharves of
the small
Hong Kong
ferries*

*Jai alai
Palace*

Grand Prix
Stand

Hydrofoil pier

Lisboa
Casino Hotel
Complex

Hong Kong ferry

Hydrofoil pier

Breakwater

*Naval
Dockyard*

*Porto
Exterior*

Old Fortress
of Barra

Saô Van Bridge

*Ponte Governador
Nobre de Varvalho*

MACAU

*Ponte de Amisade
(Friendship Bridge)*

*ILHA DA
TAIPA*

Macau
International
Airport

CHINA

*Lotus
Bridge*

*ILHA DA
COLOANE*

HENGQUIN DAO

Coloane
Alto

*SOUTH
CHINA
SEA*

0   1/2   1 Mile

0   1/2   1   1 1/2 Kilometers

Macau
(Macau Special Administrative Region)

Just 17 miles across the Pearl River estuary from Hong Kong lies the world's most densely populated territory: the former Portuguese colony of Macau (sometimes spelled Macao). Consisting of only 11 square miles of land, the peninsula and two tiny islands are home to nearly half a million people, 94 percent of whom are Chinese, with the remainder being Portuguese or Portuguese/Asian mixtures. Until December 20, 1999, when Macau reverted to China as a "Special Administrative Region" (SAR), it had been the oldest outpost of European culture in the Far East, with a 442-year history of Portuguese administration.

Macau was frequented by Portuguese traders as early as 1516, but it was not until 1557 that the Chinese agreed to Portuguese settlement of the land; unlike Hong Kong, however, China did not acknowledge Portuguese sovereignty. Indeed, the Chinese government did not recognize the Portuguese demand for "perpetual occupation" until 1887, although the administration of the territory had been in Portuguese hands for over 300 years by then.

Macau's population has varied over the years, depending on conditions in China. During the Japanese occupation of parts of China during the 1940s, for instance, Macau's Chinese population is believed to have doubled, and more refugees streamed in when the Communists took over China in 1949. Thus, like Hong Kong and Taiwan, Macau has historically served as a place of refuge for Chinese fleeing troubled conditions on the mainland of China.

In 1987, Chinese and Portuguese officials signed an agreement, effective December 20, 1999, to end European control of the first—and last—colonial outpost in China. Actually, Portugal had offered to return Macau on two earlier occasions (in 1967, during the Chinese Cultural Revolution; and in 1974, after the coup in Portugal that ended the dictatorship there), but China refused. The 1999 transition went smoothly, although Portugal's president announced that he would not attend the hand-over ceremony if China sent in troops before the official transition date. China relented, and the hand-over ceremony proceeded smoothly. When the troops arrived on December 20, they were generally cheered by crowds who hoped that they would bring some order to the gang-infested society.

The transition agreement was similar to that signed by Great Britain and China over the post-British withdrawal from Hong Kong. China agreed to allow Macau to maintain its capitalist way of life for 50 years, to permit local elections, and to allow its residents to travel freely without Chinese intervention. Unlike Hong Kong residents, who staged massive demonstrations against future Chinese rule or emigrated from Hong Kong before its return to China, Macau residents—some of whom have been openly pro-Communist—have not seemed bothered by the new arrangements. Indeed, businesses in Macau (as well as Hong Kong) have contributed to a de facto merging with the mainland by investing more than $20 billion in China since the mid-1990s.

Since it was established in the sixteenth century as a trading colony with interests in oranges, tea, tobacco, and lacquer, Macau has been heavily influenced by Roman Catholic priests of the Dominican and Jesuit

TIMELINE

A.D. 1557
A Portuguese trading colony is established at Macau

1849
Portugal declares sovereignty over Macau

1887
China signs a treaty recognizing Portuguese sovereignty over Macau

1940s
Immigrants from China flood into the colony

1967
Pro-Communist riots in Macau

1970s
Portugal begins to loosen direct administrative control over Macau

1976
Macau becomes a Chinese territory but is still administered by Portugal

1987
China and Portugal sign an agreement scheduling the return of Macau to Chinese control

1990s
Portugal sends troops to help tamp down gambling-related crime; Macau reverts to Chinese control on December 20, 1999

2000s
The reversion to Chinese control goes smoothly

Macau fights to recover from the Asian financial crisis

Macau bank is accused by United States of laundering money for North Korea

25 million tourists visited Macau in 2010, 53% from mainland China

Snapshot: MACAU

Summarized below is a quick look at the country with regard to its development, freedom, health/welfare, and achievements.

Development

The development of industries related to gambling and tourism (tourists are primarily from Hong Kong and increasingly from mainland China, where gambling is not allowed) has been very successful. Most of Macau's foods, energy, and fresh water are imported from China; Japan and Hong Kong are the main suppliers of raw materials.

Freedom

Under the "Basic Law," China agreed to maintain Macau's separate legal, political, and economic system. The Legislature is partly elected and partly appointed.

Health/Welfare

Macau has very impressive quality-of-life statistics. It has a low infant mortality rate and very high life expectancy for both males and females. Literacy is 91 percent.

Achievements

Considering its unfavorable geographical characteristics, such as negligible natural resources and a port so shallow and heavily silted that oceangoing ships must anchor offshore, Macau has had stunning economic success.

DID YOU KNOW?

Macau is the world's largest gambling area. It overtook Las Vegas in 2006 because, in 2001, the government opened up opportunities for overseas investors (including some from Las Vegas) to build or operate casinos, and billions of dollars of investment money flowed in. At about the same time, the Chinese government relaxed its travel restrictions, making it easier for mainlanders (where gambling is illegal) to travel to Macau. But Macau is not the only place ahead of Las Vegas. In 2011, Singapore took in US$6.4 billion in gambling revenues, putting it ahead of Las Vegas, too.

---

orders. Christian churches, interspersed with Buddhist temples, abound. With Buddhist immigrants from China reducing the proportion of Christians, about 15 percent of the population now claims to be Christian, and about 50 percent claims to be Buddhist. The name of Macau itself reflects its deep and enduring religious roots; until China reduced its official name to "Macau" in 1999, its name was "City of the Name of God in China, Macau, There is None More Loyal." Despite its reputation as a gambling haven, Macau has perhaps the highest density of churches and temples per square mile in the world.

▪ A HEALTHY ECONOMY

Macau's modern economy is a vigorous blend of light industry, fishing, tourism, and gambling. Revenues from the latter source—casino gambling—are impressive, accounting for almost 40 percent of the gross domestic product as well as one third of all jobs. The government imposes a 32 percent tax on casinos. Major infrastructural improvements, such as the new international airport built on reclaimed land on Taipa Island and serving passengers primarily from mainland China, Taiwan, and Singapore, are funded by gambling revenues. There are five major casinos and many other gambling

opportunities in Macau, which, along with the considerable charms of the city itself, attract more than 25 million visitors each year, most of them from Hong Kong and China (where gambling is illegal).

For 40 years, Macau's gaming industry was run by Stanley Ho and his Macau Travel and Amusement Company, which held monopoly rights on all gambling. Not everyone was pleased with that arrangement, and rival street gangs, striving to control part of the lucrative business, launched a crime spree in 1998. Military troops had to be sent from Portugal to restore order. In 2002, the government diversified control of the industry by giving one license to Ho and two to U.S. gaming interests. The three licensees had to agree to invest at least US$500 million in new casino facilities—a sum not particularly onerous for Ho, who in 2011 was listed by *Forbes* magazine as not only the wealthiest person in Macau (he actually lives in Hong Kong), but one of the richest persons in the world, with assets of US$3.1 billion (down from a high of US$6.5 billion in 2006). Ninety years old, he remains active in business and employs over 10,000 people in Macau, with operations in several other Asian countries as well.

The diversification of the gambling licenses brought another big player into Macau: U.S. billionaire Stephen Wynn. In 2006, he opened the Wynn Macau casino resort after investing US$1.2 billion. Such investment made it possible for Macau to overtake the Las Vegas strip as the world's most profitable gambling capital.

Export earnings derived from light-industry products such as textiles, fireworks, plastics, and electronics are also critical to the colony. Macau's leading export markets are Hong Kong (43%), China (16%), the United States (11%), and Germany (4%); ironically, Portugal consumes only about 3 percent of Macau's exports.

As might be expected, the general success of the economy has a downside. In Macau's case, hallmarks of modernization—crowded apartment blocks, packed casino hotels, and bustling traffic—are threatening to eclipse the remnants of the old, serene, Portuguese-style seaside town.

Other issues continue to fester. Gangs continue to intimidate business owners, and in 2006 the United States accused the Banco Delta Asia SARL of counterfeiting and of laundering illicit funds on behalf of North Korea. So clear were the offenses that even the Bank of China froze North Korea's account at that bank in support of the U.S. position.

Macau made international news briefly in 2009 when a former Tiananmen Square protest leader from the 1989 event in China tried to enter China via Macau in time for the massacre's 20th anniversary. He was denied entry and deported to Taiwan where he had fled years earlier.

Statistics

Geography

Area in Square Miles (Kilometers): 11 (28) (less than 1/6 the size of Washington, D.C.)
Capital (Population): Macau (573,003)
Environmental Concerns: air and water pollution
Geographical Features: generally flat
Climate: subtropical; marine with cool winters, warm summers

People

Population

Total: 573,003
Annual Growth Rate: 0.88%
Rural/Urban Population Ratio: 100% urban
Major Languages: Cantonese, Portuguese
Ethnic Makeup: 94% Chinese; 6% Macanese, Portuguese, and others
Religions: 50% Buddhist; 15% Roman Catholic; 35% unaffiliated or other

Health

Life Expectancy at Birth: 81.5 years (male); 87 years (female)
Infant Mortality: 3.2/1,000 live births
Physicians Available: 1/417 people

Education

Adult Literacy Rate: 91%

Communication

Telephones: 168,903 main lines (2010)
Mobile Phones: 1.1 million (2010)
Internet Users: 281,000 (2011)
Internet Penetration (% of Pop.): 49%

Transportation

Roadways in Miles (Kilometers): 257 (413)
Railroads in Miles (Kilometers): none
Usable Airfields: 1

Government

Type: Limited democracy as a Special Administrative Region of China
Independence Date: none; Special Administrative Region of China

Head of State/Government: President (of China) Hu Jintao; Chief Executive Fernando Chui Sai-on
Political Parties: no formal political parties; civic associations are used instead, including Alliance for Change; Macau Development Alliance; Macau-Guangdong Union; New Hope; Macau United Citizens' Association; New Democratic Macau Association; Union for Promoting Progress
Suffrage: direct election at 18; universal for permanent residents living in Macau for 7 years; indirect election limited to organizations registered as "corporate voters" and a 300-member Election Committee

Military

Military Expenditures (% of GDP): defense is the responsibility of China
Current Disputes: none

Economy

Currency ($ U.S. Equivalent): 8.00 pataca = $1 (tied to Hong Kong dollar)
Per Capita Income/GDP: $30,000/$18.5 billion
GDP Growth Rate: 1%
Inflation Rate: 2.8%
Unemployment Rate: 3%
Labor Force by Occupation: 13% restaurants and hotels; 13% gambling; 13% retail; 9% construction; 7% public sector; 6% transport and communications; 4% manufacturing; 35% other services and agriculture
Population Below Poverty Line: na
Natural Resources: fish
Agriculture: vegetables
Industry: clothing; textiles; toys; tourism; electronics; footwear; gambling
Exports: $870 million (primary partners Hong Kong, China, United States)
Imports: $5.5 billion (primary partners China, Hong Kong, France, Japan)

Suggested Websites

www.macautourism.gov.mo/
news.bbc.co.uk/2/hi/asia-pacific/country_profiles/4080105.stm

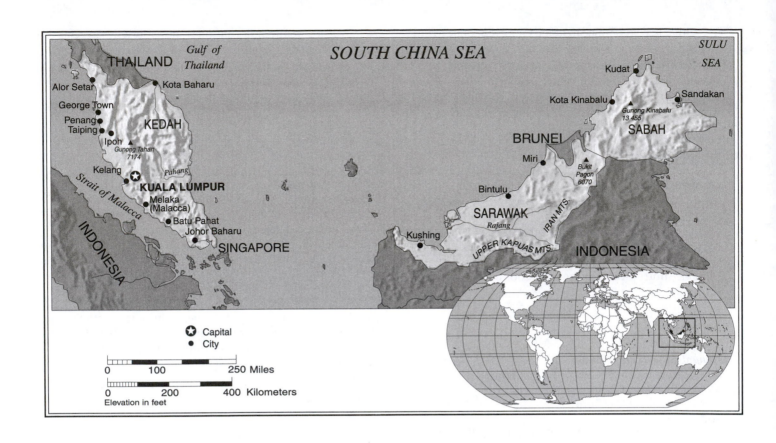

Malaysia

About the size of Japan and famous for its production of natural rubber and tin, Malaysia sounds like a true political, economic, and social entity. Although it has all the trappings of a modern nation-state, Malaysia is one of the most fragmented nations on Earth.

Consider its land. West Malaysia, wherein reside 86 percent of the population, is located on the Malay Peninsula between Singapore and Thailand; but East Malaysia, with 60 percent of the land, is located on the island of North Borneo, some 400 miles of ocean away.

Similarly, Malaysia's 29 million people are divided along racial, religious, and linguistic lines. Sixty-one percent are Malay and other indigenous peoples, many of whom adhere to the Islamic faith or animist beliefs; 24 percent are Chinese, most of whom are Buddhist, Confucian, or Taoist; 7 percent are Indians and 9 percent are Pakistanis and others, some of whom follow the Hindu faith. Bahasa Malaysia is the official language, but English, Arabic, two forms of Chinese, Tamil, and other languages are also spoken. Thus, although the country is called Malaysia (a name adopted only 35 years ago), many people living in Kuala Lumpur, the capital, or in the many villages in the countryside have a stronger identity with their ethnic group or village than with the country of Malaysia per se.

Malaysian culture is further fragmented because each ethnic group tends to replicate the architecture, social rituals, and norms of etiquette peculiar to itself. The Chinese, whose ancestors were imported in the 1800s from south China by the British to work the rubber plantations and tin mines, have become so economically powerful that their cultural influence extends far beyond their actual numbers. Like the Malays and the Hindus, they surround themselves with the trappings of their original, "mother" culture.

Malaysian history is equally fragmented. Originally controlled by numerous sultans who gave allegiance to no one, or only reluctantly to various more powerful states in surrounding regions, Malaysia first came to Western attention in A.D. 1511, when the prosperous city of Malacca, which had been founded on the west coast of the Malay Peninsula about a century earlier, was conquered by the Portuguese. The Dutch took Malacca away from the Portuguese in 1641. The British seized it from the Dutch in 1824 (the British had already acquired an island off the coast and had established the port of Singapore). By 1888, the British were in control of most of the area that is now Malaysia.

However, British hegemony did not mean total control, for each of the many sultanates—the origin of the 13 states that constitute Malaysia today—continued to act more or less independently of the British, engaging in wars with one another and maintaining an administrative apparatus apart from the British. Some groups, such as the Dayaks, an indigenous people living in the jungles of Borneo, remained more or less aloof from the various intrigues of modern state-making and developed little or no identity of themselves as citizens of any modern nation.

It is hardly surprising, then, that Malaysia has had a difficult time emerging as a nation. Indeed, it is not likely that there would have been an independent Malaysia had it not been for the Japanese, who defeated

TIMELINE

A.D. 1403
The city of Malacca is established; it becomes a center of trade and Islamic conversion

1511
The Portuguese capture Malacca

1641
The Dutch capture Malacca

1824
The British obtain Malacca from the Dutch

1941
Japan captures the Malay Peninsula

1948
The British establish the Federation of Malaya; a Communist guerrilla war begins, lasting for a decade

1957
The Federation of Malaya achieves independence under Prime Minister Tengku Abdul Rahman

1963
The Federation of Malaysia, including Singapore but not Brunei, is formed

1965
Singapore leaves the Federation of Malaysia

1980s
Malaysia attempts to build an industrial base

1990s
The NEP is replaced with Vision 2020; economic crisis

2000s
The economy rebounds; the environment suffers
Former deputy prime minister Anwar Ibrahim is arrested and convicted under questionable circumstances
The World Court awards two tiny Celebes Sea islands to Malaysia in a dispute with Indonesia
Abdullah Ahmad Badawi replaces Mahathir Mohamad as prime minister
Ex-deputy prime minister Anwar Ibrahim's conviction is overturned; he vows to push for democratic reforms
Malaysia buys US$1.1 billion of shares in Rosneft, a state-owned Russian oil firm

the British in Southeast Asia during World War II and promulgated their alluring doctrine of "Asia for Asians."

After the war, Malaysian demands for independence from European domination grew more persuasive; Great Britain attempted in 1946 to meet these demands by proposing a partly autonomous Malay Union. However, ethnic rivalries and power-sensitive sultans created such enormous tension that the plan was scrapped. In an uncharacteristic display of cooperation, some 41 different Malay groups organized the United Malay National Organization (UMNO) to oppose the British plan. In 1948, a new Federation of Malaya was attempted. It granted considerable freedom within a framework of British supervision, allowed sultans to retain power over their own regions, and placed certain restrictions on the power of the Chinese living in the country.

Opposing any agreement short of full independence, a group of Chinese Communists, with Indonesian support, began a guerrilla war against the government and against capitalist ideology. Known as "The Emergency," the war lasted more than a decade and involved some 250,000 government troops. Eventually, the insurgents withdrew.

The three main ethnic groups—Malayans, represented by UMNO; Chinese, represented by the Malayan Chinese Association, or MCA; and Indians, represented by the Malayan Indian Congress, or MIC—were able to cooperate long enough in 1953 to form a single political party under the leadership of Abdul Rahman. This party demanded and received complete independence for the Federation in 1957, although some areas, such as Brunei, refused to join. Upon independence, the Federation of Malaya (not yet called Malaysia), excluding Singapore and the territories on the island of Borneo, became a member of the British Commonwealth of Nations and was admitted to the United Nations. In 1963, a new Federation was proposed that included Singapore and the lands on Borneo. Again, Brunei refused to join. Singapore joined but withdrew in 1965. Thus, what we call Malaysia today acquired its current form in 1966.

Political troubles stemming from the deep ethnic divisions in the country, however, remain a constant feature of Malaysian life. With 9 of the 13 states controlled by independent sultans, every election is a test of the governing ability of the National Front (UMNO), a multiethnic coalition of 14 different parties. In the 2008 elections the government coalition lost its supermajority in parliament, capturing only 63 percent of the 222 House seats compared to nearly 80 percent in the 2004 elections. Furthermore, 5 state legislatures went to the opposition compared to only 1 in 2004. Part of the decline in support for the ruling party was based on ethnic Indian dissatisfaction with what they perceive as discrimination against them by Malays. Islamic party supporters rioted in one state when it was believed that university students had been paid to cast illegal votes for the ruling coalition. Police sent to quell the riot were beaten and their cars attacked. The new prime minister, Mohamed Najib bin Abdul Razak, released some political prisoners and removed bans on two newspapers, but rumors of his having an affair with a woman who was later murdered immediately dampened his popularity.

Despite some five decades in power, the ruling coalition continues to find it difficult to preserve stability in the country. In 2011, some 20,000 Malaysians defied government warnings and marched in Kuala Lumpur, the capital. They were demanding electoral reform, but the ruling government was not interested. Police blocked roads, shut down rail stations, and used water cannons, tear gas, and barbed wire to block demonstrators from gathering. In advance of the demonstration, over 200 opposition leaders were arrested, and another 1,600 were arrested during the demonstration.

Particularly troublesome has been the state of Sabah (an area claimed by the Philippines), many of whose residents have wanted independence or, at least, greater autonomy from the federal government. In recent years, the National Front was able to gain a slight majority in Sabah elections, indicating the growing confidence that people have in the federal government's economic development policies. Indeed, in recent years, the government has been running budget surpluses.

■ ECONOMIC DEVELOPMENT

For years, Malaysia's "miracle" economy kept social and political instability in check. Although it had to endure normal fluctuations in market demand for its products, the economy grew at 5 to 8 percent per year from the 1970s to the late 1990s, making it one of the world's top 20 exporters/importers. The manufacturing sector developed to such an extent that it accounted for 70 percent of exports. Then, in 1998, a financial crisis hit. Malaysia was forced to devalue its currency, the ringgit, making it more difficult for consumers to buy foreign products, and dramatically slowing the economy. The government found it necessary to deport thousands of illegal Indonesian and other workers (dozens of whom fled to foreign embassies to avoid deportation) in order to find jobs for Malaysians. In the 1980s and early 1990s, up to 20 percent of the Malaysian workforce had been foreign workers, but the downturn of the late 1990s produced "Operation Get Out," in which at least 850,000 "guest workers" were deported to their home countries. The deportations were made more urgent in 2002, when imported workers, many of them Indonesians and Filipinos working as menial construction and rubber-plantation workers, rioted in protest of a new drug-testing law. Penalties for illegal or law-breaking immigrant workers included a fine, three months in prison, or six lashes with a cane. The deportations caused consternation in Indonesia and the Philippines, especially when it was learned that 17,000 laborers were being detained in holding camps and that more than a dozen had died in detention.

Malaysia continues to tap into its abundant natural resources and has overcome the financial crisis of the late 1990s and the economic slowdown of 2009. As of 2010, with economic growth at a 10-year high, it was the

21st-largest trading nation in the world, with the lion's share of its products being sold to Singapore, China, Japan, and the United States. Moreover, the Malaysian government has a good record of active planning and support of business ventures—directly modeled after Japan's export-oriented strategy. Malaysia launched a "New Economic Policy" (NEP) in the 1970s that welcomed foreign direct investment and sought to diversify the economic base. Japan, Taiwan, and the United States invested heavily in Malaysia. So successful was this strategy that economic-growth targets set for the mid-1990s were actually achieved several years early.

In 1991, the government replaced NEP with a new plan, "Vision 2020." Its goal was to bring Malaysia into full "developed nation" status by the year 2020. Sectors targeted for growth included the aerospace industry, biotechnology, microelectronics, and information and energy technology. The government expanded universities and encouraged the creation of some 170 industrial and research parks, including "Free Zones," where export-oriented businesses were allowed duty-free imports of raw materials.

Continuing to exercise fiscal prudence, the government has scrapped some of its more ambitious plans, including a US$6 billion hydroelectric dam, and a bridge that would have connected the country with Singapore. On the other hand, a multimedia Super Corridor continues to attract high-tech companies from around the world. Some 5,000 international companies have located in Malaysia because the business infrastructure (seven major ports, five international airports, modern freeways, and an English-speaking management class) is conducive to world trade.

The careful economic planning is paying dividends in both political stability and in GDP growth, which was over 7 percent in 2010, and only fell below strong growth when the global economic meltdown affected Malaysian exports. Happily, unemployment has remained under 4 percent for several years, with ever-increasing numbers of Malaysians working in the manufacturing sector, which now accounts for 85 percent of all exports and 40 percent of GDP.

Despite Malaysia's substantial economic successes, serious social problems remain. They stem not from insufficient revenues but from inequitable distribution of wealth. The Malay portion of the population in particular continues to feel economically deprived as compared to the more affluent Chinese and Indian segments. Furthermore, most Malays are farmers, and rural areas have not benefited from Malaysia's economic boom as much as urban areas have.

In the 1960s and 1970s, riots involving thousands of college students were headlined in the Western press as having their basis in ethnicity. This was true to some degree, but the core issue was economic inequality. Included in the economic master plan of the 1970s were plans (similar to affirmative action in the United States) to change the structural barriers that prevented many Malays from fully enjoying the benefits of the economic boom. Under the leadership of Prime Minister Datuk Mahathir bin Mohamad, plans were developed that would assist Malays until they held a 30 percent interest in Malaysian businesses. In 1990, the government announced that the figure had already reached an impressive 20 percent.

Unfortunately, many Malays have insufficient capital to maintain ownership in businesses, so the government has been called upon to acquire many Malay businesses in order to prevent their being purchased by non-Malays. In addition, the system of preferential treatment for Malays has created a Malay elite, detached from the Malay poor, who now compete with the Chinese and Indian elites. Thus, interethnic and interracial goodwill remains difficult to achieve. Nonetheless, social goals have been attained to a greater extent than most observers have thought possible. Educational opportunities for the poor have been increased, farmland development has proceeded on schedule, and the poverty rate has dropped significantly.

THE LEADERSHIP

In a polity as fractured as Malaysia's, one would expect rapid turnover among political elites, but for over a decade, Malaysia was run, sometimes ruthlessly, by Malay prime minister Mahathir Mohamad of the United Malay National Organization. Most Malaysians were relieved that his aggressive nationalistic rhetoric was usually followed by more moderate behavior vis-à-vis other countries. The Chinese Democratic Action Party (DAP) was sometimes able to reduce his political strength in Parliament, but his successful economic strategies muted most critics.

Malaysia's economic success, symbolized in one of the world's tallest buildings, the mammoth Petronas Twin Towers in Luala Lumpur, has not been achieved without some questionable practices. The government under Mahathir seemed unwilling to regulate economic growth, even though strong voices were raised against industrialization's deleterious effects on the old-growth teak forests and other parts of the environment. The environmentalists' case was substantially strengthened in 1998 when forest and peat bog fires in Malaysia and Indonesia engulfed Kuala Lumpur in a thick haze for weeks. The government, unable to snuff out the fires, resorted to installing sprinklers atop the city's skyscrapers to settle the dust and lower temperatures and tempers. In 2004, a chemical plant in Prai, a town across from Penang Island, exploded, sending noxious smoke into the community and requiring the evacuation of hundreds of people. And, like other Asian nations, Malaysia has had to cope with several outbreaks of avian flu.

Blue-collar workers who are the muscle behind Malaysia's economic success are prohibited from forming labor unions, and outspoken critics have been silenced. The most outspoken critic was Anwar Ibrahim. He had been the deputy prime minister and heir-apparent

Snapshot: MALAYSIA

Summarized below is a quick look at the country with regard to its development, freedom, health/welfare, and achievements.

Development

Efforts to move the economy away from farming and toward industrial production have been very successful. Manufacturing now accounts for 30 percent of GDP, and Malaysia is the third-largest producer of semiconductors in the world. With Thailand, Malaysia will build a US$1.3 billion, 530-mile natural-gas pipeline.

Freedom

Malaysia is attempting to govern according to democratic principles. Ethnic rivalries, however, severely hamper the smooth conduct of government and limit such individual liberties as the right to form labor unions. Evidence of undemocratic tactics, such as the government's treatment and imprisonment of ex-deputy Prime Minister Anwar Ibrahim, bring out large numbers of protestors. Islamic law is applied to Muslims in matters of family law. A newspaper that reprinted Danish caricatures of the Prophet Muhammad was shut down in 2006.

Health/Welfare

City-dwellers have ready access to educational, medical, and social opportunities, but the quality of life declines precipitously in the countryside. The 2006–2010 economic growth plan emphasizes improvements in healthcare.

Achievements

Malaysia has made impressive economic advancements, and its New Economic Policy has resulted in some redistribution of wealth to the poorer classes. Malaysia has been able to recover from the Asian financial crisis of the late 1990s and now expects solid GDP growth. The country has also made impressive social and political gains.

---

to Mahathir; but when he challenged Mahathir's policies, he was fired, arrested, beaten, and eventually sent to prison for 14 years on various charges. Protesters frequently took to the streets in his defense, but they were beaten by police and sprayed with tear gas and water cannons. The largest opposition newspaper came to Anwar's defense (for which the editor was charged with sedition), and Anwar's wife started a new political party to challenge the government. The scandal severely tarnished Mahathir's reputation, and he resigned in late 2003. Another Malay, Abdullah Ahmad Badawi, whose resistance to the creation of an Islamic state won him respect from the majority of the people, including many moderate Muslims, replaced him. He was, in turn, replaced in the 2008 elections by Mohamed Najib bin Abdul Razak. Mahathir continues to unsettle the political scene, however. In 2006, when he was not elected as a delegate to the party congress, he made noisy calls for his successor's resignation, and left the party until Abdullah was replaced by Najib.

In 2004, the conviction of ex-deputy prime minister Ibrahim was overturned. Immediately upon leaving prison, he vowed to return to politics and push for greater respect for human rights. His wife and daughter won elections in 2008, and his daughter was one of the participants in the 2011 electoral reform demonstrations.

Statistics

Geography

Area in Square Miles (Kilometers): 127,355 (329,847) (slightly larger than New Mexico)
Capital (Population): Kuala Lumpur (1.49 million)
Environmental Concerns: air and water pollution; deforestation; smoke/haze from Indonesian forest fires
Geographical Features: two main land areas separated by sea; coastal plains rising to hills and mountains
Climate: tropical; annual monsoons

People

Population

Total: 28,728,607
Annual Growth Rate: 1.58%
Rural/Urban Population Ratio: 28/72
Major Languages: Bahasa Malaysia; English; Chinese (many dialects); Tamil; Telugu; Malayalam; Panjabi; Thai; other indigenous languages such as Iban and Kadazan

Ethnic Makeup: 61% Malay and other indigenous; 24% Chinese; 7% Indian; 8% others
Religions: 60% Muslim; 19% Buddhist; 9% Christian; 6% Hindu; 6% other traditional religions, unknown or none

Health

Life Expectancy at Birth: 71 years (male); 77 years (female)
Infant Mortality: 15/1,000 live births
Physicians Available: 1/1,063 people
HIV/AIDS in Adults: 0.5%

Education

Adult Literacy Rate: 89%
Compulsory (Ages): 6–16; free

Communication

Telephones: 4.3 million main lines (2009)
Mobile Phones: 30.4 million (2009)
Internet Users: 16.9 million (2009)
Internet Penetration (% of Pop.): 59%

Transportation

Roadways in Miles (Kilometers): 61,342 (98,721)
Railroads in Miles (Kilometers): 1,149 (1,849)
Usable Airfields: 118

Government

Type: constitutional monarchy
Independence Date: August 31, 1957 (from the United Kingdom)
Head of State/Government: Sultan Abdul Halim Muadzam Shar (as of Dec 2011); Prime Minister Najib Abdul Razak
Political Parties: National Front; People's Alliance; Sabah Progressive Party; others
Suffrage: universal at 21

Military

Military Expenditures (% of GDP): 2%
Current Disputes: complex dispute over the Spratly Islands; Sabah is claimed by the Philippines; various disputes with Singapore

Economy

Currency ($ U.S. Equivalent): 3.044 ringgits = $1
Per Capita Income/GDP: $14,700/$414 billion
GDP Growth Rate: 7.2%
Inflation Rate: 1.7%
Unemployment Rate: 3.5%
Labor Force by Occupation: 51% services; 36% industry; 13% agriculture
Population Below Poverty Line: 3.6%
Natural Resources: tin; petroleum; timber; natural gas; bauxite; iron ore; copper
Agriculture: Peninsular Malyasia—rubber; palm oil; cocoa; rice; Sabah—subsistence crops, coconuts, rice, rubber, timber; Sarawak—rubber; timber; pepper
Industry: rubber and palm oil processing and manufacturing; light manufacturing; pharmaceuticals; medical technology; electronics; tin mining and smelting; logging; timber processing; petroleum production and refining; agriculture processing
Exports: $210 billion (primary partners Singapore, China, Japan, United States)
Imports: $174 billion (primary partners China, Japan, Singapore, United States)

Suggested Websites

www.geographia.com/Malaysia/
www.tourism.gov.my/

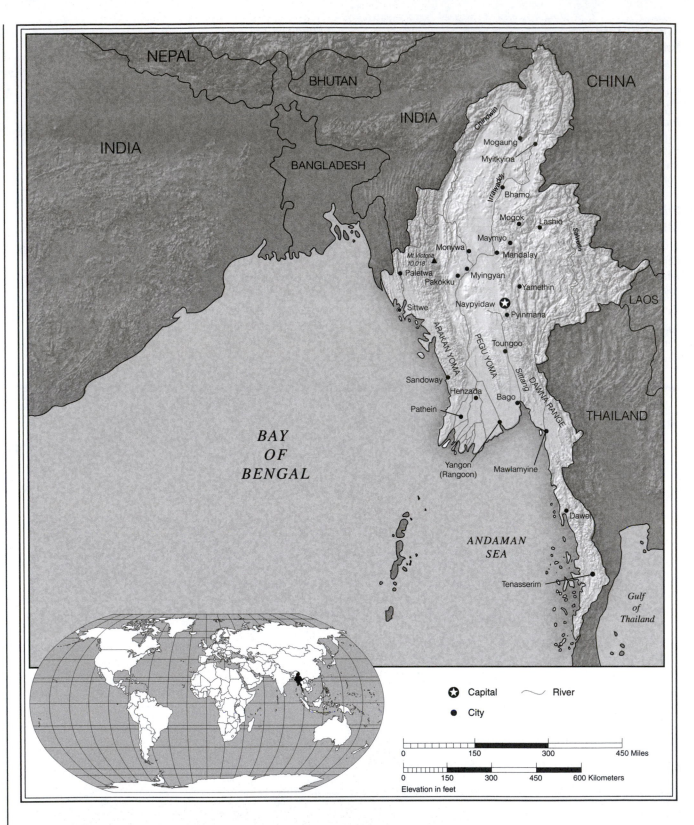

NEPAL

BHUTAN

CHINA

INDIA

INDIA

BANGLADESH

Chindwin

Mogaung

Myitkyina

Irrawaddy

Bhamo

Mogok

Lashio

Salween

Maymyo

Mt. Victoria 10,018

Monywa

Mandalay

Paletwa

Myingyan

Pakokku

Yamethin

LAOS

Sittwe

Naypyidaw

Pyinmana

ARAKAN YOMA

PEGU YOMA

Toungoo

Sittang

DAWNA RANGE

Sandoway

Henzada

Bago

THAILAND

Pathein

Mawlamyine

BAY
OF
BENGAL

Yangon
(Rangoon)

Dawei

ANDAMAN
SEA

Tenasserim

Gulf
of
Thailand

★ Capital   ∿ River

● City

| 0 | 150 | 300 | 450 Miles |

| 0 | 150 | 300 | 450 | 600 Kilometers |

Elevation in feet

118

Myanmar

(Union of Myanmar; formerly Burma)

For four decades, Myanmar (as Burma was renamed by the military junta in 1989) has been a tightly controlled society. Telephones, radio stations, railroads, and many large companies have been under the direct control of a military junta that has brutalized its opposition and forced many to flee the country. For many years, tourists were allowed to stay only two weeks (for awhile the limit was 24 hours), were permitted accommodations only at military-approved hotels, and were allowed to visit only certain parts of the country. Citizens too were highly restricted: They could not leave their country by car or train to visit nearby countries because all the roads were sealed off by government decree, and rail lines terminated at the border. Even Western-style dancing was declared illegal. Until a minor liberalization of the economy was achieved in 1989, all foreign exports—every grain of rice, every peanut, every piece of lumber—though generally owned privately, had to be sold to the government rather than directly to consumers.

Observers attribute this state of affairs to military commanders who overthrew the legitimate government in 1962, but the roots of Myanmar's political and economic dilemma actually go back to 1885, when the British overthrew the Burmese government and declared Burma to be a colony of Britain (to be administered as a province of India). In the 1930s, European-educated Burmese college students organized strikes and demonstrations against the British. Seeing that the Japanese Army had successfully toppled other European colonial governments in Asia, the students decided to assist the Japanese during their invasion of Burma in 1941. Once the British had been expelled, however, the students organized the Anti-Fascist People's Freedom League (AFPFL) to oppose Japanese rule.

When the British tried to resume control of Burma after World War II, they found that the Burmese people had given their allegiance to U Aung San, one of the original student leaders. He and the AFPFL insisted that the British grant full independence to Burma, which they reluctantly did in 1948. So determined were the Burmese to remain free of foreign domination that, unlike most former British colonies, they refused to join the British Commonwealth of Nations. This was the first of many decisions that would have the effect of isolating Burma from the global economy.

Unlike Japan, with its nearly homogeneous population and single national language, Myanmar is a multiethnic state; in fact, only about 60 percent of the people speak Burmese. The Burman people are genetically related to the Tibetans and the Chinese; the Chin are related to peoples of nearby India; the Shan are related to Thais; and the Mon migrated to Burma from Cambodia. In general, these ethnic groups live in separate political states within Myanmar—the Kachin State, the Shan State, the Karen State, and so on; and for hundreds of years, they have warred against one another for dominance.

Upon the withdrawal of the British in 1948, some ethnic groups, particularly the Kachins, the Karens, and the Shans, embraced the Communist ideology of change through violent revolution. Their rebellion against the government in the capital city of Yangon (at the time called Rangoon) had the effect of removing from government control large portions of

TIMELINE

800 B.C.
Burman people enter the Irrawaddy Valley from China and Tibet

A.D. 1500s
The Portuguese are impressed with Burmese wealth

1824–1826
The First Anglo-Burmese War

1852
The Second Anglo-Burmese War

1885
The Third Anglo-Burmese War results in the loss of Burmese sovereignty

1948
Burma gains independence from Britain

1962
General Ne Win takes control of the government in a coup

1980s
Economic crisis; the pro-democracy movement is crushed; General Saw Maung takes control of the government

1989
Burma is renamed Myanmar (though most people prefer the name Burma)

1990s
The military refuses to give up power; Aung San Suu Kyi's activities remain restricted

2000s
The world community increasingly registers its disapproval of the Myanmar junta
Military moves official capital from Yangon to Naypyidaw in 2006
A cyclone in 2008 kills 77,000 and displaces 2.5 million
Aung San Suu Kyi is released from house arrest but re-arrested again in 2003, 2006, and 2009
Protests close universities and bring condemnation from the UN, EU, and United States
Cyclone Nargis kills 140,000 people

the country. Headed by U Nu (U Aung San and several of the original government leaders having been assassinated shortly before independence), the government considered its position precarious and determined that to align itself with the Communist forces then ascendant in China and other parts of Asia would strengthen the hand of the ethnic separatists, whereas to form alliances with the capitalist world would invite a repetition of decades of Western domination. U Nu thus attempted to steer a decidedly neutral course during the cold war era and to be as tolerant as possible of separatist groups within Burma. Burma refused U.S. economic aid, had very little to do with the warfare afflicting Vietnam and the other Southeast Asian countries, and was not eager to join the Southeast Asian Treaty Organization or the Asian Development Bank.

Some factions of Burmese society were not pleased with U Nu's relatively benign treatment of separatist groups. In 1958, a political impasse allowed Ne Win, a military general, to assume temporary control of the country. National elections were held in 1962, and a democratically elected government was installed in power. Shortly thereafter, however, Ne Win staged a military coup. The military has controlled Myanmar ever since—almost 50 years. Under Ne Win, competing political parties were banned, the economy was nationalized, and the country's international isolation became even more pronounced. In what is surely a proof of the saying that "violence begets violence," the military junta arrested Ne Win, along with his daughter, son-in-law, and three grandsons, and charged them with treason—that is, with plotting the overthrow of the government. The head of the air force as well as the chief of police were also purged for their involvement in the alleged plot. Ne Win died in custody in 2002.

Years of ethnic conflict, inflexible socialism, and self-imposed isolation have severely damaged economic growth in Myanmar. In 1987, despite Burma's abundance of valuable teak and rubber trees in its forests, sizable supplies of minerals in the mountains to the north, onshore oil, rich farmland in the Irrawaddy Delta, and a reasonably well-educated population, the United Nations declared Burma one of the least-developed countries in the world (it had once been the richest country in Southeast Asia). Debt incurred in the 1970s exacerbated the country's problems, as did the government's fear of foreign investment. The per capita annual income of US$1,400 makes Myanmar the poorest country in Asia, poorer even than North Korea, Cambodia, and Papua New Guinea.

Myanmar's industrial base is still very small; nearly 70 percent of the population of 54 million makes their living by farming (rice is a major export) and by fishing. The tropical climate yields abundant forest cover, where some 250 species of valuable trees abound. Good natural harbors and substantial mineral deposits of coal, natural gas, and others also bless the land. Less than 9 percent of gross domestic product comes from the manufacturing sector (as compared to, for example, approximately 14 percent in next-door Thailand). In the absence of a strong economy, black-marketeering has increased, as have other forms of illegal economic transactions. It is estimated that 80 percent of the heroin smuggled into New York City comes from the jungles of Myanmar and northern Thailand. Methamphetamines are also a major export from the region, with profits apparently going to the army. Indeed, so heavy is the drug traffic in the area that, in response to Chinese complaints of drug activity along its border, Myanmar was forced to resettle some 120,000 Wah ethnics away from the border. Between 2003 and 2004 there was a drop in opium and heroin production, perhaps as a result of a United Nations program to control drugs, but the black market economy remains about as large as the legal economy.

Government officials have traveled to the United States in recent years in search of business opportunities for Myanmar, but most U.S. businesses refuse to trade with the country in protest of the military dictatorship and because of trade restrictions imposed by the U.S. government. The government is currently discussing a joint venture with India to develop natural gas deposits, but for the foreseeable future, Myanmar's economy will remain largely agricultural.

Over the years, the Burmese have been advised by economists to open up their country to foreign investment and to develop the private sector of the economy. They have resisted the former idea because of their deep-seated fear of foreign domination; they have similar suspicions of the private sector because it was previously controlled almost completely by ethnic minorities (the Chinese and Indians). The government has relied on the public sector to counterbalance the power of the ethnic minorities.

Beginning in 1987, however, the government began to admit publicly that the economy was in serious trouble. To counter massive unrest in the country, the military authorities agreed to permit foreign investment from countries such as Malaysia, South Korea, Singapore, and Thailand and to allow trade with China and Thailand. In 1989, the government signed oil-exploration agreements with South Korea, the United States, the Netherlands, Australia, and Japan. Both the United States and West Germany withdrew foreign aid in 1988, but Japan did not; in 1991, Japan supplied $61 million—more than any other country—in aid to Myanmar. Ironically, such aid is not welcomed by everyone. Many Burmese outside the country believe that economic assistance helped the military stay in power. One U.S.-based group, the Free Burma Coalition, worked to dissuade companies from trading with Myanmar. Their plea was heard in 2004 when the European Union approved tighter sanctions against the military dictatorship: no investment in state-run companies; no loans to Myanmar; and no visas granted to any high-level military general. Although the military released 6300 political prisoners in 2008, they subsequently re-arrested Aung San Suu Kyi. In response, U.S. President Barak Obama, in 2009, extended U.S. sanctions for another year, and U.S. Secretary of State

Hillary Rodham Clinton in 2010 expressed frustration at Myanmar's intransigence. The European Union and the United Nations, backed by both China and Russia, issued their own declarations urging the release of 2000 more political prisoners. The UN had hoped to persuade the United States to relax its restrictions and set up conditions by which the World Bank and other financial institutions could return to Myanmar, but the arrests of 2009 nixed those plans. For years, the United States government has kept Myanmar (along with China and North Korea) on its list of countries where freedom of religion is not recognized.

■ POLITICAL STALEMATE

For many years, the people of Myanmar have been in a state of turmoil caused by governmental repression. In 1988, thousands of students participated in six months of demonstrations to protest the lack of democracy in the country and to demand multiparty elections. General Saw Maung brutally suppressed the demonstrators, imprisoning many students—and killing more than 3,000 of them. He then took control of the government and reluctantly agreed to multiparty elections. About 170 political parties registered for the elections, which were held in 1990—the first elections in 30 years. Among these were the National Unity Party (a new name for the Burma Socialist Program Party, which had been the only legal party since 1974) and the National League for Democracy, a new party headed by Aung San Suu Kyi, daughter of slain national hero U Aung San.

The campaign was characterized by the same level of military control that had existed in all other aspects of life since the 1960s. Martial law, imposed in 1988, remained in effect; all schools and universities were closed; opposition-party workers were intimidated; and, most significantly, the three most popular opposition leaders were placed under house arrest and barred from campaigning. The United Nations began an investigation of civil-rights abuses during the election and, once again, students demonstrated against the military government. Several students even hijacked a Burmese airliner to demand the release of Aung San Suu Kyi, who had been placed under house arrest.

As the votes were tallied, it became apparent that the Burmese people were eager to end military rule; the National League for Democracy won 80 percent of the seats in the National Assembly. But military leaders did not want the National League for Democracy in control of the government, so they refused to allow the elected leaders to take office. Under General Than Shwe, who replaced General Saw Maung in 1992, the military organized various operations against any person or group suspected of not supporting them; Karen rebels were hunted down and some 40,000 to 60,000 Muslims were forced to flee to Bangladesh. Hundreds of students who fled the cities during the 1988 crackdown on student demonstrations joined rural guerrilla organizations, such as the Burma Communist Party and the Karen National Union, to continue the fight against the military dictatorship.

Among those most vigorously opposed to military rule have been Buddhist monks. Five months after the elections, monks in the capital city of Yangon boycotted the government by refusing to conduct religious rituals for soldiers. Tens of thousands of people joined in the boycott. The government responded by threatening to shut down monasteries in Yangon and Mandalay.

The military junta first called itself the State Law and Order Restoration Council (SLORC) and then the State Peace and Development Council. It was determined to stay in power at all costs. SLORC has kept Aung San Suu Kyi, the legally elected leader of the country, under house arrest off and on for years, watching her every move. For several years, even her husband and children were forbidden to visit her. In 1991, she was awarded the Nobel Peace Prize; in 1993, several other Nobelists gathered in nearby Thailand to call for her release from house arrest—a plea ignored by SLORC. The United Nations showed its displeasure with the military junta by substantially cutting development funds, as did the United States (which, on the basis of Myanmar's heavy illegal-drug activities, has disqualified the country from receiving most forms of economic aid). In 2002, the military released Aung San Suu Kyi from house arrest and also released some 600 other political prisoners (albeit keeping another 1,600 behind bars). These high-profile moves were intended to satisfy the United States and others that Myanmar was willing to do what was necessary to qualify for economic assistance. But shortly thereafter, Aung San Suu Kyi was put under house arrest yet again.

But perhaps the greatest pressure on the dictatorship was from within the country itself. Despite brutal suppression, the military lost control of the people. Both the Kachin and Karen ethnic groups organized guerrilla movements against the regime; in some cases, they coerced foreign lumber companies to pay them protection money, which they used to buy arms to fight the junta. Opponents of SLORC control one third of Myanmar, especially along its eastern borders with Thailand and China and in the north alongside India. In a daring move in 2006, activists gathered more than half a million signatures on a petition to demand the release of all political prisoners. But, the government once again condemned Aung San Suu Kyi to house arrest. ASEAN, of which Myanmar is a member, urged the junta to release prisoners, and the United Nations offered substantial aid

FURTHER INVESTIGATION

Aung San Suu Kyi was elected to be the Prime Minster of Myanmar but was never allowed to take office and spent years in prison or under house arrest instead. To learn more about the most famous person in Myanmar, see: http://www.nobelprize.org/nobel_prizes/peace/laureates/1991/kyi-bio.html or http://www.independent.co.uk/arts-entertainment/books/reviews/the-lady-and-the-peacock-the-life-of-aung-san-suu-kyi-by-peter-popham-6263578.html in return for some movement toward democracy and the release of some 1,000 political detainees. But the junta ignored these initiatives for months and continued its oppression of democracy and of the Karens, some 2,000 of whom fled to Thailand. Finally, in early 2007, the junta released about 3,000 prison inmates; most were criminals, but a few were political prisoners.

With the economy in shambles, the military got involved with the heroin trade as a way of acquiring needed funds. It engaged in bitter battles with drug lords for control of the trade. Eventually, to ease economic pressure, the military rulers ended their monopoly of certain businesses and legalized the black market, making products from China, India, and Thailand available on the street.

Pressure from the European Union and the United States occasionally brought positive results; in 2004, SLORC released nearly 10,000 prisoners, including some who were high-profile members of the National League for Democracy (although party leader Aung San Suu Kyi remained under house arrest). The National League for Democracy was allowed to reopen its party headquarters, and the junta even proposed a multi-party national constitutional convention. The NLD, however, announced it would boycott the meetings until their leader was released from house arrest. The brief political softening suffered a setback when the relatively moderate premier, Khin Nyunt (along with some of his supporters) was suddenly removed from his position, held in house arrest, and replaced by the more hard-line Lt. Gen. Soe Win.

In 2009, an American citizen claiming he was there to protect Aung San Suu Kyi, illegally entered the country and stayed overnight in Aung San Suu Kyi's home where she was under house arrest. Claiming that she was complicit in the illegal entry, Aung San Suu Kyi was re-arrested and taken to jail. With national elections planned for 2010, many people believed the arrest was designed to keep her in jail until after the elections. There was a worldwide outcry against the arrest, and even ASEAN, which usually refrains from criticizing its member countries complained of Suu Kyi's inhumane treatment.

In 2010, the military junta reluctantly allowed national elections to be held. Elected was a former general whose party, the Union Solidarity and Development Party, was strongly backed by the military. General Thein Sein had resigned his military position just in time to run for the office of Prime Minister. Surprisingly, since his election (he took office in March 2011), he has made efforts to restore lost liberties and to open up a dialogue with outside countries such as the United States. President Barack Obama praised the new tone in Myanmar and promised to send Secretary of State Hillary Clinton to discuss assistance—the first such visit in 50 years. Further, the new government released more than 6,000 prisoners, including many political prisoners. The United Nations started considering ways to open the door for a return of international financing, and ASEAN agreed to allow Myanmar to chair the 2014 ASEAN meetings. The National League for Democracy boycotted the 2010 elections, but Aung San Suu Kyi was later released from arrest and says she will likely stand for election to the parliament. Thus, at long last, a light is appearing at the end of a long tunnel. Many problems remain, including illicit opium production, military conscription of child soldiers, repression of civil liberties, and many, many others, but at least Myanmar appears to be heading in a positive direction for the first time since the 1960s.

For the average Burman, especially those in the countryside, life is anything but pleasant. A 1994 human-rights study found that as many as 20,000 women and girls living in Myanmar near the Thai border had been abducted to work as prostitutes in Thailand. For several years, SLORC has carried out an "ethnic-cleansing" policy against villagers who have opposed their rule; thousands of people have been carried off to relocation camps, forced to work as slaves or prostitutes for the soldiers, or simply killed. Some 400,000 members of ethnic groups have fled the country, including 300,000 Arakans who escaped to Bangladesh and 5,000 Karenni, 12,000 Mon, and 52,000 Karens who fled to Thailand. Food shortages plague certain regions of the country, and many young children are forced to serve in the various competing armies rather than acquire an education or otherwise enjoy a normal childhood. Indeed, warfare and violence are the only reality many youth know. In 1999, a group of "Burmese Student Warriors" seized the Myanmar Embassy in Thailand, taking hostages and demanding talks between the military junta and Aung San Suu Kyi. The next year, a youth group calling itself God's Army and led by 12-year-old twin brothers, seized a hospital in Thailand and held 800 patients and staff hostage. The boys were from the Karen people, a Christian subculture long persecuted in Burma. The boys attacked Thai residents to protest their villages having been shelled by the Thai military, which is increasingly uneasy with the large number of Burmese refugees inside and along its border. In 2006, the United States agreed to resettle hundreds of Karens living in Thailand.

■ THE CULTURE OF BUDDHA

Although Myanmar's most famous religious buildings are of Hindu origin, Buddhism, representing the beliefs of 89 percent of the population, has been the dominant religion for decades. In fact, for a brief period in the 1960s, Buddhism was the official state religion of Burma. The government quickly repealed this appellation in order to weaken the power of the Buddhist leadership, or *Sangha*, vis-à-vis the polity. Still, Buddhism remains the single most important cultural force in the country. Even the Burmese alphabet is based, in part, on Pali, the sacred language of Buddhism. Buddhist monks joined with college students after World War II to pressure the British government to withdraw from Burma, and they have brought continual pressure to bear on the current military junta.

Historically, so powerful has been the Buddhist Sangha in Burma that four major dynasties have fallen because of it. This has not been the result of ideological antagonism between church and state (indeed, Burmese rulers have usually been quite supportive of Buddhism)

Snapshot: MYANMAR

Summarized below is a quick look at the country with regard to its development, freedom, health/welfare, and achievements.

Development

Primarily an agricultural nation, Myanmar has a poorly developed industrial sector. For years, the military government forbade foreign investment and severely restricted tourism. In 1989, recognizing that the economy was on the brink of collapse, the government permitted foreign investment and signed contracts with Japan and others for oil exploration.

Freedom

For almost 50 years, Myanmar has suffered under a repressive military dictatorship. When the National League for Democracy won the 1990 elections, the military refused to allow the elected government to take office. The military restricted the activities of Buddhist monks, carried out ethnic cleansing against minorities, and maintained a socialist economy that

led to an impoverished nation. In 2011, elections were held, and a military-backed former general, who seems to be in favor of democratic change, was elected president.

Health/Welfare

The Myanmar government provides free health care and pensions to citizens, but the quality and availability of these services are erratic, to say the least. Malnourishment and preventable diseases are common, and infant mortality is high. Overpopulation is not a problem; Myanmar is one of the most sparsely populated nations in Asia.

Achievements

Myanmar is known for the beauty of its Buddhist architecture. Pagodas and other Buddhist monuments and temples dot many of the cities, especially Pagan, one of Burma's earliest cities. Politically, it is notable that the country was able to remain free of the warfare that engulfed much of Indochina during the 1960s and 1970s. Imprisoned leader Aung San Suu Kyi received the Nobel Peace Prize in 1991 and was released from arrest in 2011.

---

but, rather, because Buddhism soaks up resources that might otherwise go to the government or to economic development. Believers are willing to give money, land, and other resources to the religion, because they believe that such donations will bring them spiritual merit; the more merit one acquires, the better one's next life will be. Thus, all over Myanmar, but especially in older cities such as Pagan, one can find large, elaborate Buddhist temples, monuments, or monasteries, some of them built by kings and other royals on huge, untaxed parcels of land. These monuments drained resources from the government but brought to the donor unusual amounts of spiritual merit. As Burmese scholar Michael Aung-Thwin explained it: "One built the largest temple because one was spiritually superior, and one was spiritually superior because one built the largest temple."

The Buddhist Sangha was at the forefront of opposition to military rule. This was a rather unusual position for Buddhists who generally prefer a more passive approach to "worldly" issues. Some monks joined college students in peaceful-turned-violent demonstrations

against the junta. Others staged spiritual boycotts against the soldiers by refusing to accept merit-bringing alms from them or to perform weddings and funerals. The junta retaliated by banning some Buddhist groups altogether, purging others of rebellious leaders, and closing schools and universities.

Eventually, the regime relaxed its intimidation of the Buddhists and reopened universities. They also started paying attention to the economy by inviting outside investment and promoting tourism. Wary of these moves, all 15 European Union members as well as the national U.S. Chamber of Commerce supported a 2000 Massachusetts state law that would have penalized companies for doing business with Myanmar; the law was invalidated by the Supreme Court, but the sentiment of broad opposition to the military junta remained. Although the Japanese have invested in Myanmar throughout the military dictatorship (most cars on the roads are Japanese-made), many potential investors from other countries have refused to invest in the regime. Perhaps under the new government, things will change.

Statistics

Geography

Area in Square Miles (Kilometers): 261,228 (676,578) (slightly smaller than Texas)

Capital (Population): Yangon (4.3 million; Nay Pyi Taw is the administrative capital)

Environmental Concerns: deforestation; air, soil, and water pollution; inadequate sanitation and water treatment

Geographical Features: central lowlands ringed by steep, rugged highlands

Climate: tropical monsoon

People

Population

Total: 53,999,804

Annual Growth Rate: 1.08%

Rural/Urban Population Ratio: 66/34

Major Languages: Burmese; various minority languages

Ethnic Makeup: 68% Burman; 9% Shan; 7% Karen; 4% Rakhine; 3% Chinese; 9% Mon, Indian, and others

Religions: 89% Buddhist; 4% Muslim; 4% Christian; 1% animist; 2% others

Health

Life Expectancy at Birth: 63 years (male); 67 years (female)

Infant Mortality: 49/1,000 live births

Physicians Available: 1/2,188 people

HIV/AIDS Rate in Adults: 0.6%

Education

Adult Literacy Rate: 90% (official)

Compulsory (Ages): 5–10; free

Communication

Telephones: 812,200 main lines (2009)

Mobile Phones: 502,000 (2009)

Internet Users: 110,000 (2010)

Internet Penetration (% of Pop.): 0.2%

Transportation

Roadways in Miles (Kilometers): 16,777 (27,000)

Railroads in Miles (Kilometers): 3,126 (5,031)

Usable Airfields: 76

Government

Type: military regime

Independence Date: January 4, 1948 (from the United Kingdom)

Head of State/Government: President Thein Sein; Minister of Defense Sein Than Shwe

Political Parties: All Mon Region Democracy Party; National Democratic Force; National League for Democracy; National Unity Party; Rakhine Nationalities Development Party; Shan Nationalities Democratic Party; Shan Nationalities League for Democracy; Union Solidarity and Development Party; many others

Suffrage: universal at 18

Military

Military Expenditures (% of GDP): 2.1%

Current Disputes: internal strife; border and boundary conflicts with Thailand, Bangladesh, and India

Economy

Currency ($ U.S. Equivalent): 966 kyat = $1

Per Capita Income/GDP: $1,400/$76.5 billion

GDP Growth Rate: 5.3%

Inflation Rate: 7.7%

Unemployment Rate: 5.7%

Labor Force by Occupation: 70% agriculture; 23% services; 7% industry

Population Below Poverty Line: 33%

Natural Resources: petroleum; tin; timber; antimony; zinc; copper; tungsten; lead; coal; marble; limestone; precious stones; natural gas

Agriculture: rice; pulses; beans; sesame; groundnuts; sugarcane; hardwood; fish and fish products

Industry: agricultural processing; textiles; footwear; wood and wood products; petroleum refining; mining; construction materials; pharmaceuticals; fertilizer

Exports: $8.8 billion (primary partners Thailand, India, China)

Imports: $4.3 billion (primary partners China, Thailand, Singapore)

Suggested Websites

www.myanmar.com/

www.myanmars.net/

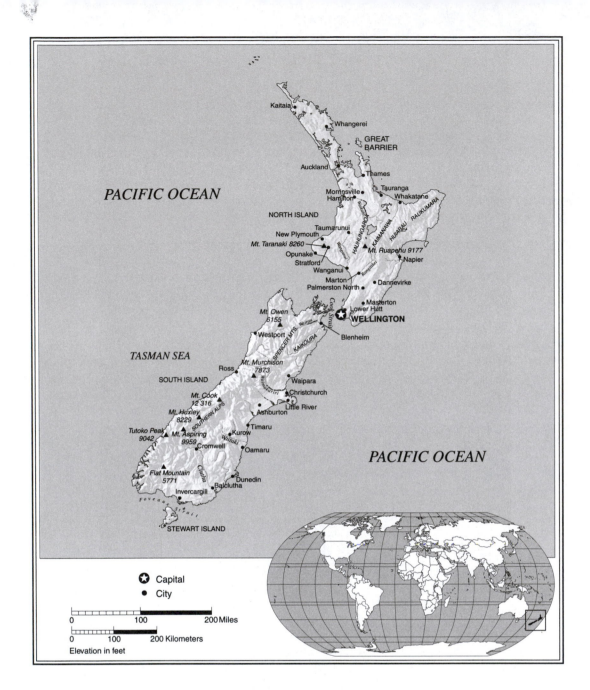

PACIFIC OCEAN

Kaitaia
Whangerei
GREAT BARRIER
Auckland
Thames
Tauranga
Morrinsville
Whakatane
Hamilton
NORTH ISLAND
Taumarunui
New Plymouth
Mt. Taranaki 8260
Mt. Ruapehu 9177
Opunake
Napier
Stratford
Wanganui
Marton
Dannevirke
Palmerston North
HAUHUNGAROA
KAIMANAWA
HUIARAU
RAUKUMARA
Masterton
Mt. Owen 6155
Lower Hutt
WELLINGTON
Westport
SPENCER MTS.
KAIKOURA
Cook Strait
Blenheim

TASMAN SEA

Ross
Mt. Murchison 7873
Waipara
SOUTH ISLAND
Christchurch
Mt. Cook 12 316
Little River
Mt. Huxley 8229
Ashburton
SOUTHERN ALPS
Timaru
Tutoko Peak 9042
Mt. Aspiring 9959
Kurow
Cromwell
Oamaru

PACIFIC OCEAN

Flat Mountain 5771
Dunedin
Balclutha
Invercargill
Foveaux Strait
STEWART ISLAND

★ Capital
● City

0   100   200 Miles
0   100   200 Kilometers
Elevation in feet

New Zealand

Thanks to the filming in New Zealand of the blockbuster movie series *Lord of the Rings,* many people have had an opportunity to see some of the natural beauty of the country; but few have visited there, owing, no doubt, to its geographic isolation from the rest of the world. No doubt, the isolation, along with a sparse population, has helped preserve the incredible natural beauty of the country. Glacier-carved snow-capped mountains, lush green valleys, crystal-clear rivers, waterfalls (including Sutherland Falls, the world's fifth highest), and white plumes of hot steam from numerous underground lava fields make New Zealand a country-size version of North America's Yellowstone National Park. Rare animals, such as the yellow-eyed penguin, the Kea or alpine parrot, and the royal albatross, add to the dazzling beauty of the country, as do the miles of spectacular coastline; one can never get more than 80 miles from the ocean. Urban areas, both large and small, are also unusually clean and tidy.

The cleanliness of New Zealand's cities is not the only quality that differentiates it from some of the more populated megacities of the Pacific Rim. Ethnicity is another difference. As with Australia, the majority of the people (93 percent in Australia and 69 percent in New Zealand) are of European descent (primarily British and from rural England), English is the official language, and most people, even many of the original Maori inhabitants, are Christians. Britain claimed the beautiful, mountainous islands officially in 1840, after agreeing to respect the property rights of Maoris, most of whom lived on the North Island. Blessed with a temperate climate and excellent soils for crop and dairy farming, New Zealand—divided into two main islands, North Island and South Island—became an important part of the British Empire.

Although largely self-governing since 1907, and independent as of 1947 (with Queen Elizabeth II as head of state), the country has always maintained very close ties with the United Kingdom and is a member of the Commonwealth of Nations. It has, in fact, attempted to re-create British culture—customs, architecture, even vegetation—in the Pacific. English sheep, which literally transformed the economy, were first imported to the South Island in 1773. Some towns seem very, very English, such as Christchurch, the largest city on the South Island, through which flows the Avon River and which was founded in part by the Church of England. Others seem more Scottish, such as Dunedin. Everywhere one can find town names derived from British life: Wellington, Nelson, Queenstown, Stirling. So close were the links with Great Britain in the 1940s that England purchased fully 88 percent of New Zealand's exports (mostly agricultural and dairy products), while 60 percent of New Zealand's imports came from Britain. As a part of the British Empire, New Zealand always sided with the Western nations on matters of military defense.

Efforts to maintain a close cultural link with Great Britain do not stem entirely from the common ethnicity of the two nations; they also arise from New Zealand's extreme geographical isolation from the centers of European and North American activity. Even Australia is more than 1,200 miles away. Therefore, New Zealand's policy—until the 1940s—was

TIMELINE

A.D. 1300s
Maoris, probably from Tahiti, settle the islands

1642
New Zealand is "discovered" by Dutch navigator A. J. Tasman

1769
Captain James Cook explores the islands

1840
Britain declares sovereignty over New Zealand with Treaty of Waitangi

1865
A gold rush attracts new immigrants

1907
New Zealand becomes an almost independent dominion of Great Britain

1941
Socialized medicine is implemented

1947
New Zealand becomes fully independent within the British Commonwealth of Nations
New Zealand backs creation of the South Pacific Commission

1950s
Restructuring of export markets. Joined ANZUS

1970s
The National Party takes power; New Zealand forges foreign policy more independent of traditional allies

1980s
The Labour Party regains power; New Zealand withdraws from ANZUS

1990s
New Zealanders consider withdrawing from the Commonwealth; Maoris and white New Zealanders face economic challenges from other Pacific Rimmers

2000s
New Zealand is led by its second woman prime minister, Helen Clark
New Zealand champions cultural and environmental goals
Parliament debates the merits of liberal social legislation such as a parent-leave law, a euthanasia law, and a civil union law for homosexuals.
First Maori political party formed.

Maoris occupied New Zealand long before the European settlers moved there. The Maoris quickly realized that the newcomers were intent on depriving them of their land, but it was not until the 1920s that the government finally regulated unscrupulous land-grabbing practices. However, there remains an ongoing dispute over how many acres of land were guaranteed to the Maoris in the Treaty of Waitangi. Today, the Maoris pursue a lifestyle that preserves key parts of their traditional culture while incorporating skills necessary for survival in the modern world.

to encourage the British presence in Asia and the Pacific, by acquiring more lands or building up naval bases, to make it more likely that Britain would be willing and able to defend New Zealand in a time of crisis. New Zealand had involved itself somewhat in the affairs of some nearby islands in the late 1800s and early 1900s, but its purpose was not to provide development assistance or defense. Rather, its aim was to extend the power of the British Empire and put New Zealand in the middle of a mini-empire of its own. To that end, New Zealand annexed the Cook Islands in 1901 and took over German Samoa in 1914. In 1925, it assumed formal control over the atoll group known as the Tokelau Islands.

■ REGIONAL RELIANCE

During World War II (or, as the Japanese call it, the Pacific War), Japan's rapid conquest of the Malay Archipelago, its seizure of the British colony of Hong Kong and of many Pacific Islands, as well as its plan to attack Australia demonstrated to New Zealanders the futility of relying on the British to guarantee their security. After the war, and for the first time in its history, New Zealand

? DID YOU KNOW?

In February 2011, a massive 6.1 earthquake hit South Island. This quake followed just 6 months after an even more powerful 7.1 quake damaged buildings all over ChristChurch, New Zealand's second largest city. The February quake took the lives of over 180 people, knocked out power to 80 percent of ChristChurch, and damaged about 2,000 central city buildings, including ChristChurch Cathedral, a television station, and a 26-story hotel, and some 10,000 homes. Liquefaction and other ground damage meant that some areas would no longer be suitable for buildings. Powerful aftershocks (some 7,500 of them) continued for eight months, putting residents continually on edge. So many people were displaced that the national census, required by law to be taken in 2011, was postponed until 2012. The cost of the damage is estimated to be more than US$12 billion, yet it was a disaster that few outside of New Zealand remember. Why? Because two weeks later, the massive earthquake, tsunami, and nuclear disaster hit Japan, and that seemed to capture the world's attention.

began to pay serious attention to the real needs and ambitions of the peoples nearby rather than to focus on Great Britain. In 1944 and again in 1947, New Zealand joined with Australia and other colonial nations to create regional associations on behalf of the Pacific islands. One of the organizations, the South Pacific Commission, has itself spawned many regional subassociations dealing with trade, education, migration, and cultural and economic development. Although it had neglected the islands that it controlled during its imperial phase, in the early 1900s, New Zealand cooperated fully with the United Nations in the islands' decolonization during the 1960s, while at the same time increasing development assistance. Some islands chose to remain dependencies of New Zealand. The Ross Dependency is one of those, as is Tokelau, which receives 80 percent of its budget from New Zealand and which voted against independence in 2006.

New Zealand's first alliance with Asian nations came in 1954, when it joined the Southeast Asian Treaty Organization. New Zealand's continuing involvement in regional affairs is demonstrated by its sending of troops to help maintain the peace in East Timor and its agreement to take in residents ("environmental refugees") from the island of Tuvalu, an island that is being literally engulfed by the Pacific as global warming produces higher sea levels.

New Zealand's new international focus certainly did not mean the end of cooperation with its traditional allies, however. In fact, the common threat of the Japanese during World War II strengthened cooperation with the United States to the extent that, in 1951, New Zealand joined a three-way, regional security agreement known as ANZUS (for Australia, New Zealand, and the United States). Moreover, because the United States was, at war's end, a Pacific/Asian power, any agreement with the United States was likely to bring New Zealand into more, rather than less, contact with Asia and the Pacific. Indeed, New Zealand sent troops to assist in all of the United States' military involvements in Asia: the occupation of Japan in 1945, the Korean War in 1950, and the Vietnam War in the 1960s. And, as a member of the British Commonwealth, New Zealand sent troops in the 1950s and 1960s to fight Malaysian Communists and Indonesian insurgents.

◼ A NEW INTERNATIONALISM

Beginning in the 1970s, especially when the Labour Party of Prime Minister Norman Kirk was in power, New Zealand's orientation shifted even more markedly toward its own region. Under the Labour Party, New Zealand defined its sphere of interest and responsibility as the Pacific, where it hoped to be seen as a protector and benefactor of smaller states. Of immediate concern to many island nations was the issue of nuclear testing in the Pacific. Both the United States and France had undertaken tests by exploding nuclear devices on tiny Pacific atolls. In the 1960s, the United States ceased these tests, but France continued. On behalf of the smaller islands, New Zealand argued before the United Nations against testing, but France still did not stop. Eventually, the desire to end testing congealed into the more comprehensive position that the entire Pacific should be declared a nuclear-free zone. Not only testing but also the transport of nuclear weapons through the area would be prohibited under the plan.

New Zealand's Labour government issued a ban on the docking of ships with nuclear weapons in New Zealand, despite the fact that such ships were a part of the ANZUS agreement. When the National Party regained control of the government in the late 1970s, the nuclear ban was revoked, and the foreign policy of New Zealand tipped again toward its traditional allies. The National Party government argued that, as a signatory to ANZUS, New Zealand was obligated to open its docks to U.S. nuclear ships. However, under the subsequent Labour government of Prime Minister David Lange, New Zealand once again began to flex its muscles over the nuclear issue. Lange, like his Labour Party predecessors, was determined to create a foreign policy based on moral rather than legal rationales. In 1985, a U.S. destroyer was denied permission to call at a New Zealand port, even though its presence there was due to joint ANZUS military exercises. Because the United States refused to say whether or not its ship carried nuclear weapons, New Zealand insisted that the ship could not dock. Diplomatic efforts to resolve the standoff were unsuccessful; in 1986, New Zealand, claiming that it was not fearful of foreign attack, formally withdrew from ANZUS, although its bilateral military agreement with Australia, via ANZUS, remains in force.

The issue of use of the Pacific for nuclear-weapons testing by superpowers is still of major concern to the New Zealand government. The nuclear test ban treaty signed by the United States in 1963 has limited U.S. involvement in that regard, but France has continued to test atmospheric weapons, and at times both the United States and Japan have proposed using uninhabited Pacific atolls to dispose of nuclear waste. In 1995, when France ignored the condemnation of world leaders and detonated a nuclear device in French Polynesia, New Zealand recalled its ambassador to France out of protest.

In the early 1990s, a new issue came to the fore: nerve-gas disposal. With the end of the cold war, the U.S. military proposed disposing of most of its European stockpile of nerve gas on an atoll in the Pacific. The atoll is located within the trust territory granted to the United States at the conclusion of World War II. The plan is to burn the gas away from areas of human habitation, but those islanders living closest (albeit hundreds of miles away) worry that residues from the process could contaminate the air and damage humans, plants, and animals. The religious leaders of Melanesia, Micronesia, and Polynesia have condemned the plan, not only on environmental grounds but also on grounds that outside powers should not be permitted to use the Pacific region without the consent of the inhabitants there—a position with which the Labour government of New Zealand strongly concurs.

In addition to a strong stand on the nuclear issue, the Labour government successfully implemented laws that have made New Zealand more socially "liberal" than even some European nations, and certainly more so than the increasingly conservative United States. Abortion is not only legal, it is paid for by the National Health Service, and young girls, even pre-teens, are not required to inform their parents if they have an abortion. The use of birth control has kept the fertility rate per woman steady for the past 20 years at almost the perfect replacement level of 2.1 births per woman. Prostitution has been legalized, and bills that would allow euthanasia for the elderly and civil unions for homosexuals have been seriously considered by Parliament. Some religious leaders are speaking out against such proposals, but many New Zealanders seem comfortable with the trend and continue to elect liberal candidates. In fact, New Zealand was the first country in the world, in the 1970s, to have an environmentalist Green Party, and the Labour/Green coalition constantly reveals itself in policies such as New Zealand's signing of a fishing-regulations treaty to slow depletion of world fish stocks.

■ ECONOMIC CHALLENGES

The New Zealand government's new foreign-policy orientation has caught the attention of observers around the world, but more urgent to New Zealanders themselves is the state of their own economy. Until the 1970s, New Zealand had been able to count on a nearly guaranteed export market in Britain for its dairy and agricultural products. Moreover, cheap local energy supplies as well as inexpensive oil from the Middle East had produced several decades of steady improvement in the standard of living. Whenever the economy showed signs of being sluggish, the government would artificially protect certain industries to ensure full employment.

This comfortable situation changed in 1973, when Britain joined the European Union (then called the European Economic Community) and when the Organization of Petroleum Exporting Countries sent the world into its first oil shock. New Zealand actually has the potential of near self-sufficiency in oil, but the easy availability of Middle East oil over the years has prevented the full development of local oil and natural-gas reserves. As for exports, New Zealand had to find new outlets for its agricultural products, which it did by contracting with various countries throughout the Pacific Rim. Currently, about one third of New Zealand's trade occurs within the Pacific Rim.

In the transition to these new markets, farmers complained that the manufacturing sector—intentionally protected by the government as a way of diversifying New Zealand's reliance on agriculture—was getting

FURTHER INVESTIGATION
To learn more about Maori culture, see www .maori.org.nz/

unfair, favorable treatment. Subsequent changes in government policy toward industry resulted in a new phenomenon for New Zealand: high unemployment—over 8 percent in 2006 (but moderated to about 6.5 percent in 2011). Moreover, New Zealand had constructed a rather elaborate social-welfare system since World War II, so, regardless of whether economic growth was high or low, social-welfare checks still had to be sent. This untenable position has made for a difficult political situation, for, when the National Party cut some welfare benefits and social services, it lost the support of many voters. The welfare issue, along with a change to a mixed member proportional voting system that enhanced the influence of smaller parties, threatened the National Party's political power. Thus, in order to remain politically dominant, in 1996 the National Party was forced to form a coalition with the United Party—the first such coalition government in more than 60 years.

In the 1970s, for the first time in its history, New Zealand's standard of living began to drop when compared to other Pacific Rim nations. Still high by world standards, New Zealand's per capita gross domestic product of about US$33,000 in 2010 put it below eight other Pacific Rim countries, meaning that the average person in New Zealand was getting less of the national wealth than the average person in Singapore, Japan, and Brunei. The government realized that the country's dependence on agriculture to support the economy was no longer feasible and began a program of economic diversification. Food processing, wood and paper, and many other industries were encouraged. The effort seems to be paying dividends; the country's annual economic growth rate was above 3 percent in the early 2000s (although it had dropped substantially to under one percent in 2008 and remained there through 2010 due to the global economic crisis). Before that time, government revenue surpluses were prompting calls for tax cuts. Still, in 2000, the Labour Party prime minister, Helen Clark, had to cancel a large purchase of F-16 fighter jets from the United States because she said the budget could not handle it. The bankruptcy of Air New Zealand's Australian subsidiary also caused concern.

New Zealanders are well aware of the economic strength of countries in North Asia such as Japan and China, and they see the potential for benefiting their own economy through joint ventures, loans, and trade. Yet they also worry that Japanese and Chinese wealth may come to constitute a symbol of New Zealand's declining strength as a culture. For instance, in the 1980s, as Japanese tourists began traveling en masse to New Zealand, complaints were raised about the quality of New Zealand's hotels. Unable to find the funds for a massive upgrading of the hotel industry, New Zealand agreed to allow Japan to build its own hotels; it reasoned that the local construction industry could use an economic boost, and that the better hotels would encourage well-heeled Japanese to spend even more tourist dollars in the country. However, they also worried that, with the Japanese

Snapshot: NEW ZEALAND

Summarized below is a quick look at the country with regard to its development, freedom, health/welfare, and achievements.

Development

Government protection of manufacturing has allowed this sector to grow at the expense of agriculture. Nevertheless, New Zealand continues to export large quantities of dairy products, wool, meat, fruits, and wheat. Full development of the country's oil and natural-gas deposits could alleviate New Zealand's dependence on foreign oil.

Freedom

New Zealand partakes of the democratic heritage of English common law and subscribes to all the human-rights protections that other Western nations have endorsed. Maoris, originally deprived of much of their land, are now guaranteed the same legal rights as whites. Social discrimination against Maoris is much milder than with many other colonized peoples. Tensions remain, but the government is taking steps to resolve long-standing disputes. Debates, sometimes violent, continue over the interpretation of the Treaty of Waitangi.

Health/Welfare

New Zealand established pensions for the elderly as early as 1898. Child-welfare programs were started in 1907, followed by the Social Security Act of 1938, which augmented the earlier benefits and added a minimum-wage requirement and a 40-hour work week. A national health program was begun in 1941. The government began dispensing free birth-control pills to all women in 1996 in an attempt to reduce the number of abortions, although, recently, the abortion rate has been rising.

Achievements

New Zealand is notable for its efforts on behalf of the smaller Islands of the Pacific. In addition to advocating a nuclear-free Pacific, New Zealand has promoted interisland trade and has established free-trade agreements with Western Samoa, the Cook Islands, and Niue. It provides educational and employment opportunities to Pacific Islanders who reside within its borders.

---

owning the hotels, New Zealanders might be relegated to low-level jobs.

Historically, concern about their status vis-à-vis nonwhites had never been much of an issue to many Anglo-Saxon New Zealanders; they always simply assumed that nonwhites were inferior. Many settlers of the 1800s believed in the Social Darwinistic philosophy that the Maori and other brown- and black-skinned peoples would gradually succumb to their European "betters." It did not take long for the Maoris to realize that, land guarantees notwithstanding, the whites intended to deprive them of their land and culture. Violent resistance to these intentions occurred in the 1800s, but Maori landholdings continued to be gobbled up, usually deceptively, by white farmers and sheep herders. Government control of these unscrupulous practices was lax until the 1920s. Since that time, many Maoris (whose population has increased to about 660,000) have intentionally sought to create a lifestyle that preserves key parts of traditional culture while incorporating the skills necessary for survival in a white world. In 1999, Maoris on the South Island, using land-loss funds provided by the government, made such a large land purchase (nearly 300,000 acres) that they became the largest land owner on the island. In 2004, Maori's started their own television station, broadcasting 50 percent in English and 50 percent in Maori. Maoris continue to insist that the Treaty of Waitangi, which they signed with the British, and which guaranteed them rights to their own lands, be honored in full. Some members of Parliament, such as those affiliated with the New Zealand First Party and the National Party, want to end what they see as "special treatment" of Maoris, i.e., entitlements such as health care and government paid education. But the Maoris are, literally, fighting back, throwing mud at the National Party leader and marching 10,000-strong, in what sometimes turn into violent demonstrations. The Maoris have generally accommodated themselves to those who rule over them, but they are increasingly educated and increasingly vocal. They recently had a stand-off with the Japanese over net fishing in the oceans and the damage the Japanese were inflicting on the Maori fishing industry, and in 2004, they created their own political party—the first ever Maori party.

More so than the Australians, New Zealanders are attempting to rectify the historic discrimination against the country's indigenous people. The configuration of one recent government cabinet, for example, included four Maoris and a member of Pacific Island descent (as well as 11 women and 1 Muslim). The new Labour/Green Party coalition government has also attempted to eliminate some of the colonial cobwebs from their society by abolishing knighthoods bestowed by the British Crown. Local honors are awarded instead. Outspoken former Prime Minister Helen Clark, defining the queen of England's role in New Zealand's affairs as "absurd," once said it was only a matter of time until New Zealand would make a clean break from its old British Empire orientation and become a republic. Naturally, opinions on such matters differ; Clark's Labour Party never won enough seats in Parliament to govern without allying itself with others; and evidence of a right-wing tilt is found in recent elections where the far-right, anti-immigration "New Zealand First" party was supported by 11 percent of the voters, giving them more seats in Parliament than before.

In November 2008, John Key, a wealthy businessperson with Jewish roots, took his conservative National Party to success at the polls after nine years of Labour government. In 2009, Key was physically assaulted by two Maoris. Unhappy with the National Party's traditional stance toward Maoris, many Maoris consider Prime Minister Key to be a "smiling snake," charismatic and pleasant but not out to do them any good.

New Zealand will have to assert itself with vigor if it wants to prevent a slide to mediocrity when compared to the economic strength of some of its Asian neighbors. Still, given the favorable land-to-population ratio, the beauty of the land, the high quality of the infrastructure, and the general ease of life, New Zealanders can, in the main, be said to have a lifestyle that is the envy of most other peoples in the Pacific Rim—indeed, in the world.

Statistics

Geography

Area in Square Miles (Kilometers): 103,363 (267,710) (about the size of Colorado)

Capital (Population): Wellington (391,000)

Environmental Concerns: deforestation; soil erosion; damage to native flora and fauna from outside species

Geographical Features: mainly mountainous with some large coastal plains

Climate: temperate; sharp regional contrasts

People

Population

Total: 4,290,347

Annual Growth Rate: 0.88%

Rural/Urban Population (Ratio): 14/86

Major Languages: English; Maori; Samoan; French; Hindi; Yue; Northern Chinese; others

Ethnic Makeup: 57% European; 8% Asian; 7% Maori; 5% Pacific Islander; 10% mixed; 13% other

Religions: 63% Christian; 2% Hindu; 1% Buddhist; 2% other; 32% unspecified or none

Health

Life Expectancy at Birth: 79 years (male); 83 years (female)

Infant Mortality: 4.8/1000 live births

Physicians Available: 1/419 people

HIV/AIDS Rate in Adults: 0.1%

Education

Adult Literacy Rate: 99%

Compulsory (Ages): 6–16; free

Communication

Telephones: 1.9 million main lines (2009)

Mobile Phones: 4.7 million (2009)

Internet Users: 3.4 million (2010)

Internet Penetration (% of Pop.): 84%

Transportation

Roadways in Miles (Kilometers): 58,354 (93,911)

Railroads in Miles (Kilometers): 2,850 (4,128)

Usable Airfields: 122

Government

Type: parliamentary democracy

Independence Date: September 26, 1907 (from the United Kingdom)

Head of Sate/Government: Queen Elizabeth II; Prime Minister John Key

Political Parties: ACT New Zealand; Green Party; Maori Party; National Party; New Zealand First Party; New Zealand Labor Party; Progressive Party; United Future

Suffrage: universal at 18

Military

Military Expenditures (% of GDP): 1%

Current Disputes: disputed territorial claim in Antarctica

Economy

Currency ($ U.S. Equivalent): 1.39 New Zealand dollars = $1

Per Capita Income/GDP: $27,700/$118 billion

GDP Growth Rate: 1.5%

Inflation Rate: 2.3%

Unemployment Rate: 6.5%

Labor Force by Occupation: 74% services; 19% industry; 7% agriculture

Natural Resources: natural gas; iron ore; sand; coal; timber; hydropower; gold; limestone

Agriculture: dairy products; lamb and mutton; wheat; barley; potatoes; pulses; fruits; vegetables; wool; beef; fish

Industry: food processing; wood and paper products; textiles; machinery; transportation equipment; banking; insurance; tourism; mining

Exports: $32 billion (primary partners Australia, China, United States, Japan)

Imports: $30 billion (primary partners Australia, China, United States, Japan)

Suggested Websites

www.newzealand.govt.nz

www.newzealand.com

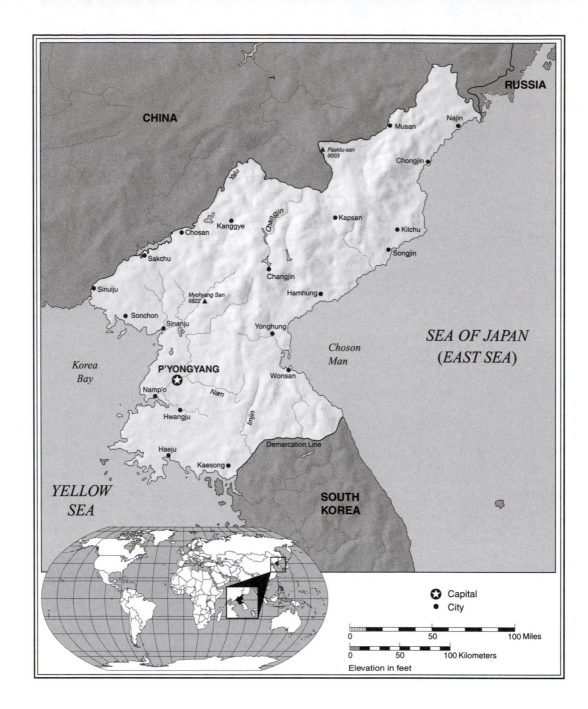

CHINA

RUSSIA

Musan

Najin

▲ Paektu-san
9003

Chongjin

Chosan

Kanggye

Kapsan

Kilchu

Sakchu

Songjin

Changjin

Sinuiju

Myohyang San
6822 ▲

Hamhung

Sonchon

Sinanju

Yonghung

Choson
Man

SEA OF JAPAN
(EAST SEA)

Korea
Bay

P'YONGYANG

Wonsan

Namp'o

Nam

Hwangju

Imjin

Haeju

Demarcation Line

Kaesong

SOUTH
KOREA

YELLOW
SEA

⭐ Capital
● City

0   50   100 Miles
0   50   100 Kilometers
Elevation in feet

North Korea
(Democratic People's Republic of Korea)

The area that we now call North Korea has, at various times in Korea's long history, been separated from the South. In the fifth century A.D., the Koguryo Kingdom in the North was distinct from the Shilla, Paekche, and Kaya Kingdoms of the South. Later, the Parhae Kingdom in the North remained separate from the expanded Shilla Kingdom in the South. Thus, the division of Korea in 1945 into two unequal parts was not without precedent. Yet this time, the very different paths of development that the North and South chose rendered the division more poignant and, to those separated with little hope of reunion, more emotionally painful.

Beginning in 1945, Kim Il-Song, with the strong backing of the Soviet Union, pursued a hard-line Communist policy for both the political and economic development of North Korea. The Soviet Union's involvement on the Korean Peninsula arose from its opportunistic entry into the war against Japan, just eight days before Japan's surrender to the Allies. Thus, when Japan withdrew from its long colonial rule over Korea, the Soviets, an allied power, were in a position to be one of the occupying armies. Reluctantly, the United States allowed the Soviet Union to move troops into position above the 38th Parallel, a temporary dividing line for the respective occupying forces. It was the Soviet Union's intention to establish a Communist buffer state between itself and the capitalist West. Therefore, it moved quickly to establish the area north of the 38th Parallel as a separate political entity. The northern city of P'yongyang was declared to be the capital.

When United Nations representatives arrived in 1948 to oversee elections and ease the transition from military occupation and years of Japanese rule to an independent Korea, the Soviets would not cooperate. Kim Il-Song took over the reins of power in the North. Separate elections were held in the South, and the beginning of separate political systems got underway. The 38th Parallel came to represent not only the division of the Korean Peninsula but also the boundary between the worlds of capitalism and communism.

■ THE KOREAN WAR (1950–1953)

Although not pleased with the idea of division, the South, without a strong army, resigned itself to the reality of the moment. In the North, a well-trained military, with Soviet and Chinese help, began preparations for a full-scale invasion of the South. The North attacked in June 1950, a year after U.S. troops had vacated the South, and quickly overran most of the Korean Peninsula. The South Korean government requested help from the United Nations, which dispatched personnel from 19 nations, under the command of U.S. general Douglas MacArthur. U.S. president Harry Truman ordered American troops to Korea just a few days after the North attacked.

MacArthur's troops advanced against the North's armies and by October were in control of most of the peninsula. However, with massive Chinese help, the North once again moved south. In response, UN troops bombed the North, inflicting heavy destruction. Whereas South Korea was primarily agricultural, North Korea was the industrialized sector of

TIMELINE

A.D. 1945
Kim Il-Song comes to power

1948
The People's Democratic Republic of Korea is created

1950
The Korean War begins

1953
A truce is arranged between North Korea and UN troops

1968
A U.S. spy boat, the *Pueblo,* is seized by North Korea

1969
A U.S. spy plane is shot down over North Korea

1971
Reunification talks begin

1990s
A nonaggression pact is signed with the South; North and South are granted seats in the UN; the world fears that North Korea is developing nuclear weapons; Kim Il-Song dies and is succeeded by his son, Kim Jong-Il

2000s
Famine causes thousands to flee the country
North and South Korea meet in a dramatic summit in 2000; North and South Koreans march together in the 2000 Summer Olympics
Tensions increase with the Bush administration
The government admits that it is still developing a nuclear-weapons program
A massive train explosion kills nearly 200 and destroys thousands of homes
Six-nation talks to resolve the nuclear issue continue intermittently
The government detonates a small nuclear bomb
In 2010, North Korea sank a South Korean warship with a torpedo
Kim Jong Il dies in 2011 at age 69

Socks Delivered by Balloon

For years, South Koreans have been sending pamphlets with political messages inside balloons floated across the border to the North. The pamphlets have encouraged some North Koreans to defect to the South. One of them, Lee Ju-sung, who defected in 2005, believes that the balloons can also help keep North Koreans alive in the cold winters. He fills dozens of helium balloons and attaches socks—socks for babies, children, and adults—to a box with a timer set to release the socks in three hours, and then he hopes the wind will carry the balloons anywhere northward. He tries to send as many as 1,000 pairs of socks each month. Not only are the socks used to keep people warm, but they can also be sold to buy food.

the peninsula. Bombing of the North's industrial targets severely damaged the economy, forcing several million North Koreans to flee south to escape the war, the economic hardships, and the Communist dictatorship under which they found themselves.

Eventually, U.S. troops recaptured South Korea's capital, Seoul. Realizing that further fighting would lead to an expanded Asian war, the two sides agreed to cease-fire talks. They signed a truce that established a 2.5-mile-wide "demilitarized zone" (DMZ) for 155 miles across the peninsula and more or less along the former 38th Parallel division. The Korean War took the lives of more than 54,000 American soldiers, 58,000 South Koreans, and 500,000 North Koreans—but when it was over, both sides occupied about the same territory as they had at the beginning. Yet because neither side has ever declared peace, the two countries remain officially in a state of war.

The border between North and South has been one of the most volatile places in Asia. The North staged military exercises along the border in 1996, firing into the DMZ, and thus breaking the cease-fire of 1953 (over the years there have many shooting and other incidents across the border, including in 2006 when South Korean soldiers fired on soldiers from the North that were entering the DMZ). The South responded by raising its intelligence-monitoring activities and by requesting U.S. AWACS surveillance planes to monitor military movements in the North.

Scholars are still debating whether the Korean War should be called the United States' first losing war and whether or not the bloodshed was really necessary. To understand the Korean War, one must remember that in the eyes of the world, it was more than a civil war among different kinds of Koreans. The United Nations, and particularly the United States, saw North Korea's aggression against the South as the first step in the eventual communization of the whole of Asia. Just a few months before North Korea attacked, China had fallen to the Communist forces of Mao Zedong, and Communist guerrilla activity was being reported throughout Southeast Asia. The "Red Scare" frightened many Americans, and witch-hunting for suspected Communist sympathizers—a college professor who might have taught about Karl Marx in class or a news reporter who might have praised the educational reforms of a Communist country—became the everyday preoccupation of such groups as the John Birch Society and the supporters of U.S. senator Joseph McCarthy.

In this highly charged atmosphere, it was relatively easy for the U.S. military to promote a war whose aim it was to contain communism. Containment rather than defeat of the enemy was the policy of choice, because the West was weary after battling Germany and Japan in World War II. The containment policy also underlay the United States' approach to Vietnam. Practical though it may have been, this policy denied Americans the opportunity of feeling satisfied in victory, since there was to be no victory, just a stalemate. Thus, the roots of the United States' dissatisfaction with the conduct of the Vietnam War actually began in the policies shaping the response to North Korea's offensive in 1950. North Korea was indeed contained, but the communizing impulse of the North remained. In a way, the war continues, not only because a peace accord has never been signed, but also because remnants of the war keep coming back, as in the case of the remains of an American soldier that were found as recently as 2004 in North Korea and repatriated to the United States.

▪ COLLECTIVE CULTURE

With Soviet backing, North Korean leaders moved quickly to repair war damage and establish a Communist culture. The school curriculum was rewritten to emphasize nationalism and equality of the social classes. Traditional Korean culture, based on Confucianism, had stressed strict class divisions, but the Communist

FURTHER INVESTIGATION

The Korean language is related to Japanese, Hungarian, Finnish, Estonian, and Mongolian. To learn more about it and its *hangul* script, see: www.omniglot.com/writing/korean.htm

authorities refused to allow any one class to claim privileges over another (although eventually the families of party leaders came to constitute a new elite). Higher education at the more than 600 colleges and training schools was redirected to technical rather than analytical subjects. Industries were nationalized; farms were collectivized into some 3,000 communes; and the communes were invested with much of the judicial and executive powers that other countries grant to cities, counties, and states. To overcome labor shortages, nearly all women were brought into the workforce, and the economy slowly returned to prewar levels. Of course, while the North was economically normalizing, the South was creating itself into an economic dynamo that soon eclipsed the North.

Today, many young North Koreans bypass formal higher education in favor of service in the military. Although North Korea has not published economic statistics for nearly 30 years, it is estimated that military expenses consume one quarter to one third of the entire national budget—this despite near-starvation conditions in many parts of the country.

With China and the former Communist-bloc nations constituting natural markets for North Korean products, and with substantial financial aid from both China and the former Soviet Union in the early years, North Korea was able to regain much of its former economic, and especially industrial, strength. Today, North Korea successfully mines iron and other minerals and exports such products as cement and cereals. After the collapse of the Soviet/Communist bloc, China has remained North Korea's only reliable ally; trade between the two countries is substantial, with China accepting about 50 percent of North Korea's exports. In one Chinese province, more than two thirds of the people are ethnic Koreans, most of whom take the side of the North in any dispute with the South.

Tensions with the South have remained high since the war. Sporadic violence along the border has left patrolling soldiers dead, and the assassination of former South Korean president Park Chung Hee and attempts on the lives of other members of the South Korean government have been attributed to North Korea, as was the bombing of a Korean Airlines flight in 1987. Both sides have periodically accused each other of attempted sabotage. In 1996, North Korea tried to send spies to the South via a small submarine; the attempt failed, and most of the spies were killed. More worrisome have been clashes with the North Korean Navy. Dozens of North Korean sailors were killed in 1999 while attempting to penetrate the South; and in 2002, two North Korean patrol boats fired on and sank a South Korean ship in the Yellow Sea. Thirty North Korean sailors were killed in the attack. To most observers, such attacks seem pointless and irrational.

The North has long criticized the South for its suppression of dissidents. Although the North's argument is bitterly ironic, given its own brutal suppression of human rights, it is nonetheless accurate in its view that the government in the South has been blatantly dictatorial. To suppress opponents, the South Korean government has, among other things, abducted its own students from Europe, abducted opposition leader Kim Dae Jung from Japan, tortured dissidents, and violently silenced demonstrators. All of this was said to be necessary because of the need for unity in the face of the threat from the North; as pointed out by scholar Gavan McCormack, the South used the North's threat as an excuse for maintaining a rigid dictatorial system. Recent South Korean leaders have responded to the will of the people and have slowly implemented significant democratic reforms. No such progress is evident in the North, however.

Under these circumstances, it is not surprising that the formal reunification talks, begun in 1971 with much fanfare, have only recently shown any promise. Visits of residents separated by the war were approved in 1985—the first time in 40 years that an opening of the border had even been considered. In 1990, in what many saw as an overture to the United States, North Korea returned the remains of five American soldiers killed during the Korean War. But real progress came in late 1991, when North Korean premier Yon Hyong Muk and South Korean premier Chung Won Shik signed a nonaggression and reconciliation pact, whose goal was the eventual declaration of a formal peace treaty between the two governments. In 1992, the governments established air, sea, and land links and set up mechanisms for scientific and environmental cooperation. North Korea also signed the nuclear nonproliferation agreement with the International Atomic Energy Agency. This move placated growing concerns about North Korea's rumored development of nuclear weapons and opened the way for investment by such countries as Japan, which had refused to invest until they received assurances on the nuclear question.

■ THE NUCLEAR ISSUE FLARES UP

The goodwill deteriorated quickly in 1993 and 1994, when North Korea refused to allow inspectors from the International Atomic Energy Agency (IAEA) to inspect its nuclear facilities, raising fears in the United States, Japan, and South Korea that the North was developing a nuclear bomb, despite their promise not to. When pressured to allow inspections, the North responded by threatening to withdraw from the IAEA and expel the inspectors. Tensions mounted, with all parties engaging in military threats and posturing and the United States, South Korea, and Japan (whose shores could be reached in minutes by the North's new ballistic missiles) threatening economic sanctions. Troops in both Koreas were put on high alert. Former U.S. president Jimmy Carter helped to defuse the issue by making a private goodwill visit to Kim Il-Song in P'yongyang, the unexpected result of which was a promise by the North to hold a first-ever summit meeting with the South. Then, in a near-theatrical turn of events, Kim Il-Song, at five decades the longest national office-holder in the world, died, apparently of natural causes. The summit was canceled and international diplomacy was frozen while the North Korean government mourned the loss of its "Great Leader" and informally selected a new one, "Dear Leader" Kim Jong-Il, Kim Il-Song's son. Eventually, the North agreed to resume talks, a move interpreted as evidence that, for all its bravado, the North wanted to establish closer ties with the West. In the 1994 Agreed Framework with the United States, North Korea agreed to dismantle its graphite-moder-ated nuclear reactors in exchange for oil and help in building two light-water reactors. The world commu-nity breathed easier for a while. Unfortunately, tensions increased when North Korea launched a missile over Japan in 1998 and promised to keep doing so.

In the midst of these developments, a most amaz-ing breakthrough occurred: the North agreed to a summit. In June 2000, President Kim Dae Jung of the South flew to P'yongyang in the North for talks with President Kim Jung-Il. Unlike anything the South had expected, the South Korean president was greeted with cheers from the crowds lining the streets, and was feted at a state banquet. An agreement was reached that seemed to pave the way for peace and eventual reunification. To reward the North for this dramatic improvement in relations, the United States immedi-ately lifted trade sanctions that had been in place for 50 years. Both the North and the South suspended pro-paganda broadcasts and began plans for reuniting fami-lies long separated by the fortified DMZ. The North even went so far as to establish diplomatic ties with Australia and Italy.

Yet the West's trust of North Korea was soundly weakened when, in late 2006, the North Koreans deto-nated a small nuclear bomb near a military base in the northeast. This was solid evidence that the North had broken its promises to halt nuclear development. The test was denounced by the United States, South Korea, Japan, and even China, which is the North's only sub-stantial ally. The nuclear test, along with ballistic mis-sile tests, prompted most people to worry about North Korea's real intentions toward its neighbors. Substan-tial offers of financial assistance, and even an offer to allow North Korea to join the Asia Pacific Economic Cooperation forum, have not altered the North's mili-taristic posture. Likewise, the North's motives have been questioned when, after requesting food relief for its starving peoples, it refused to allow observers from the United Nations World Food Program to monitor the distribution of aid. Moreover, despite saying it will cooperate with the six-nation talks that would bring resolution to many of its problems, the North always seems to find a reason to halt talks with the West once they have started.

■ THE CHANGING INTERNATIONAL LANDSCAPE

North Korea has good reason to promote better rela-tions with the West, because the world of the 1990s is not the world of the 1950s. In 1989, for instance, several former Soviet-bloc countries cut into the North's eco-nomic monopoly by welcoming trade initiatives from South Korea; some even established diplomatic rela-tions. At the same time, the disintegration of the Soviet Union meant that North Korea lost its primary political and military ally. Perhaps most alarming to the North is its declining economy; it has suffered negative growth for several years. Severe flooding in 1995 destroyed much of the rice harvest and forced the North to do the unthinkable: accept rice donations from the South. More flooding in 2006 left some 50,000 people dead and 2.5 million homeless. Hundreds of North Koreans have defected to the South (over 7,000 between 2002 and 2006), all of them complaining of near-famine condi-tions. Sometimes, defectors attempt to flee to the South in boats but are carried elsewhere, such as the 9 people found in 2011 by Japan officials when their boat was carried to Japan. As many as many as 2,000 per month have fled to China. With the South's economy consis-tently booming and the example of the failed economies of Eastern Europe as a danger signal, the North appears to understand that it must break out of its decades of isolation or lose its ability to govern. Nevertheless, it is not likely that North Koreans will quickly retreat from the Communist model of development that they have espoused for so long.

Kim Il-Song, who controlled North Korea for nearly 50 years, promoted the development of heavy industries, the collectivization of agriculture, and strong linkages with the then–Communist bloc. Governing with an iron hand, Kim denied basic civil rights to his people and for-bade any tendency of the people to dress or behave like the "decadent" West. He kept tensions high by asserting his intention of communizing the South. His son, Kim Jong-Il, who had headed the North Korean military but was barely known outside his country, was eventually named successor to his father—the first dynastic power transfer in the Communist world. An enigmatic leader, the younger Kim approved actions that increased ten-sions with the South while doing little to improve the stumbling economy. In 2006, Kim canceled the family reunion policy that had allowed a few people each year to receive family members from the South. He did this because South Korea halted some financial assistance programs.

When communism was introduced in North Korea in 1945, the government nationalized major compa-nies and steered economic development toward heavy industry. In contrast, the South concentrated on heavy industry to balance its agricultural sector until the late 1970s but then geared the economy toward meeting con-sumer demand. Thus, the standard of living in the North for the average resident remains far behind that of the South. Indeed, Red Cross, United Nations, and other observers have documented widespread malnutrition and starvation in North Korea, conditions that are likely to continue well into the twenty-first century unless the North dramatically alters its current economic policies.

Conditions are so bad in some areas that there is no electricity nor chlorine to run water-treatment plants, resulting in contaminated water supplies for about 60 percent of the population.

Even before the June 2000 summit, North Korea exhibited signs of liberalization. The government agreed to allow foreign companies to establish joint ventures inside the country, tourism was being promoted as a way of earning foreign currency, and two small Christian churches were allowed to be established. In an unusual move, North Korea openly requested aid from several foreign countries after nearly 200 people were killed and some 8,000 homes destroyed or damaged when a train exploded in the city of Ryongchon near the Chinese border in 2004. About the same time, the North even stopped broadcasting propaganda across the DMZ from large loudspeakers it had once erected to spread the philosophy of communism to Southerners across the border. Nevertheless, years of a totally controlled economy in the North and shifting international alliances indicate many difficult years ahead for North Korea.

For instance, when North Korea admitted in 2002 that it had not dismantled its nuclear-bomb-making program, U.S. president George W. Bush immediately cut off shipments of oil and other supplies. The North responded by evicting the United Nations inspectors who had been monitoring the nuclear program. In 2007, North Korea at last agreed to "shut down and seal" its Yongbyon and other nuclear facilities in exchange for fuel oil and humanitarian economic assistance from China, Russia, the United States and South Korea. Japan supported the agreement in principle but would not provide fuel oil until the matter of abducted Japanese citizens had been fully addressed by North Korea.

In early 2010, the North sank a South Korean warship with a torpedo. The unprovoked attack caused a storm of protests, and the South recast the North as its "principal enemy." To many, the expenditure of large amounts of the national budget on military ventures such as that seem incomprehensible, especially given the general poverty of the country. It is estimated that 25 percent of the GDP is spent on the military. Moreover, the country continues to engage in activities that frustrate the entire world: additional nuclear tests, building a facility to enrich uranium, exchanging artillery fire with the South over a disputed island, secretly selling military equipment to the Congo Middle Eastern countries in violation of United Nation sanctions, and so on. In early 2011, former U.S. Secretary of Defense Robert Gates warned that the North is within five years of being able to strike North America with ballistic missiles. Despite (or because of) these things, the U.S. decided in late 2011 to resume suspended talks with the North, this after the North agreed to consider a moratorium on nuclear weapons testing and production. The disarmament-for-aid talks had been stalled for two years when the North walked out after the U.N. Security Council imposed sanctions because the North tested a nuclear weapon in 2009.

■ RECENT TRENDS

Although political reunification still seems to be years away, social changes are becoming evident everywhere as a new generation, unfamiliar with war, comes to adulthood, and as North Koreans are being exposed to outside sources of news and ideas. Many North Koreans now own radios that receive signals from other countries. South Korean stations are now heard in the North, as are news programs from the Voice of America. Modern North Korean history, however, is one of repression and control, first by the Japanese and then by the Kim governments, who used the same police surveillance apparatus as did the Japanese during their occupation of the Korean Peninsula. It is not likely, therefore, that a massive push for democracy will be forthcoming soon from a people long accustomed to dictatorship.

Most worrisome is North Korea's policy volatility. On one day the government seems to want some kind of reconciliation with the South or with the West; on the next it intentionally provokes everyone. For instance, after their surprise nuclear test in 2006, the North was supposed to have stopped making nuclear bombs, but they did not due to a dispute with the United States over verification. Then they started launching long-range missiles, further provoking the United States and Japan. After additional verification agreements, the United States removed the North from its list of terrorists only to be chagrined when, one month later, the North launched a second nuclear test.

At other times, they have appeared to soften their position on the South, only to reverse course and declare an all-out effort to smash South Korea for placing new conditions on economic cooperation. In 2009 they temporarily canceled all contracts regarding the Kaesong Industrial Park—the only bright spot in the North's economy.

After the latest missile launches and nuclear tests, the UN Security Council, including China, banned the sale of arms to or from North Korea—a major blow to the always enfeebled economy. The North declared that any attempt by the South to interdict vessels carrying arms would be considered "an act of war." Yet, once again, the North did an about face when, after arresting three U.S. journalists who had crossed the North's border and sentencing them to twelve years of hard labor, the North released them with little comment when former President Bill Clinton went to P'yongyang seeking their release.

What the future holds is unclear, in part because it is not certain who will be leading the country. In late 2011, Kim Jong-Il died at age 69. He is said to have named his youngest son, Kim Jong-Un (25) as his successor, but Kim Jong-Un is a virtual unknown both inside and outside North Korea.

Snapshot: NORTH KOREA

Summarized below is a quick look at the country with regard to its development, freedom, health/welfare, and achievements.

Development

Already more industrialized than South Korea at the time of the Korean War, North Korea built on this foundation with massive assistance from China and the Soviet Union. Heavy industry was emphasized to the detriment of consumer goods. Economic isolation presages more negative growth ahead. North Korea is one of the poorest countries in the Pacific Rim, but the Kaesong Industrial Park with 120 factories, is flourishing.

Freedom

The mainline Communist approach has meant that the human rights commonplace in the West have never been enjoyed by North Koreans. Through suppression of dissidents, a controlled press, and restrictions on travel, the regime has kept North Koreans isolated from the world. As many as 200,000 political prisoners are thought to be held in prison camps. North Koreans are not free to choose their own employment or change jobs at will.

Health/Welfare

Under the Kim Il-Song government, illiteracy was greatly reduced. Government housing is available at low cost, but shoppers are often confronted with empty shelves and low-quality goods. Malnutrition is widespread, and mass starvation has been reported in some regions. Life expectancy has dropped 3.4 years (66 years for men and 72 for women), and infant mortality has risen to 19 per every 1,000 births.

Achievements

North Korea has developed its resources into solid industries for the production of tools and machinery. However, the country is not able to feed many of its people. In 2006, North Korea detonated a small nuclear bomb but few in the world regarded that as an achievement.

Statistics

Geography

Area in Square Miles (Kilometers): 44,358 (120,540) (about the size of Mississippi)

Capital (Population): P'yongyang (3.3 million (2008))

Environmental Concerns: water pollution and water-borne disease; insufficient potable water; deforestation; soil erosion and degradation

Geographical Features: mostly hills and mountains separated by deep, narrow valleys; coastal plains

Climate: temperate with summer rainfall

People*

Population

Total: 24,457,492

Annual Growth Rate: 0.54%

Rural/Urban Population (Ratio): 40/60

Major Language: Korean

Ethnic Makeup: homogeneous Korean

Religions: mainly Buddhist and Confucianist, some Christian and Chondogyo (autonomous religious activities mostly nonexistent; government-sponsored religions give the appearance of freedom)

Health

Life Expectancy at Birth: 65 years (male); 73 years (female)

Infant Mortality: 27/1,000 live births

Physicians Available: 1/304 people (2003)

Education

Adult Literacy Rate: 99%

Compulsory (Ages): 6–17; free

Communication

Telephones: 1.18 million

Transportation

Roadways in Miles (Kilometers): 15,879 (25,554)

Railroads in Miles (Kilometers): 3,257 (5,242)

Usable Airfields: 79

Government

Type: authoritarian socialist; one-man dictatorship

Independence Date: August 15, 1945 (from Japan)

Head of State/Government: President Kim Jong-Il; Premier Choe Yong Rim

Political Parties: Korean Workers' Party (KWP); Korean Social Democratic Party; Chondoist Chongu Party (the latter two are under KWP control)

Suffrage: universal at 17

Military

Military Expenditures (% of GDP): N/A

Current Disputes: Demarcation Line with South Korea; unclear border with China

Economy

Currency ($ U.S. Equivalent): 1,800 won = $1

Per Capita Income/GDP: $1,800/$40 billion

GDP Growth Rate: −0.9%

Labor Force by Occupation: 65% nonagricultural; 35% agricultural

Natural Resources: hydropower; iron ore; copper; lead; salt; zinc; coal; magnesite; gold; tungsten; graphite; pyrites; fluorspar

Agriculture: rice; corn; potatoes; soybeans; pulses; livestock; eggs

Industry: machinery; military products; electric power; chemicals; mining; metallurgy; textiles; food processing; tourism

Exports: $2 billion (primary partners South Korea, China, Thailand)

Imports: $3.1 billion (primary partners China, South Korea, Thailand)

Suggested Websites

www.topics.nytimes.com/top/news/international/countriesandterritories/northkorea/index

*Note: Statistics for North Korea are generally estimated due to unreliable official information.

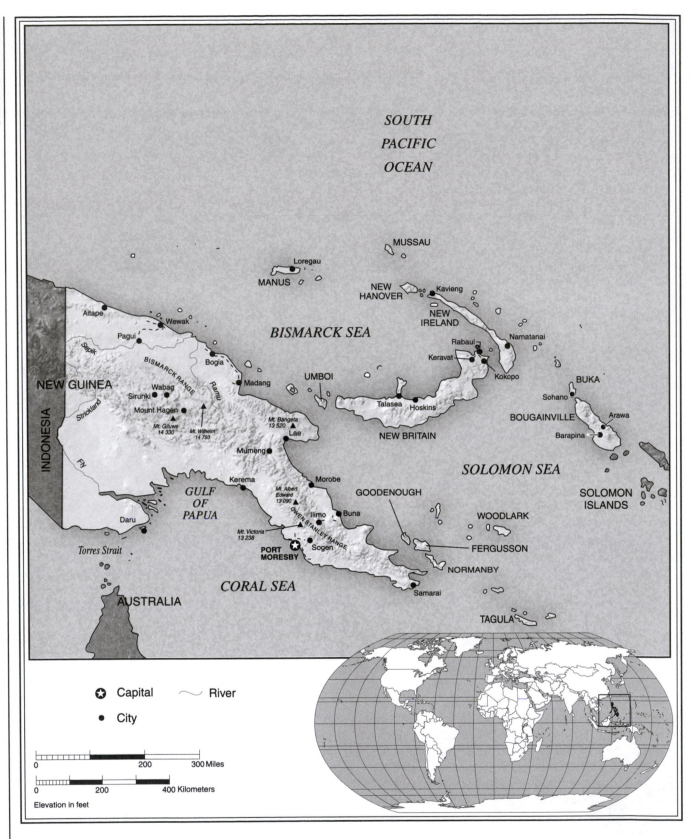

SOUTH

PACIFIC

OCEAN

MUSSAU

Loregau

MANUS

NEW
HANOVER

Kavieng

NEW
IRELAND

BISMARCK SEA

Aitape

Wewak

Pagui

Namatanai

Rabaul

Keravat

Bogia

UMBOI

Kokopo

BUKA

NEW GUINEA

Madang

Sepik

Wabag

Sirunki

Ramu

Talasea

Hoskins

Sohano

INDONESIA

Mount Hagen

Mt. Giluwe
14 330

Mt. Wilhelm
14 793

Bismarck Range

Strickland

Mt. Bangeta
13 520

Lae

NEW BRITAIN

BOUGAINVILLE

Arawa

Barapina

Mumeng

SOLOMON SEA

SOLOMON
ISLANDS

Fly

Kerema

Morobe

GOODENOUGH

WOODLARK

Mt. Albert
Edward
13 090

Ilimo

Buna

Daru

Owen Stanley Range

GULF
OF
PAPUA

Mt. Victoria
13 238

PORT
MORESBY

Sogen

FERGUSSON

NORMANBY

Torres Strait

CORAL SEA

Samarai

AUSTRALIA

TAGULA

⊛ Capital   ∿ River

● City

0   200   300 Miles

0   200   400 Kilometers

Elevation in feet

Papua New Guinea

(Independent State of Papua New Guinea)

Papua New Guinea is an independent nation and a member of the British Commonwealth. The capital is Port Moresby, where, in addition to English, the Motu language and a hybrid language known as New Guinea Pidgin are spoken. Occupying the eastern half of New Guinea (the second-largest island in the world) and many outlying islands, Papua New Guinea is perhaps one of the most overlooked of all the nations in the Pacific Rim.

It was not always overlooked, however. Spain claimed the vast land in the mid-sixteenth century, followed by Britain's East India Company in 1793. The Netherlands laid claim to part of the island in the 1800s and eventually came to control the western half (known for many years as Irian Jaya but now called Papua, a province of the Republic of Indonesia). In the 1880s, German settlers occupied the northeastern part of the island; and in 1884, Britain signed a treaty with Germany, which gave the British about half of what is now Papua New Guinea. In 1906, Britain gave its part of the island to Australia. Australia invaded and quickly captured the German area in 1914. Eventually, the League of Nations and, later, the United Nations gave the captured area to Australia to administer as a trust territory.

During World War II, the northern part of New Guinea was the scene of bitter fighting between a large Japanese force and Australian and U.S. troops. The Japanese had apparently intended to use New Guinea as a base for the military conquest of Australia. Australia resumed control of the eastern half of the island after Japan's defeat, and it continued to administer Papua New Guinea's affairs until 1975, when it granted independence. Australia continues to support Papua New Guinea in many ways, not the least of which is financial aid. In 2006, Australia was supplying 20 percent of the country's operating budget (as well as other assistance such as patrol boats). It also helps the Papuans interdict drug traffickers along its Indonesian border.

■ STONE-AGE PEOPLES

Early Western explorers found the island's resources difficult to reach. The coastline and some of the interior are swampy and mosquito- and tick-infested, and the high, snow-capped mountainous regions are densely forested and hard to traverse. But perhaps most daunting to early would-be settlers and traders were the local inhabitants. Consisting of hundreds of sometimes warring tribes with totally different languages and customs, the New Guinea populace was determined to prevent outsiders from settling the island. Many adventurers were killed, their heads displayed in villages as victory trophies. The origins of the Papuan people are unknown, but some tribes share common practices with Melanesian islanders. Others appear to be Negritos, and some may be related to the Australian Aborigines. More than 860 languages are spoken in Papua New Guinea.

Australians and other Europeans found it beneficial to engage in trade with coastal tribes who supplied them with unique tropical lumbers,

TIMELINE

A.D. 1511
The main island is sighted by Portuguese explorers; it was claimed first by Spain and later by Britain

1828
The Dutch annex the west half of the island

1884
A British protectorate over part of the eastern half of the island; the Germans control the northeast

1890
Gold is discovered in Papua New Guinea

1906–1914
Australia assumes control of the island

1920
Australia is given the former German areas as a UN trust territory

1940s
Japan captures the northern part of the island; Australia resumes control in 1945

1975
Australia grants independence to Papua New Guinea

1988
A revolt against the government begins on the island of Bougainville

1990s
An economic blockade of Bougainville is lifted, but violence continues, claiming 3,000 lives; country joined APEC in 1993

2000s
Bougainville independence referendum is set
The government moves to privatize industries
100 women run for parliamentary seats; 1 is elected
With cooperation of 35 indigenous communities, Papua New Guinea creates its first nature preserve where 190,000 acres will be off-limits to logging and hunting

(United Nations Photo Library/Ray Witlin)

The interior of Papua New Guinea is very difficult to reach. Achieving easier access to the country's valuable minerals and exotic timber have caused a push for the development of transportation services. The island has over 550 airstrips, some in very isolated areas, as well as a primitive but expanding road network. The negative impact of the development of roads on the environment is of great concern.

such as sandalwood and bamboo, and foodstuffs such as sugarcane, coconut, and nutmeg. Rubber and tobacco were also traded. Tea, which grows well in the highland regions, is an important cash crop.

But the resource that was most important for the economic development of Papua New Guinea was gold. It was discovered there in 1890 and produced two major gold rushes, in 1896 and 1926. Most prospectors were not Papuans; rather, they came from outside,

most often Australia, and their presence antagonized the locals. Some prospectors were killed and cannibalized. A large number of airstrips in the otherwise undeveloped interior eventually were built by miners who needed a safe and efficient way to receive supplies. Today, copper is more important than gold—copper is, in fact, the largest single earner of export income for Papua New Guinea.

Meanwhile, pollution from mining is increasingly of concern to environmentalists, as is deforestation of Papua New Guinea's spectacular rain forests. A diplomatic flap occurred in 1992, when Australian environmentalists protested that a copper and gold mine in Papua New Guinea was causing enormous environmental damage. They called for the mine to be shut down. The Papuan government strongly resented the verbal intrusion into its sovereignty and reminded the protesters and the Australian government that it alone would establish environmental standards for companies operating inside its borders. The Papuan government holds a 20 percent interest in the mining company.

DID YOU KNOW?

Like other countries in the "ring of fire" earthquake zone, Papua New Guinea is often hit by earthquakes. In 1998, a 7.0 earthquake hit the northern coast of the country causing a massive underwater landslide. That, in turn, created a 50 ft (15 m) tsunami that wiped out or severely damaged many coastal villages. Over 2,000 people died and 9,000 were left homeless.

FURTHER INVESTIGATION

In 2011, a dozen people were killed and thousands of homes were burned down in rioting in Papua New Guinea's second-largest city, Lae. To find out why, see www.pngindustrynews.net/storyview.asp?storyid=2491543

A similar flap occurred in the late 1990s and early 2000s over proposed logging of the forests near Collingwood Bay. The Prime Minister wanted to allow clearcutting of the forests, sell the logs to China, replace the forests with coconut sap plantations, and create 50,000 jobs in the process. The Maisin people who live in the area issued a formal declaration opposing the logging, but the timber companies, including one from the Philippines, proceeded anyway. The Maisins believed that the development of painted tapa cloth industries would be better suited to their culture and protect the forests at the same time. Eventually, the matter ended up in the courts, and in late 2002, the local inhabitants won the battle. A four-day celebration ensued, during which the victors, in a traditional practice, wiped pork fat on those who helped them win the case.

The tropical climate that predominates in all areas except the highest mountain peaks produces an impressive variety of plant and animal life. For decades, botanists and other naturalists have been attracted to the island for scientific study. Despite extensive contacts with these and other outsiders over the past century, and despite the establishment of schools and a university by the Australian government, some inland mountain tribes continue to live much as they probably did in the Stone Age. Thus, the country lures not only miners and naturalists but also anthropologists and archaelogists looking for clues to humankind's early lifestyles. One of the most famous of these was Bronislaw Malinowski, the Polish-born founder of the field of social anthropology. In the early 1900s, he spent several years studying the cultural practices of the tribes of Papua New Guinea, particularly those of the Trobriand Islands.

Most of the 6 million Papuans live by subsistence farming. Agriculture for commercial trade is limited by the absence of a good transportation network, because most roads are unpaved, and there is no railway system. Travel on tiny aircraft and helicopters is common, however; New Guinea boasts some 562 airstrips, most of them unpaved and dangerously situated in mountain valleys. The harsh conditions of New Guinea life have produced some unique ironies. For instance, Papuans who have never ridden in a car or truck may have flown in a plane dozens of times.

In 1998, 23-foot-high tidal waves caused by offshore, undersea earthquakes inundated dozens of villages along the coast and drowned 6,000 people. A social earthquake, of sorts, occurred in the late 1980s when secessionist rebels on Bougainville Island, located about 500 miles from the capital of Port Moresby, launched a violent campaign to gain independence. Before the violence ended, some 20,000 had been killed, some with such armaments as stone-age spears and clubs. In 2001, the government signed a peace agreement with the rebels that granted them autonomy, including the right to operate their own police force. The government also agreed that residents could establish a provincial government and vote on independence within 10 to 15 years.

The government has had to handle other touchy issues in recent years. For instance, Prime Minister Bill Skate established diplomatic ties with Taiwan—which, as usual, raised a political storm from the mainland Chinese. The controversy eventually caused Skate's resignation as prime minister and the defection of several members of his party to the opposition. His successor, Sir Mekere Moruata, immediately reversed the decision over Taiwan, but barely had he been in office when he fired three of his cabinet ministers, claiming that they were plotting against him. Nevertheless, he moved quickly to privatize almost all state-owned enterprises, including the airlines, telecommunications, and others.

Moruata was replaced in the next election by Sir Michael Somare, the person most responsible for promoting independence from Australia and often called the nation's "founding father." It was Somare's third time as prime minister. But the election that brought him back to office was nothing short of chaotic. Forty-three political parties running 3,000 candidates contended for the right to govern the island nation. During the campaign, some 30 people were killed; ballot boxes were stolen and plundered, and gangs hired by tribal leaders intimidated voters in many outlying villages. There were so many voting irregularities that a final count was delayed for weeks, and several regions had to redo the vote. In the end, Somare's National Alliance had to ally with five other parties in order to gain a majority and constitute a government. Somare found himself in the middle of an international flap when Papuan officials helped an alleged child sex abuser avoid extradition to Australia to be tried. The man, Julian Moti, an Australian citizen and lawyer, had been appointed by the Prime Minister of the Solomon Islands to be attorney general. He took refuge in Papua New Guinea, and the Papuans eventually took him to the Solomons in a military plane, rather than turning him over to the Australians. He was arrested in the Solomons, but that government also refused to send him to Australia, claiming the charges were unfounded. In the fallout to this fiasco, the Solomons expelled the Australian High Commissioner (Ambassador), and the Australians temporarily suspended some diplomatic ties with Papua New Guinea.

In the 2007 elections, Michael Somare was returned as prime minister, but the event that captured the news was the candidacy of James Yali, a provincial governor and member of parliament. The year before, he had been convicted of rape and sent to prison for 12 years,

Snapshot: PAPUA NEW GUINEA

Summarized below is a quick look at the country with regard to its development, freedom, health/welfare, and achievements.

Development

Agriculture (especially coffee and copra) is the mainstay of Papua New Guinea's economy. Copper, gold, and silver mining are also important, but large-scale development of other industries is inhibited by rough terrain, illiteracy, and a huge array of spoken languages—more than 860. There are substantial reserves of untapped oil.

Freedom

Papua New Guinea is a member of the British Commonwealth and officially follows the English heritage of law.

However, in the country's numerous, isolated small villages, effective control is wielded by village elites with personal charisma; tribal customs take precedence over national law—of which many inhabitants are virtually unaware.

Health/Welfare

Just over half the population can read and write, and even those who can have had very little formal education. Daily nutritional intake falls far short of recommended minimums, and tuberculosis and malaria are common diseases.

Achievements

Papua New Guinea, lying just below the equator, is world-famous for its astoundingly varied and beautiful flora and fauna, including orchids, birds of paradise, butterflies, and parrots. Dense forests cover 70 percent of the country. Some regions receive as much as 350 inches of rain a year.

---

but his supporters found a way to help him escape and proceeded with him to the electoral office where he applied to run for his old seat. He solidly won the seat, despite his rape conviction, because, it was said, of fear of him as a sorcerer. The courts eventually overturned his victory.

Given the differences in socialization of the Papuan peoples and the difficult conditions of life on their island, it will likely be many decades before Papua New Guinea, which joined the Asia Pacific Economic Cooperation group in 1993, is able to participate fully in the Pacific Rim community.

Statistics

Geography

Area in Square Miles (Kilometers): 178,704 (462,840) (about the size of California)
Capital (Population): Port Moresby (314,000)
Environmental Concerns: deforestation; pollution from mining projects; drought
Geographical Features: mostly mountains; coastal lowlands and rolling foothills
Climate: tropical monsoon

People

Population

Total: 6,187,591
Annual Growth Rate: 1.99%
Rural/Urban Population Ratio: 87/13
Major Languages: Tok Pisin; English; Hiri Motu; 860 indigenous languages
Ethnic Makeup: predominantly Melanesian and Papuan; some Negrito, Micronesian, and Polynesian
Religions: 96% Christian

Health

Life Expectancy at Birth: 64 years (male); 68.5 years (female)
Infant Mortality: 43/1,000 live births
Physicians Available: 1/18,868 people
HIV/AIDS Rate in Adults: 0.9%

Education

Adult Literacy Rate: 57%

Communication

Telephones: 60,000 main lines
Mobile Phones: 900,000 (2009)
Internet Users: 125,000 (2010)
Internet Penetration (% of Pop.): 2%

Transportation

Roadways in Miles (Kilometers): 5,809 (9,349)
Railroads in Miles (Kilometers): none
Usable Airfields: 562

Government

Type: parliamentary democratic/monarchy
Independence Date: September 16, 1975 (from the Australian-administered UN trusteeship)
Head of State/Government: Queen Elizabeth II (represented by Governor Michael Ogio as of February 2011); Prime Minister Peter Paire O'Neill
Political Parties: National Alliance Party; Papua and Niugini Union Party; Papua New Guinea Party; People's Democratic Movement; People's Action Party; United Resources Party
Suffrage: universal at 18

Military

Military Expenditures (% of GDP): 1.4%
Current Disputes: illegal cross-border activities mainly from Indonesia

Economy

Currency ($ U.S. Equivalent): 2.75 kina = $1
Per Capita Income/GDP: $2,500/14.95 billion
GDP Growth Rate: 7%
Inflation Rate: 16%
Labor Force by Occupation: 85% agriculture; 15% industry & services
Population Below Poverty Line: 37%
Natural Resources: gold; copper; silver; natural gas; timber; petroleum; fisheries

Agriculture: coffee; cocoa; copra; palm kernels; tea; sugar; rubber; sweet potatoes; fruit; vegetables; vanilla; shellfish; poultry; pork
Industry: copra crushing; palm oil processing; wood processing and production; mining; construction; tourism
Exports: $6.2 billion (primary partners Australia, Japan, China)
Imports: $3.75 billion (primary partners Australia, Singapore, China)

Suggested Websites

www.pngonline.gov.pg/
www.pngtourism.org.pg/

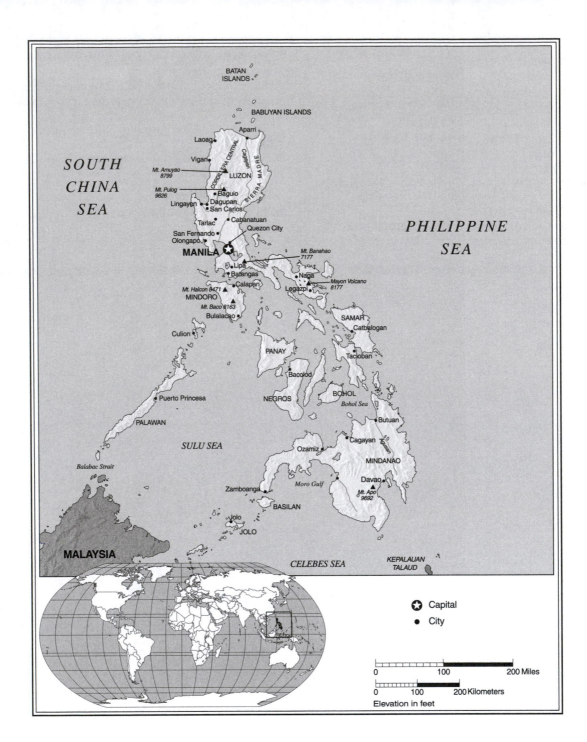

SOUTH
CHINA
SEA

PHILIPPINE
SEA

BATAN
ISLANDS

BABUYAN ISLANDS

Aparri

Laoag

Vigan

Mt. Amuyao
8799

LUZON

Baguio

Mt. Pulog
9626

Lingayen
Dagupan
San Carlos

Tarlac

Cabanatuan

San Fernando
Olongapo

Quezon City

MANILA

Mt. Banahao
7177

Lipa

Batangas
Naga

Calapan
Legazpi

Mt. Halcon 8471

Mayon Volcano
8177

MINDORO

Mt. Baco 8163

SAMAR

Bulalacao

Catbalogan

Culion

PANAY

Tacloban

Bacolod

Puerto Princesa

NEGROS

BOHOL

Bohol Sea

Butuan

PALAWAN

Cagayan

SULU SEA

Ozamiz

MINDANAO

Balabac Strait

Davao

Moro Gulf

Mt. Apo
9692

Zamboanga

BASILAN

MALAYSIA

Jolo

JOLO

CELEBES SEA

KEPALAUAN
TALAUD

⊛ Capital

● City

0   100   200 Miles

0   100   200 Kilometers

Elevation in feet

Philippines
(Republic of the Philippines)

The Philippines has close historic ties to the West. Eighty-one percent of Filipinos are Roman Catholics, and most speak at least some English. Many use English daily in business and government; in fact, English is the language of instruction at school. Moreover, when they discuss their history as a nation, Filipinos will mention Spain, Mexico, the Spanish-American War, the United States, and cooperative Filipino–American attempts to defeat the Japanese in World War II. The country was even named after a European, King Philip II of Spain. (Currently, 3.2 million Filipinos live in the United States, making them the second-largest Asian ethnic group after the Chinese, and some 300,000 Americans live in the Philippines). If this does not sound like a typical Asian nation, it is because Philippine nationhood essentially began with the arrival of Westerners. That influence continues to dominate the political and cultural life of the country.

Yet the history of the islands certainly did not begin with European contact; indeed, there is evidence of human habitation in the area as early as 25,000 B.C. Beginning about 2,000 B.C., Austronesians, Negritos, Malays, and other tribal peoples settled many of the 7,107 islands that constitute the present-day Philippines. Although engaged to varying degrees in trade with China and Southeast Asia, each of these ethnic groups (nearly 60 distinct groups still exist) lived in relative isolation from one another, speaking different languages, adhering to different religions, and, for good or ill, knowing nothing of the concept of national identity.

Although 5 million ethnic peoples remain marginated from the mainstream today, for most islanders the world changed forever in the mid-1500s, when soldiers and Roman Catholic priests from Spain began conquering and converting the population. Eventually, the disparate ethnic groups came to see themselves as one entity, Filipinos, a people whose lives were controlled indirectly by Spain from Mexico—a fact that, unique among Asian countries, linked the Philippines with the Americas. Thus, the process of national-identity formation for Filipinos actually began in Europe and North America.

Members of some ethnic groups assimilated rather quickly, marrying Spanish soldiers and administrators and acquiring the language and cultural outlook of the West. The descendants of these mestizos ("mixed" peoples, including local/Chinese mixes) have become the cultural, economic, and political elite of the country. Others, particularly among the Islamic communities on the island of Mindanao, resisted assimilation right from the start and continue to challenge the authority of Manila. Indeed, the Communist insurgency, reported so often in the news, does not seem to go away. Several groups, among them the Moro Islamic Liberation Front, the Moro National Liberation Front, and the Islamic Abu Sayyaf Group, continue to take tourists and others as hostages, set off bombs, and inflict violence on villagers. For example, a bomb explosion at a basketball game in the southern Philippines in 2004 killed 11 people and wounded 40, including the town mayor, who was probably the target of the attack. About the same time, an American businessperson was abducted and chained by his neck and feet for nearly a month. It is believed that

TIMELINE

25,000 B.C.
Negritos and others begin settling the islands

2,000 B.C.
Malays arrive in the islands

A.D. 400–1400
Chinese, Arabs, and Indians control parts of the economy and land

1542
The islands are named for the Spanish king Philip II

1890s
Local resistance to Spanish rule

1898
A treaty ends the Spanish-American War Independence from Spain but now controlled by the United States

1941
The Japanese attack and occupy the country

1944
General Douglas MacArthur makes a triumphant return to Manila, re-taking the islands from Japan

1946
The United States grants complete independence to the Philippines

1947
Military-base agreements are signed with the United States

1965
Ferdinand Marcos is elected president

1972
Marcos declares martial law

1980s
Martial law is lifted; Corazon Aquino and her People Power movement drive Marcos into exile

1990s
The United States closes its military bases in the Philippines; economic crisis

2000s
The crippling "Love Bug" computer virus emanates from the Philippines
President Estrada is ousted for "economic plunder"
The Philippines withdraws more troops from Iraq in an effort to save the life of a Filipino hostage
After finishing out the term for her predecessor, Gloria Macapagal-Arroyo is elected president in 2004 elections
Mudslides from typhoons bury entire villages and kill thousands
Arroyo signs a ban on the use of the death penalty
Former president and "people power" leader Corazon Aquino dies
Benigno (NoyNoy) Aquino elected President

The Philippines has suffered from the misuse of funds entrusted to the government over the past several decades. The result has been a polarity of wealth, with the elites living lives of luxury while other citizens live in crushing poverty. Slums, such as Tondo in Manila, pictured above, are a common sight in many of the urban areas of the Philippines.

some of these groups capture hostages (such as a Christian missionary from Kansas in 2002 and three International Red Cross workers in 2008) in order to send the ransom money to support the al-Qaeda terrorist network. On the day that nearly 3,000 U.S. troops arrived in the Philippines to help the local military in its efforts to eliminate the Abu Sayyaf rebels, bombs exploded at stores, a radio station, and a bus terminal, killing many innocent passersby. Such violence is, in part, connected to the worldwide rise of radical Islamism, but also it is an attempt by marginated ethnics and others to regain the cultural independence that their peoples lost some 400 years ago. Talks with the rebels resumed in 2009 after a cease-fire with the military went into effect, and the United States is attempting to facilitate the process by promising to remove Philippine communist rebels from a list of terrorist organizations if they conclude a peace agreement and resume talks in Oslo, Norway.

Unfortunately, the violence in or emanating from the Islamic Mindanao region continues. The government has accused the powerful Ampatuan family, whose members wield much power in Mindanao, of plotting the 2009 massacre of 197 people, including 65 police and some top provincial leaders. In 2011, the U.S. government renewed its travel warning, advising Americans of possible indiscriminate attacks at such places as airports and malls.

As elsewhere in Asia, the Chinese community has played an important but controversial role in Philippine life. Dominating trade for centuries, the Philippine Chinese have acquired clout (and enemies) that far exceeds their numbers (fewer than 1 million). Former president Corazon Aquino was of part-Chinese ancestry, and some of the resistance to her presidency stemmed from her ethnic lineage. The Chinese-Philippine community, in particular, has been the target of ethnic violence—kidnappings

? DID YOU KNOW?

Under the South China Sea are such large deposits of oil and other valuable resources that half a dozen countries claim the space as their own. The disputants claim ownership of some 190 islands, reefs, and sand banks, but the most aggressive claimant, by far, is China. In 2011, when a Filipino government official accused China of illegal intrusions in the Manila-claimed areas, a Chinese embassy official became so enraged and so "undiplomatic" that the Philippine government banned him from taking part in any further meetings. The United States, aware that small countries such as the Philippines cannot compete with China militarily, has intensified military assistance.

and abductions—because their wealth (relative to other Filipino groups) makes them easy prey.

■ FOREIGN INTERESTS

Filipinos occupy a resource-rich, beautiful land. Monsoon clouds dump as much as 200 inches of rain on the fertile, volcanic islands. Rice and corn grow well, as do hemp, coconut, sugarcane, and tobacco. Tuna, sponges, shrimp, and hundreds of other kinds of marine life flourish in the surrounding ocean waters. Part of the country is covered with dense tropical forests yielding bamboo and lumber and serving as habitat to thousands of species of plant and animal life. The northern part of Luzon Island is famous for its terraced rice paddies.

Given this abundance, it is not surprising that, historically, several foreign powers took a serious interest in the archipelago. The Dutch held military bases in the country in the 1600s, the British briefly controlled Manila in the 1800s, and the Japanese overran the entire country in the 1940s. But it was Spain, in control of most of the country for more than 300 years (1565–1898), that established the cultural base for the modern Philippines. Spain's interest in the islands—its only colony in Asia— was primarily material and secondarily spiritual. It wanted to take part in the lucrative spice trade and fill its galleon ships each year with products from Asia for the benefit of the Spanish Crown. It also wanted (or, at least, Rome wanted) to convert the so-called heathens (that is, nonbelievers) to Christianity. The friars were particularly successful in winning converts to Roman Catholicism because, despite some local resistance, there were no competing Christian denominations in the Philippines, and because the Church quickly gained control of the resources of the island, which it used to entice converts. Eventually, a Church-dominated society was established that mirrored in structure—social class divisions as well as religious and social values—the mother cultures of Spain and particularly Mexico.

Resisting conversion were the Muslims of the island of Mindanao, a group that continues to remain on the fringe of Philippine society but which signed a cease-fire with the government in 1994, after 20 years of guerrilla warfare. Another cease-fire was agreed to in 2009. The cease-fire has been only partially observed, with violence flaring periodically, including immediately after the agreement, when 200 armed Muslims attacked and burned the town of Ipil on Mindanao Island, and off and on since then.

Spanish rule in the Philippines came to an inglorious end in 1898, at the end of the Spanish-American War. Spain granted independence to Cuba and ceded the Philippines, Guam, and Puerto Rico to the United States. Filipinos hoping for independence were disappointed to learn that yet another foreign power had assumed control of their lives. Resistance to American rule cost several thousand lives in the early years, and the American public became outraged at the brutality exhibited by some of the U.S. soldiers against the Philippine people. Eventually, the majority of the people accepted their fate and even began to realize that, despite the deception and loss of independence, the U.S. presence was fundamentally different from that of Spain. The United States was interested in trade, and it certainly could see the advantage of having a military presence in Asia, but it viewed its primary role as one of tutelage. American officials believed that the Philippines should be granted independence, but only when the nation was sufficiently schooled in the process of democracy. Unlike Spain, the United States encouraged political parties and attempted to place Filipinos in positions of governmental authority.

Preparations were under way for independence when World War II broke out. The war and the occupation of the country by the Japanese undermined the economy, devastated the capital city of Manila, caused divisions among the political elite, and delayed independence. After Japan's defeat, the country was, at last, granted independence, on July 4, 1946. Manuel Roxas, a well-known politician, was elected president. Despite armed opposition from Communist groups, the country, after several elections, seemed to be maintaining a grasp on democracy.

■ MARCOS AND HIS AFTERMATH

Then, in 1965, Ferdinand E. Marcos, a Philippines senator and former guerrilla fighter with the U.S. armed forces, was elected president. He was reelected in 1969. Rather than addressing the serious problems of agrarian reform and trade, Marcos maintained people's loyalty through an elaborate system of patronage, whereby his friends and relatives profited from the misuse of government power and money. Opposition to his corrupt rule manifested itself in violent demonstrations and in a growing Communist insurgency. In 1972, Marcos declared martial law, arrested some 30,000 opponents, and shut down newspapers as well as the National Congress. Marcos continued to rule the country by personal proclamation until 1981. After the lifting of martial law, Marcos remained in power for another five years, and he and his wife, Imelda, and their extended family and friends increasingly came under public criticism for corruption. Finally, in 1986, after nearly a quarter-century of his rule, an uprising of thousands of dissatisfied Filipinos overthrew Marcos, who fled to Hawaii. He died there in 1990.

FURTHER INVESTIGATION

The President of the Philippines has said he would push forward a bill to allow contraceptive devices to be used in the country, even if it meant excommunication from the Catholic Church. To learn more: http://blogs.reuters.com/faith -world/2011/05/10/philippine-catholic-bishops-clash-with -aquino-over-contraception-bill/

Taking on the formidable job of president was Corazon Aquino, the widow of murdered opposition leader Benigno Aquino. Aquino's People Power revolution had a heady beginning. Many observers believed that at last Filipinos had found a democratic leader around whom they could unite and who would end corruption and put the persistent Communist insurgency to rest. Aquino, however, was immediately beset by overwhelming economic, social, and political problems.

Opportunists and factions of the Filipino military and political elite still loyal to Marcos attempted numerous coups d'état in the years of Aquino's administration. Much of the unrest came from within the military, which had become accustomed to direct involvement in government during Marcos's martial-law era. A few Communist separatists, lured by Aquino's overtures of peace, turned in their arms, but most continued to plot violence against the government. Thus, the sense of security and stability that Filipinos needed in order to attract more substantial foreign investment and to reestablish the habits of democracy continued to elude them.

During the Aquino years, the anemic economy showed tentative signs of improvement. Some countries, particularly Japan and the United States and, more recently, Hong Kong, invested heavily in the Philippines, as did half a dozen international organizations. In fact, some groups complained that further investment was unwarranted, because already-allocated funds had not yet been fully utilized. Moreover, misuse of funds entrusted to the government—a serious problem during the Marcos era—continued, despite Aquino's promise to eradicate corruption.

A 1987 law, enacted after Corazon Aquino assumed the presidency, limited the president to one term in office. Half a dozen contenders vied for the presidency in 1992, including Imelda Marcos and other relatives of former presidents Marcos and Aquino. U.S. West Point graduate General Fidel Ramos, who had thwarted several coup attempts against Aquino and who thus had her endorsement, won the election. It was the first peaceful transfer of power in more than 25 years (although campaign violence claimed the lives of more than 80 people).

Six years later, in the 1998 presidential campaign, 83 candidates filed with the election commission, including Imelda Marcos. Despite the deaths of nearly 30 people during the campaign and the occurrence of some bizarre moments, such as when a mayoral candidate launched a mortar attack on his opponents, the election was the most orderly in years. Former movie star Joseph Estrada won by a landslide. However, when he attempted to repeal the law limiting presidents to one six-year term, some 100,000 people took to the streets in protest.

Estrada had to handle such problems as the Mayon Volcano eruption, which forced 66,000 people from their homes, as well as the sometimes violent dispute over ownership of the Spratly Islands in the South China Sea. During the Estrada presidency, the Philippine Navy shot at and sank Chinese fishing vessels that were operating near the Islands. The Philippines demanded that China remove a pier that it had built on Mischief Reef. The dispute was not resolved until 2002, when (under Estrada's successor) the half-dozen claimants to the Spratlys reached agreement on a code of conduct for use of the resources in and under the sea. The Philippines is sure to watch the islands carefully, however, given that China appears to be building up its military capabilities there.

In 2001, growing evidence of "economic plunder" by the once-adored actor-turned-president led to popular demonstrations to remove Estrada from office. Investigators believe that Estrada siphoned away as much as $300 million in government money, hiding the funds in banks under various aliases. An attempt at impeachment failed, but public pressure drove Estrada from office. He was replaced by his vice-president, Gloria Macapagal-Arroyo, who filled out the remaining years of Estrada's term and then decided to stand for the presidency in her own right in 2004, which she won by narrow margins. Arroyo wanted to adopt a parliamentary system for the government, but the plan was rejected by the Supreme Court. Arroyo's influence was also weakened in 2006 when she was accused of vote-rigging and corruption by the opposition—the second such impeachment charge against her. Even former president Aquino urged her to resign. The House dismissed the charges, but Arroyo's hold on office always appeared to be tenuous. For instance, she declared a state of emergency when she learned that an army general and some members of the opposition were plotting a coup against her. Faced with the prospect of civil unrest if she tried to extend her stay in office by removing the term limits rule, Arroyo announced in 2009 that she would follow the constitution and step down when her term of office expired. In 2010, the son of Corazon Acquino, Benigno (NoyNoy) Acquino was elected President in a peaceful transfer of power. Macapagal-Arroyo, rather than retiring from politics, ran for office in the House of Representatives and won election, as did Imelda Marcos.

■ SOCIAL PROBLEMS

Much of the foreign capital coming into the Philippines in the 1990s was invested in stock and real-estate speculation rather than in agriculture or manufacturing. Thus, with the financial collapse of 1997, there was little of substance to fall back on. Even prior to the financial crisis, inflation had been above 8 percent (in 2011 about 4 percent) per year and unemployment was nearing 9 percent (in 2011 about 7 percent).

And one problem never seemed to go away: extreme social inequality. As in nearby Malaysia, where ethnic Malays have constituted a seemingly permanent class of poor peasants, Philippine society is fractured by distinct classes. Chinese and mestizos constitute the top of the hierarchy, while Muslims and most rural dwellers form the bottom. A little over a third of the Filipino population of 101 million make their living in agriculture and fishing; but even in Manila, where the economy is stronger than anywhere else, thousands of residents live in abject poverty as urban squatters. Officially, one third of Filipinos live below the poverty line. Disparities of wealth are striking. Worker discontent is so widespread that the Philippines often ranks near the top of Asian countries in days of work lost due to strikes and other protests. Economic leaders, once proud of the Philippines' economic strength, now see the country slipping further and further behind the rest of Asia. The per capita annual income is only US $3,500. China's, by comparison, is US $7,600.

Adding to the country's financial woes was the sudden loss of income from the six U.S. military bases that closed in 1991 and 1992. The government had wanted the United States to maintain a presence in the country, but in 1991, the Philippine Legislature, bowing to nationalist sentiment, refused to renew the land-lease agreements that had been in effect since 1947. Occupying many acres of valuable land and bringing as many as 40,000 Americans at one time into the Philippines, the bases had come to be seen as visible symbols of American colonialism or imperialism. But they had also been a huge boon to the economy. Subic Bay Naval Base alone had provided jobs for 32,000 Filipinos on base and, indirectly, to 200,000 more. Moreover, the United States paid nearly $390 million each year to lease the land and another $128 million for base-related expenses. Base-related monies entering the country amounted to 3 percent of the entire Philippines economy.

After the base closures, the U.S. Congress cut other aid to the Philippines, from $200 million in 1992 to $48 million in 1993. To counterbalance the losses, the Philippines accepted a $60 million loan from Taiwan to develop 740 acres of the former Subic Bay Naval Base into an industrial park. The International Monetary Fund also loaned the country $683 million—funds that have been successfully used to transform the former military facilities into commercial zones.

Leaders have made other efforts to revitalize the economy. For example, in 1999, the Legislature passed a law allowing more foreign investment in the retail sector; and, in the most surprising development, the government approved resumption of large-scale military exercises with U.S. forces in 2001 to speed the hunt for terrorists. Currently, some 4,000 U.S. troops are operating in the Philippines. U.S. troops joined with the Philippine military in 2006 to attack Abu Sayyat strongholds. In 2011, on the predominantly Muslim island of Basilan, dozens of Philippine soldiers engaged in a battle with Abu Sayyaf, ultimately freeing a businesswoman but leaving three rebels dead.

Adding to the economic stress is the constant devastation from natural disasters. In recent years, Filipinos have suffered the massive explosion of the Mayon

Volcano, which forced 66,000 people from their homes; Typhoon Nida which triggered landslides, destroyed houses, killed 19 people and left hundreds homeless; monsoon landslides and huge waves that killed 87 people in the eastern Philippines; a 6.4 magnitude earthquake in Manila; a storm and typhoon in the north that killed 650 people; and several other "acts of God." Some of the landslides appear to be the result of logging of the natural forests. One such case was that of mudslides on Leyte Island in 2006 that buried an entire village of 1,500 people under 115 feet of mud. Other villages were wiped out when tons of volcanic mud, loosened by rains from Typhoon Durian, cascaded down the slopes of the Mayon Volcano. The typhoons of 2009 that destroyed millions of tons of rice were followed in 2010 by a severe drought. Back-to-back typhoons and flooding in 2011 killed 59 people and left hundreds of villages stranded for days on their rooftops. During some typhoons, the rainfall is so heavy that an entire month of rain is dumped in just 12 hours or so. US$61 million of crops were damaged, the water supply was reduced, and hydroelectric dams were so low that there were electricity blackouts. The Philippines is already the worlds largest importer of rice, and the typhoons and drought greatly exacerbated the situation.

■ CULTURE

Philippine culture is a rich amalgam of Asian and European customs. Family life is valued, and few people have to spend their old age in nursing homes. Divorce is frowned upon. Women have traditionally involved themselves in the worlds of politics and business to a greater degree than have women in other Asian countries. Educational opportunities for women are about the same as those for men; adult literacy in the Philippines is estimated at 93 percent, and many people are bi- or trilingual, with well over 50 percent of the population said to understand English. Evidence of gender discrimination continues, however, in this Latin-based culture. For example, so many women are regularly groped by men in the tightly packed commuter rail cars in Manila that the rail line has established women-only coaches for their safety. Another serious problem is the inability of college-educated men and women to find employment befitting their skills. Discontent among these young workers continues to grow, as it does among the many rural and urban poor.

Nevertheless, many Filipinos take a rather relaxed attitude toward work and daily life. They enjoy hours of sports and folk dancing or spend their free time in conversation with neighbors and friends, with whom they construct patron/client relationships. In recent years, the growing nationalism has been expressed in the gradual replacement of the English language with Filipino, a version of the Malay-based Tagalog language.

Snapshot: PHILIPPINES

Summarized below is a quick look at the country with regard to its development, freedom, health/welfare, and achievements.

Development

The Philippines labors under more than US$63.75 billion in foreign debt. Returns on investment in development projects have been so slow that about half of the earnings from all exports have to be used to service the debt. Still, the successful development of the former U.S. military bases into industrial zones is a bright spot in the economy, as is an annual economic growth rate of about 7 percent.

Freedom

The Philippine Constitution is similar in many ways to that of the United States. However, under Marcos, both the substance and structure of democracy were ignored. Corazon Aquino, who came to office in a "people power" revolution against Marcos, attempted to adhere to democratic principles, as have her successors. The Communist Party was legalized in 1992. The *Da Vinci Code* movie was banned in 2006. In 2010, electronic voting machines were used for the first time.

Health/Welfare

Quality of life varies considerably between the city and the countryside. Except for the numerous urban squatters, city residents generally have better access to health care and education. Safe drinking water is now available to over 90% of the population. The gap between the upper-class elite and the poor is hugely pronounced, and growing. Controversy over government-funded access to contraceptives continues, with the Catholic Church blocking family planning bills in the Legislature. A new reproductive health bill was introduced in 2011.

Achievements

Filipino women often run businesses or hold important positions in government. Two of the presidents since independence have been women. Folk dancing is very popular, as is the *kundiman*, a unique blend of music and words found only in the Philippines.

Statistics

Geography

Area in Square Miles (Kilometers): 110,400 (300,000) (about the size of Arizona)

Capital (Population): Greater Manila (11.4 million)

Environmental Concerns: deforestation; air and water pollution; soil erosion; pollution of mangrove swamps; coral reef degradation

Geographical Features: mostly mountainous archipelago; coastal lowlands

Climate: tropical marine; monsoonal

People

Population

Total: 101,833,938

Annual Growth Rate: 1.9%

Rural/Urban Population Ratio: 51/49

Major Languages: Filipino (based on Tagalog); English; eight major dialects—Tagalog, Cebuano, Nocano, Hiligaynon or Ilonggo, Bicol, Waray, Pampango, Pangasinan

Ethnic Makeup: 28% Tagalog; 13% Cebuano; 9% Ilocano; 8% Bisaya/Binisaya; 7.5% Hiligaynon Ilonggo; 6% Bikol; 3.5% Waray; 25% other

Religions: 81% Roman Catholic; 5% Muslim; 3% Evangelical; 2% Iglesia ni Kristo; 2% Aglipayan; 4.5% other Christian; 2.5% none or other

Health

Life Expectancy at Birth: 69 years (male); 75 years (female)

Infant Mortality: 19/1,000 live births

Physicians Available: 1/654 people (2004)

HIV/AIDS Rate in Adults: Less than 0.1%

Education

Adult Literacy Rate: 93%

Compulsory (Ages): 7–12; free

Communication

Telephones: 6.8 million main lines (2010)

Mobile Phones: 92 million (2010)

Internet Users: 29.7 million (2010)

Internet Penetration (% of Pop.): 20%

Transportation

Roadways in Miles (Kilometers): 132,446 (213,151)

Railroads in Miles (Kilometers): 618 (995)

Usable Airfields: 254

Government

Type: republic

Independence Date: July 4, 1946 (from the United States)

Head of State/Government: President Benigno Aquino III is both head of state and head of government

Political Parties: Struggle of Filipino Democrats; Christian Muslim Democrats; Liberal Party; Nacionalista Party; Nationalist People's Coalition; People's Reform Party; Force of the Philippine Masses

Suffrage: universal at 18

Military

Military Expenditures (% of GDP): 0.9%

Current Disputes: internal conflicts with Moro insurgencies; territorial disputes with China, Malaysia, Taiwan, and Vietnam

Economy

Currency ($ U.S. Equivalent): 45.11 Philippine pesos = $1

Per Capita Income/GDP: $3,500/$351 billion

GDP Growth Rate: 7.3%

Inflation Rate: 3.8%

Unemployment Rate: 7.3%

Labor Force by Occupation: 52% services; 33% agriculture; 15% industry

Population Below Poverty Line: 33%

Natural Resources: timber; petroleum; nickel; cobalt; silver; gold; salt; copper

Agriculture: rice; coconuts; corn; sugarcane; fruit; animal products; fish

Industry: food processing; chemicals; textiles; pharmaceuticals; wood products; electronics assembly; petroleum refining; fishing

Exports: $51 billion (primary partners China, United States, Singapore, Japan)

Imports: $61 billion (primary partners Japan, China, United States, Singapore)

Suggested Websites

www.tourism.gov.ph/

www.philippines.hvu.nl/

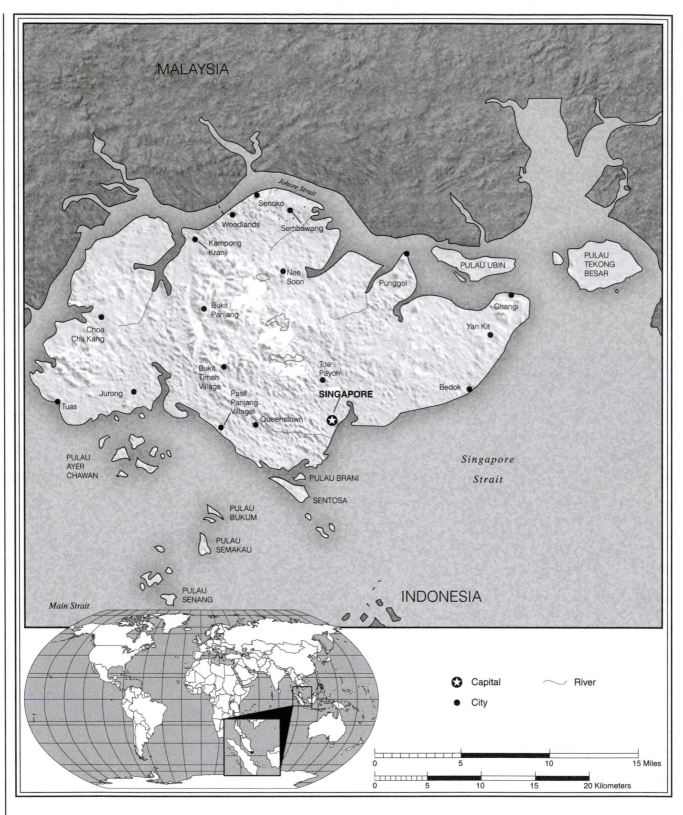

MALAYSIA

Johore Strait

Senoko

Woodlands

Sembawang

Kampong
Kranji

Nee
Soon

Punggol

PULAU UBIN

PULAU
TEKONG
BESAR

Bukit
Panjang

Choa
Chu Kang

Changi

Yan Kit

Bukit
Timah
Village

Toa
Payoh

SINGAPORE

Bedok

Jurong

Pasil
Panjang
Village

Tuas

Queenstown

Singapore

Strait

PULAU
AYER
CHAWAN

PULAU BRANI

SENTOSA

PULAU
BUKUM

PULAU
SEMAKAU

PULAU
SENANG

INDONESIA

Main Strait

★ Capital   〜 River

● City

0   5   10   15 Miles

0   5   10   15   20 Kilometers

Singapore
(Republic of Singapore)

It is often said that North Americans are well off because they inhabit a huge continent that abounds with natural resources. How then can one explain the phenomenal prosperity of land- and resource-poor Singapore? The inhabitants of this tiny, flat, humid, tropical island, located near the equator off the tip of the Malay Peninsula, have so few resources that they must import even their own drinking water. With only 250 square miles (648 sq. kilometers) of land (including 58 mostly uninhabited islets), Singapore is half the size of tiny Hong Kong, yet in recent years, it has had the highest per capita income in all of the Pacific Rim. Singapore's economic success has benefited everyone living there. For instance, with one of the highest population densities in the region, Singapore might be expected to have the horrific slums that characterize parts of other crowded urban areas. But almost 85 percent of its 4.7 million people live in spacious and well-equipped, albeit government-controlled, apartments.

Imperialism, geography, and racism help to explain Singapore's unique characteristics. For most of its recorded history, beginning in the thirteenth century A.D., Singapore was controlled variously by the rulers of Thailand, Java, Indonesia, and even India. In the early 1800s, the British were determined to wrest control of parts of Southeast Asia from the Dutch and expand their growing empire. Facilitating their imperialistic aims was Sir Stamford Raffles, a Malay-speaking British administrator who not only helped defeat the Dutch in Java but also diminished the power of local elites in order to fortify his position as lieutenant-governor.

Arriving in Singapore in 1819, Raffles found it to be a small, neglected settlement with an economy based on fishing. Yet he believed that the island's geographic location endowed it with great potential as a transshipment port. He established policies that facilitated its development just in time to benefit from the British exports of tin, rubber, and timber leaving Malaya. Perhaps most important was his declaration of Singapore as a free port. Skilled Chinese merchants and traders, escaping racist discrimination against them by Malays on the Malay Peninsula, flocked to Singapore, where they prospered in the free-trade atmosphere.

In 1924, the British began construction of a naval base on the island, the largest in Southeast Asia, which was nonetheless overcome by the Japanese in 1942. Returning in 1945, the British continued to build Singapore into a major maritime center. Today, oil supertankers from Saudi Arabia must exit the Indian Ocean through the Strait of Malacca and skirt Singapore to enter the South China Sea for deliveries to Japan and other Asian nations. Thus, Singapore has found itself in the enviable position of helping to refine and transship millions of barrels of Middle Eastern oil. Singapore's oil-refining capacities have been ranked the world's third largest since 1973.

Singapore's container port has now surpassed Rotterdam in the Netherlands as the busiest port in the world. It has become the largest shipbuilding and repair port in the region, as well as a major shipping-related financial center. During the 1990s, Singapore's economy grew at the astounding rates of between 6 and 12 percent a year, making it one of the fastest-growing economies in the world. The Asian financial crisis

TIMELINE

A.D. 1200–1400
Singapore is controlled by several different nearby nations, including Thailand, Java, India, and Indonesia

1800s
British take control of the island

1942
The Japanese capture Singapore

1945
The British return to Singapore

1959
Full elections and self-government; Lee Kuan Yew comes to power

1963
Singapore, now unofficially independent of Britain, briefly joins the Malaysia Federation

1965
Singapore becomes an independent republic

1980s
Singapore becomes the second-busiest port in the world and achieves one of the highest per capita incomes in the Pacific Rim

1990s
The U.S. Navy moves some of its operations from the Philippines to Singapore; the Asian financial crisis briefly slows Singapore's economic growth

2000s
Singapore tries to position itself to become a regional financial center by liberalizing foreign investment in the banking sector

The People's Action Party continues to win virtually all seats in Parliament of the late 1990s slashed growth to between 1.1 and 1.5 percent. The economy had started to rebound when the global economic meltdown occurred, returning Singapore to slow growth for a while. But the downturn was short-lived. By 2010, growth was an amazing 14.7 percent.

In recent years, the government has aggressively sought out investment from non-shipping–related industries in order to diversify the economy. For example, Singapore is investing half a billion dollars to build up the cyber-games-development industry, holding the World Cyber Games Asian Championships in 2006. Tourism is also being promoted, with some US$6 billion being added to the economy as a result of more hotel and resort development that attracted nearly 12 million tourists in 2010. Some 32 million people passed through Singapore's modern Changi Airport in 2005.

In 1992, Singapore hosted a summit of the Association of Southeast Asian Nations in which a decision was made to create a regional common market by the year 2008. In order to compete with the emerging European and North American regional trading blocs, it was decided that tariffs on products traded within the ASEAN region would be cut to 5 percent or less.

■ A UNIQUE CULTURE

Britain maintained an active interest in Singapore throughout its empire period. At its peak, some 100,000 British military men and their dependents were stationed on the island. The British military remained until 1971. (After the closure of U.S. military bases in the Philippines, logistics operations of the U.S. Navy's Seventh Fleet were transferred to Singapore, thereby increasing the number of U.S. military personnel in Singapore to about 300 persons.) Thus, British culture, from the architecture of the buildings, to the leisure of a cricket match, to the prevalence of the English language, is everywhere present in Singapore. Yet, because of the heterogeneity of the population (77 percent Chinese, 14 percent Malay, and 7 percent Indian), Singapore accommodates many philosophies and belief systems, including Confucianism, Buddhism, Islam, Hinduism, and Christianity. In recent years, the government has attempted to promote the Confucian ethic of hard work and respect for law, as well as the Mandarin Chinese language, in order to develop a greater Asian consciousness among the people. But most Singaporeans seem content to avoid extreme ideology in favor of pragmatism; they prefer to believe in whatever approach works—that is, whatever allows them to make money and have a higher standard of living.

Their great material success has come with a price. The government keeps a firm hand on the people. For example, citizens can be fined as much as $250 for dropping a candy wrapper on the street or for driving without a seat belt. Worse offenses, such as importing or selling chewing gum, carry fines of $6,000 and $1,200 respectively. The chewing gum restriction—originally implemented because the former prime minister did not like discarded gum fouling sidewalks and train station platforms—was softened a little in 2004 when the government decided to allow gum to be imported for "medicinal" reasons. Thus, gum to help nicotine addicts break the smoking habit is now allowed, as are "dental" gum products. But there is still a catch: gum buyers must submit their names and government identification card numbers to the pharmacies (the only place gum can be sold) before they can buy a stick of gum, and if the pharmacists fail to record that information, they could be jailed for up to two years and fined nearly $3000. If gum buyers are treated with such Draconian measures, what about those who commit more serious crimes? Death by hanging is the punishment for murder, drug trafficking, and kidnapping; while lashing is inflicted on attempted murderers, robbers, rapists, and vandals. Being struck with a cane is the punishment for crimes such as malicious damage, as an American teenager, in a case that became a brief international cause célèbre in 1994, found out when he allegedly sprayed graffiti on cars in Singapore. Later that year, a Dutch businessperson was executed for alleged possession of heroin. The death penalty is required when one is convicted of using a gun in Singapore. The United Nations regularly cites Singapore for a variety of human-rights violations, and the world press frequently makes fun of the Singapore government for such practices as giving prizes for the cleanest public toilet and for the incongruity of some laws. For example, a person has to register to buy a pack of nicotine gum but not to buy a pack of cigarettes or to visit a prostitute (which practice is legal in parts of Singapore).

If restrictions are the norm for ordinary citizens, they are even more severe for politicians, especially those who challenge the government. In recent years, opposition politicians have been fined and sent to jail for speaking in public without police permission, or for selling political pamphlets on the street without permission. Such actions can potentially bar would-be politicians from running for office for up to five years. Foreign media coverage of opposition parties is allowed only under strict guidelines, and in 1998, Parliament banned all political advertising on television. Government leaders argue that order and hard work are necessities since, being a tiny island, Singapore could easily be overtaken by the envious and more politically unstable countries nearby; with few natural resources, Singapore must instead develop its people into disciplined, educated workers. Few deny that Singapore is an amazingly clean and efficient city-state; yet in recent years, younger residents have begun to clamor for more flexibility in their social lives—and their desires are producing changes in Singapore's image. Beginning in 2003, the Singapore government began allowing dance clubs to stay open all night, and young people now find they can get away with such things as body piercings, tattoos, and even tabletop dancing in clubs. Yet, government influence in people's private lives remains substantial; Singaporeans are regularly reminded via government-sponsored advertising campaigns to wash their hands, smile, speak properly, and never litter.

The law-and-order tone exists largely because after its separation from Malaysia in 1965, Singapore was controlled by one man and his personal hard-work ethic, Prime Minister Lee Kuan Yew, along with his Political Action Party (PAP). He remained in office for some 25 years, resigning in 1990, but his continuing role as "minister-mentor" gives him considerable clout. In 2000 and 2005, for example, he was able to engineer the selection process for president in such a way that no election took place at all; his personal choice, S. R. Nathan, was simply appointed by the election board. Similarly, Lee saw to it that his personal preference for prime minister, Goh Chok Tong, would, in fact, become his chosen successor. Goh had been the deputy prime minister and had been the designated successor-in-waiting since 1984. The transition was smooth, and the PAP's hold on the government remained intact. The strength of Lee's continuing influence on the government was revealed again when Prime Minister Goh announced that he would give up his seat to his deputy prime minister who was none other than Lee Hsin Loong, the son of Lee Kuan Yew. The transfer happened in 2004. Parliament is composed of 84 elected seats (and 9 appointed seats), and the People's Action Party (PAP) has continuously controlled the majority of seats. However, in 2011, it lost ground to the opposition parties, reducing its hold to 60 percent of the seats, with the Worker's Party gaining the attention of the voters. In the past, so few seats were contested that the PAP was automatically declared the winner without actually having an election. But in 2011, 94 percent of the seats were contested by various opposition parties. In 2011, the largely ceremonial post of President was also contested for the first time since 1993. In recent years,

some restrictions that hampered opposition parties have been removed. For example, despite its policy of "selective liberalization," the prime minister has now allowed the use of political podcasting and videocasting and opened a park for public protests. Still, foreigners who live in Singapore and provide needed talent for the high-tech economy, find the restrictions on human rights to be annoying at best; some of them leave for countries with fewer roadblocks to expression.

The PAP originally came into prominence in 1959, when the issue of the day was whether Singapore should join the proposed Federation of Malaysia. Singapore joined Malaysia in 1963, but serious differences persuaded Singaporeans to declare independence two years later. Lee Kuan Yew, a Cambridge-educated, ardent anti-Communist with old roots in Singapore, gained such strong support as prime minister that not a single opposition-party member was elected for more than 20 years. The opposition is comprised of 22 parties but holds only 2 seats.

The two main goals of the administration have been to fully utilize Singapore's primary resource—its deep-water port—and to develop a strong Singaporean identity. The first goal has been achieved in a way that few would have thought possible; the question of national identity, however, continues to be problematic. Creating a Singaporean identity has been difficult because of the heterogeneity of the population, a situation that is only likely to increase as foreign workers are imported to fill gaps in the labor supply resulting from a very successful birth-control campaign started in the 1960s. Identity formation has also been difficult because of Singapore's seesaw history in modern times. First, Singapore was a colony of Britain, then it became an outpost of the Japanese empire, followed by a return to Britain. Next Malaysia drew Singapore into its fold, and finally, in 1965, Singapore became independent. All these changes transpired within the lifetime of many contemporary Singaporeans, so their confusion regarding national identity is understandable. They are aware that English as an international language connects them with the rest of the world; about half of Singaporeans speak English fluently.

Many still have a sense that their existence as a nation is tenuous, and they look for opportunities to build stabilizing networks. In 1996, Singapore reaffirmed its support for a five-nation defense agreement among itself, Australia, Malaysia, New Zealand, and Great Britain. It also strengthened its economic agreements with Australia. But the most controversial governmental decision was that of sending troops to support the U.S.-led

FURTHER INVESTIGATION

To learn why it is illegal to chew gum in Singapore, see: www.wisegeek.com/what-is-the-penalty-for-chewing-gum-in-singapore.htm

DID YOU KNOW?

Singapore has the highest percentage of millionaires of any country in the world. One in six households have assets over US$1 million.

Snapshot: SINGAPORE

Summarized below is a quick look at the country with regard to its development, freedom, health/welfare, and achievements.

Development

Development of the deepwater Port of Singapore has been so successful that at any single time, 400 ships are in port, making it the busiest container port in the world. (Shanghai, China is number two). Singapore has also become a base for fleets engaged in offshore oil exploration and a major financial center, the "Switzerland of Southeast Asia." Singapore has key attributes of a developed country. Singapore has the third-largest oil refinery in the world, and claims to have the most efficient airport—Changi Airport—in Asia. Based on tonnage, Singapore is one of the busiest ports in the world, with 28 million units of cargo imported or exported each year.

Freedom

Under former Prime Minister Lee Kuan Yew, Singaporeans had to adjust to a strict regimen of behavior involving both political and personal freedoms. Citizens want more freedoms but also realize that law and order have helped produce their high quality of life. Political-opposition voices have largely been silenced since 1968, when the People's Action Party captured all the seats in the Parliament. In May 2011, 5 seats were won by opposition parties, the most ever, but still small considering there are 22 opposition parties.

Health/Welfare

About 85 percent of Singaporeans live in government-built dwellings. A government-created pension fund, the Central Provident Fund, takes up to one quarter of workers' paychecks; some of this goes into a compulsory savings account that can be used to finance the purchase of a residence. Other forms of social welfare are not condoned. Care of the elderly is the duty of the family, not the government.

Achievements

Housing remains a serious problem for many Asian countries, but virtually every Singaporean has access to adequate housing. Replacing swamplands with industrial parks has helped to lessen Singapore's reliance on its deepwater port. Singapore successfully overcame a Communist challenge in the 1950s to become a solid home for free enterprise in the region.

---

invasion of Iraq. Singapore's most recent action in this regard was to send a troop landing ship with a crew of 180 to the Persian Gulf to conduct patrols and provide logistics support.

In Asia, Singapore has always admired the success of Japan. In 2006, the Singaporean ambassador pointed out that Japan had always been the "spark plug" of economic growth in Southeast Asia, and he looked forward to the day when the spark would return. Culturally too, Singapore has found it valuable to implement some of the methods used in Japanese higher education, and former prime minister Lee was unabashed in his admiration for the Japanese work ethic. The stunning economic success of Japan, despite its limited resources, presents a logical model for the resource-poor Singapore.

Statistics

Geography

Area in Square Miles (Kilometers): 269 (697) (about 3 1/2 times the size of Washington, D.C.)
Capital (Population): Singapore (4.2 million)
Environmental Concerns: air and industrial pollution; limited fresh water; waste-disposal problems
Geographical Features: lowlands; central plateau; focal point for sea routes; many small islands
Climate: tropical; hot, humid, rainy; two monsoon seasons

People

Population

Total: 4,740,737
Annual Growth Rate: 0.82%
Rural/Urban Population (Ratio): 100% urban
Major Languages: Mandarin Chinese; English; Malay; Tamil; Hokkien; other Chinese dialects
Ethnic Makeup: 77% Chinese; 14% Malay; 8% Indian; 1% others

Religions: 42.5% Buddhist; 15% Muslim; 8.5% Taoist; 15% Christian; 4% Hindu; 15% none or other

Health

Life Expectancy at Birth: 79.5 years (male); 85 years (female)
Infant Mortality: 2/1,000 live births
Physicians Available: 1/546 people (2009)
HIV/AIDS Rate in Adults: 0.1%

Education

Adult Literacy Rate: 93%

Communication

Telephones: 1.85 million (2009)
Mobile Phones: 6.65 million (2009)
Internet Users: 3.66 million (2010)
Internet Penetration (% of Pop.): 77%

Transportation

Roadways in Miles (Kilometers): 2,085 (3,356)
Railroads in Miles (Kilometers): 23 (38)

Usable Airfields: 8

Government

Type: parliamentary republic

Independence Date: August 9, 1965 (from Malaysian Federation)

Head of State/Government: President Tony Tan Keng Yam; Prime Minister Lee Hsien Loong

Political Parties: National Solidarity Party; People's Action Party; Reform Party; Singapore Democratic Alliance (includes Singapore People's Party and Singapore Justice Party); Singapore Democratic Party; Worker's Party

Suffrage: universal and compulsory at 21

Military

Military Expenditures (% of GDP): 4.9%

Current Disputes: boundary, bridge construction, and fresh water delivery disputes with Malaysia; boundary disputes with Indonesia; piracy in the Malacca Strait

Economy

Currency ($ U.S. Equivalent): 1.37 Singapore dollars = $1

Per Capita Income/GDP: $62,100/$292 billion

GDP Growth Rate: 14.5%

Inflation Rate: 2.8%

Unemployment Rate: 2.2%

Labor Force by Occupation: 70% services; 30% industry; 0.1% agriculture

Natural Resources: fish; deepwater ports

Agriculture: orchids; vegetables; poultry; eggs; fish; ornamental fish

Industry: electronics; chemicals; financial services; oil drilling equipment; petroleum refining; rubber processing and products; processed food and beverages; ship repair; offshore platform construction; life sciences; commercial center trade

Exports: $358 billion (primary partners Malaysia, Hong Kong; China; Indonesia)

Imports: $310 billion (primary partners Malaysia; United States; China; Japan)

Suggested Websites

www.gov.sg/

www.yoursingapore.com

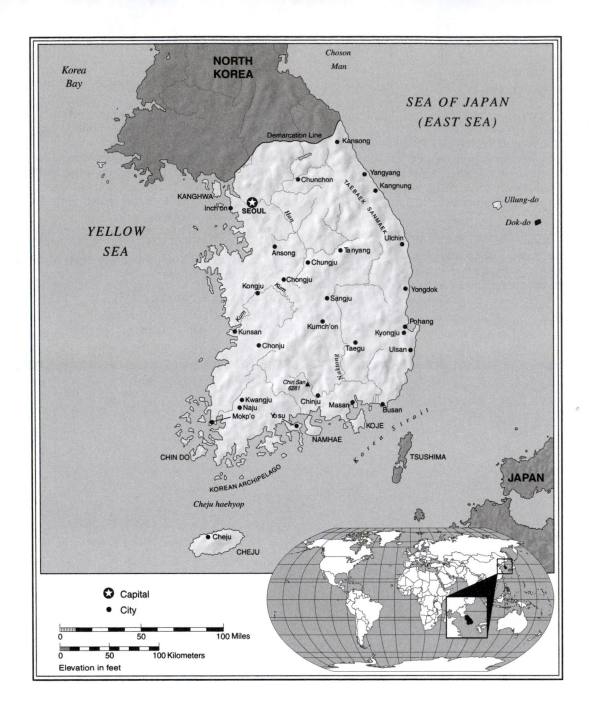

Korea
Bay

**NORTH
KOREA**

*Choson
Man*

*SEA OF JAPAN
(EAST SEA)*

Demarcation Line

● Kansong

● Chunchon

● Yangyang
● Kangnung

Ullung-do

KANGHWA

Dok-do ◗

Inch'on ● ⊛ SEOUL

● Ulchin

*YELLOW
SEA*

● Ansong
● Tanyang

● Chungju

● Chongju

● Kongju

● Sangju

● Yongdok

● Kunsan

● Kumch'on

● Pohang

● Kyongju

● Chonju

● Taegu

● Ulsan

*Chiri San
6281* ▲

● Kwangju
● Naju

● Chinju

● Masan

● Busan

Mokp'o ●

Yo su

KOJE

Korea Strait

NAMHAE

CHIN DO

TSUSHIMA

JAPAN

KOREAN ARCHIPELAGO

Cheju haehyop

● Cheju

CHEJU

⊛ Capital

● City

0   50   100 Miles

0   50   100 Kilometers

Elevation in feet

162

South Korea

(Republic of Korea)

Korea was inhabited for thousands of years by an early people who may or may not have been related to the Ainus of northern Japan, the inhabitants of Sakhalin Island, and the Siberian Eskimos. Distinct from this early civilization are today's Koreans whose ancestors migrated to the Korean Peninsula from Central Asia and, benefiting from close contact with the culture of China, established prosperous kingdoms as early as 1000 B.C.(legends put the date as early as 2333 B.C.). By A.D. 935, the Koryo Dynasty had established itself in Korea (the word *Korea* comes from the Koryo Dynasty). It was famous for, among other things, the invention of the world's first movable type printer.

The era of King Sejong (1418–1450) of the subsequent Choson or Yi Dynasty, is also notable for its many scientific and humanistic accomplishments. Ruling his subjects according to neo-Confucian thought, Sejong taught improved agricultural methods; published books on astronomy, history, religion, and medicine; and was instrumental in the invention of sundials, rain gauges to improve farming, and various musical instruments. Of singular importance was his invention of *hangul,* a simplified writing system that even uneducated peasants could easily learn. Before hangul, Koreans had to use the more complicated Chinese characters to represent sounds in the Korean language. Today, hangul is used in both North and South Korea, although some Chinese characters remain in use in newspapers and in family names.

■ REGIONAL RELATIONS

For most of its long history, Korea has remained at least nominally independent of foreign powers. China, however, always wielded tremendous cultural influence and at times politically dominated the Korean Peninsula. Likewise, Japan frequently cast longing eyes toward Korea and controlled it directly for 40 years from 1905 to 1945.

Korean influence on Japanese culture was pronounced in the 1400s and 1500s, when, through peaceful trade as well as forced labor, Korean artisans and technicians taught advanced skills in ceramics, textiles, painting, and other arts to the Japanese. (Historically, the Japanese received most of their cultural influence from China via Korea.)

In this century, the direction of influence reversed—to the current Japan-to-Korea flow—with the result that the two cultures share numerous qualities. Ironically, cultural closeness has not eradicated emotional distance: Some Japanese people today discriminate against Koreans who reside in Japan, while Japanese brutality during the four decades of Japan's occupation of Korea remains a frequent topic of conversation among Koreans. The Japanese emperor Akihito appears to want warmer ties between the two nations. To the surprise of many, in 2002 he announced that one of his royal ancestors was, in fact, a Korean.

Japan achieved its desire to rule Korea in 1905, when Russia's military, along with its imperialistic designs on Korea, was soundly defeated by the Japanese in the Russo–Japanese War; Korea was granted to Japan as part of the peace settlement. Unlike other expansionist nations, Japan

TIMELINE

A.D. 935
The Yi dynasty begins a 518-year reign over Korea

1637
Korea pays tribute to Mongol rulers in China

1876
Korea opens its ports to outside trade

1910
Japan formally annexes Korea at the end of the Russo–Japanese War

1945
Korea is divided into North and South

1950
North Korea invades South Korea: the Korean War begins

1953
Cease-fire agreement; the DMZ is established

1980s
Democratization movement; the 1988 Summer Olympic Games are held in Seoul

1990s
Reunification talks; a nonaggression pact is signed with North Korea; cross-border exchanges begin; the economy suffers a major setback

2000s
The first ever summit between South and North Korea in 2000 holds promise for improved peace on the peninsula

South and North Korean athletes march together in the 2000 Summer Olympics in Sydney

South Korea and Japan, often bitter rivals, successfully co-hosted the World Soccer Cup in both countries

The workweek is officially reduced from 6 to 5 days in 2003

South and North Korea agree to build an industrial park in Kaesong, North Korea

Liberal Roh Moo-hyun is re-elected president in 2004

Diplomat Ban Ki-moon is selected Secretary-General of the United Nations

Businessman Lee Myung-Bak is elected president

The economy is so strong that South Korea becomes a donor nation, sending US$1 billion in aid to poor countries

The 2018 Winter Olympics

The 2018 Winter Olympics will be held at a new resort village in Pyeongchang county, about 3 hours from Seoul. The Alpensia resort is situated at only 6,000 feet (1,800 meters) above sea level, and the snow is not nearly as plentiful as higher locations would be in other places. Indeed, the mountains are not particularly Alps-like, the snow is usually wet and not powdery, and some wonder if the Olympics will really work at such a place. There is much to do before 2018: there is no bobsled course, no media village, and no high-speed train (yet) to take people from venue to venue. But South Koreans know what to do, having hosted the 1988 summer games in Seoul.

was not content to rule Korea as a colony but, rather, attempted the complete cultural and political annexation of the Korean Peninsula. Koreans had to adopt Japanese names, serve in the Japanese Army, and pay homage to the Japanese emperor. (The current President, Lee Myung-Bak, was born in Osaka Japan and originally had a Japanese name). Some 1.3 million Koreans were forcibly sent to Japan to work in coal mines or to serve in the military. The Korean language ceased to be taught in school, and more than 200,000 books on Korean history and culture were burned.

Many Koreans joined clandestine resistance organizations. In 1919, a "Declaration of Korean Independence" was announced in Seoul by resistance leaders, but the brutally efficient Japanese police and military crushed the movement. They killed thousands of demonstrators, tortured and executed the leaders, and set fire to the homes of those suspected of cooperating with the movement. Despite suppression of this kind throughout the 40 years of Japanese colonial rule, a provisional government was established by the resistance in 1919, with branches in Korea, China, and Russia. However, a very large police force—one Japanese for every 40 Koreans—kept active resistance inside Korea in check.

One resistance leader, Syngman Rhee, vigorously promoted the cause of Korean independence to government leaders in the United States and Europe. Rhee, supported by the United States, became the president of South Korea after the defeat of the Japanese in 1945.

Upon the surrender of Japan, the victorious Allied nations decided to divide Korea into two zones of temporary occupation for the purposes of overseeing the orderly dismantling of Japanese rule and establishing a new Korean government. The United States was to occupy all of Korea south of the 38th Parallel of latitude (a demarcation running east and west across the peninsula, north of the capital city of Seoul), while the Soviet Union was to occupy Korea north of that line. The United States was uneasy about permitting the Soviets to move troops into Korea, as the Soviet Union had entered the war against Japan just eight days before Japan surrendered, and its commitment to the democratic intentions

of the Allies was questionable. Nevertheless, it was granted occupation rights.

Later, the United Nations attempted to enter the zone occupied by the Soviet Union in order to oversee democratic elections for all of Korea. Denied entry, UN advisers proceeded with elections in the South, which brought Syngman Rhee to the presidency. The North created its own government, with Kim Il-Song at the head. Tensions between the two governments resulted in the elimination of trade and other contacts across the new border. This was difficult for each side, because the Japanese had developed industries in the North while the South had remained primarily agricultural. Each side needed the other's resources; in their absence, considerable civil unrest occurred. Rhee's government responded by suppressing dissent, rigging elections, and using strong-arm tactics on critics. Rhee was forced to resign in 1960. Autocratic rule, not unlike that imposed on Koreans by the colonial Japanese, remained the norm in South Korea for several decades after the Japanese withdrew. Citizens, particularly university students, have frequently taken to the streets—often risking their lives and safety—to protest human rights violations by the various South Korean governments. Even more draconian measures were instituted by the Communist government in the North, meaning that despite six decades as independent nations, the repressive legacy of the Japanese police state remains firmly in place in the North, and only began to give way to democracy in the South in the late 1980s.

■ AN ECONOMIC POWERHOUSE

Upon the establishment of two separate political entities in Korea, the North pursued a Communist model of economic restructuring. South Korea, bolstered by massive infusions of economic and military aid from the United States, pursued a decidedly capitalist strategy. The result of these choices is dramatic, with South Korea having emerged as a powerhouse economy with a rapidly modernizing lifestyle, and the North stagnating under the weight of starvation and despair. Before the Asian financial crisis of the late 1990s, it had been predicted that South Korea's per capita income would rival some of the stronger European economies by the year 2010. The predictions have been revised downward, but as the 15th-largest economy in the world (based on GDP purchasing power parity), South Korea has long since eclipsed the smaller economies of Europe such as Belgium, Portugal, Finland, and others. Korea's success in improving the living standards of its people has been phenomenal. The current government's goal is to raise the per capita GDP to US$40,000 and make South Korea the seventh-largest economy in the world (the US per capita GDP in 2009 was US$48,000).

About 81 percent of South Korean people live in urban centers, where they have access to good education and jobs. Manufacturing accounts for about 30 percent of the gross domestic product. Economic success and recent improvements in the political climate seem to be slowing the rate of outward migration. In recent years, some Koreans have even returned home after spending years abroad.

Imbued with the Confucian work ethic, and following the Japanese business model, South Korean businesspeople work long hours (a six-day work week was the law until 2003, when it was reduced to five days) and have it as their goal to capture market share rather than to gain immediate profit—that is, they are willing to sell their products at or below cost for several years in order to gain the confidence of consumers, even if they make no profit. Once a sizable proportion of consumers buy their product and trust its reliability, the price is raised to a profitable level.

During the 1980s and much of the 1990s, South Korean businesses began investing in other countries, and South Korea became a creditor rather than a debtor member of the Asian Development Bank, putting it in a position to loan money to other countries. There was even talk that Japan (which is separated from Korea by only 150 miles of ocean) was worried that the two Koreas would soon unify and thus present an even more formidable challenge to its own economy—a situation not unlike some Europeans' concern about the economic strength of a reunified Germany.

The magic ended, however, in late 1997, when the world financial community would no longer provide money to Korean banks. This happened because Korean banks had been making questionable loans to Korean *chaebol,* or business conglomerates, for so long, that the banks' creditworthiness came into question. With companies unable to get loans, with stocks at an 11-year low, and with workers eager to take to the streets in mass demonstrations against industry cutbacks, many businesses went under. One of the more well-known firms that went into receivership in 1998 was Kia Motors Corporation. In 2000, the French automaker Renault bought the failed Samsung Motors company. With unemployment reaching 7 percent, the government applied for a financial bailout—with all of its restrictions and forced closures of unprofitable businesses—from the International Monetary Fund. Workers deeply resented the belt-tightening required by the IMF (shouting "No to layoffs!" thousands of them threw rocks at police, who responded with tear gas and arrests), but IMF funding probably prevented the entire economy from collapsing.

Despite these problems and the additional stress caused by the global business meltdown of 2008 and 2009, the South Korean economy remains incredibly strong. In fact, by 2010, South Korea had become the first former recipient of international aid to move into the ranks of the donor "class" of countries. Already sending US$1 billion in aid to poor countries, the government intends to triple that amount by 2015. When compared to North Korea, the differences are staggering. For the North, life is unbearable. Thousands of North Koreans have defected, some via a "safe house" system through China (there are large numbers of Koreans living on the China side of the border), which is similar to the famous "Underground Railroad" that allowed blacks in the United States to escape slavery in the 1860s. Some military pilots have flown their jets across the border to South Korea seeking freedom. Food shortages are increasingly evident, and some reports indicate that as many as 2,000 hungry North Koreans attempt to flee into China each month.

The South provides financial assistance to these refugees, but having thousands of poor North Koreans fleeing to the South is not tenable in the long run. A better solution is being attempted just one hour north of Seoul, across the border in North Korea. There, the Kaesong Industrial Park has been opened (2005). Some 250 South Korean companies have been given permission to operate manufacturing and assembly plants there, taking advantage of low labor costs. Eventually, over 1,000 companies are expected to participate, thus giving employment to some 700,000 North Koreans. The park will help the economies of both countries, and it represents a major breakthrough in the generally chilly relations between North and South. In 2009, however, the North suddenly demanded huge pay increases for its workers and unilaterally abrogated the wage and rental agreements it had made with South Korean companies.

South Korea is using some of its economic strength to make major infrastructure improvements. It will soon start construction of a rapid transit rail system, the Korea Train Express, and it is developing a plan to restore polluted rivers (part of a plan called the Green New Deal) and reclaim an additional 10 percent of rainwater. In addition to the 2018 Winter Games, the country will also host the 2012 World Expo and the 2014 Incheon Asian Games.

SOCIAL PROBLEMS

Despite some recent setbacks, the South Korean economic recovery is well on its way, and South Korea will continue to be an impressive showcase for the fruits of capitalism. Politically, however, the country has been wracked with problems. Under Presidents Syngman Rhee (1948–1960), Park Chung Hee (1964–1979), and Chun Doo Hwan (1981–1987), South Korean government was so centralized as to constitute a virtual dictatorship. Human-rights violations, suppression of workers, and other acts incompatible with the tenets of democracy were frequent occurrences. Student uprisings, military revolutions, and political assassinations became more influential than the ballot box in forcing a change of government. Ex-general Noh Tae-woo took power in 1987, but mass protests forced him to agree to democratic elections, which he won, thus making him the first democratically elected leader in years. The 1992 elections, also accompanied by mass demonstrations, brought to office the first civilian president in 30 years, Kim Young-sam. Kim was once a dissident himself and was victimized by government policies against free speech; once elected, he promised to make major democratic reforms. The reforms, however, were not good enough for thousands of striking subway workers, farmers, or students whose demonstrations against low pay, foreign rice imports, or the deployment of U.S. Patriot missiles in South Korea sometimes had to be broken up by riot police.

Replacing Kim Young-sam as president in 1998 was opposition leader Kim Dae Jung. Like his predecessor, Kim had once been a political prisoner. Convicted of sedition by a corrupt government and sentenced to die, Kim had spent 13 years in prison or house arrest, and then, like Nelson Mandela of South Africa, rose to defeat the system that had abused him. That the Korean people were ready for real democratic reform was also revealed in the 1996 trial of former president Chun Doo Hwan, who was sentenced to death (later commuted) for his role in a 1979 coup. In 2003, a former human-rights lawyer, Roh Moo-hyun, occupied the Blue House (the presidential residence). Promising to continue Kim's "sunshine policy" toward North Korea while establishing a more equal relationship with the United States, Roh won the support of many younger voters.

But Roh, a liberal with a labor-friendly platform, found himself outnumbered in the National Assembly by conservatives who tried to have him impeached over election law violations and who successfully overrode his veto of legislation. So intense was the acrimony that one Roh supporter set himself afire with gasoline just outside the Assembly building. Legislators got into fistfights on the floor of the Assembly during the debate over impeachment. The impeachment was dismissed after Roh's Uri Party was able to gain a slight majority in the one-chamber National Assembly in the 2004 elections, with many of those having voted for his impeachment losing their

seats. In 2008, conservative Lee Myung-Bak was elected president. A businessman, Lee has gained both supporters and detractors for his plans for a canal system, educational reforms stressing the English language, and openness to trade deals with the United States.

A primary focus of the South Korean government's attention at the moment is the several U.S. military bases in South Korea, currently home to approximately 40,000 U.S. troops. The government (and apparently most of the 48 million South Korean people), although not always happy with the military presence, believes that the U.S. forces are useful in deterring possible aggression from North Korea, which, despite an enfeebled economy, still invests massive amounts of its budget in its military. Many university students, however, are offended by the presence of these troops. They claim that the Americans have suppressed the growth of democracy by propping up authoritarian regimes—a claim readily admitted by the United States, which believed during the cold war era that the containment of communism was a higher priority. Strong feelings against U.S. involvement in South Korean affairs have precipitated hundreds of violent demonstrations, sometimes involving as many as 100,000 protesters. The United States' refusal to withdraw its forces from South Korea leaves the impression with many Koreans that Americans are hard-line, cold war ideologues who are unwilling to bend in the face of changing international alignments. This image was bolstered by U.S. president George W. Bush's harsh rhetoric toward North Korea and his decision to invade Iraq (which the South Korean government supported by sending several thousand troops, but which many Koreans strongly opposed). After denouncing North Korea for developing nuclear capabilities, the U.S. government was chagrined to learn that South Korea had done the same. The South, however, did not receive the criticism by the U.S. government that the North received.

In 1990, U.S. officials announced that in an effort to reduce its military costs in the post–cold war era, the United States would pull out several thousand of its troops from South Korea and close three of its five air bases. When the Iraq War was launched, the redeployment of troops from South Korea to Iraq was speeded up. The United States also declared that it expected South Korea to pay more of the cost of the U.S. military presence, in part as a way to reduce the unfavorable trade balance between the two countries. The South Korean government agreed to build a new U.S. military base about 50 miles south of the capital city of Seoul, where current operations would be relocated. South Korea would pay all construction costs—estimated at about $1 billion—and the United States would be able to reduce its presence within the Seoul metropolitan area, where many of the anti-U.S. demonstrations take place. Despite the effort to minimize the "footprint" of the U.S. military in Seoul, anti-American sentiment seems to be on the rise. When U.S. military courts acquitted American soldiers who had hit and killed two Korean teenagers in 2002, protests erupted again. Students broke into military installations, a white-robed Protestant minister led a march to the U.S. Embassy holding a wooden cross, and some 20 Catholic priests launched a hunger strike. Protestors also fought police in Pyeongtaek to prevent the razing of villages for construction of the new U.S. base. As is often the case, the most strident protests come from those members of South Korean society who are the most Westernized—Christians and students educated overseas.

The issue of the South's relationship with North Korea has occupied the attention of every government since the 1950s. The division of the Korean Peninsula left many families unable to visit or even communicate with relatives on the opposite side. Moreover, the threat of military incursion from the North forced South Korea to spend huge sums on defense. Both sides engaged in spying, counter-spying, and other forms of subversive activities. Most worrisome of all was that the two antagonists would not sign a peace treaty, meaning that, since the 1950s (!), they remain technically still at war. In the South, a unification ministry exists to further peaceful relations with the North, but the North often does not cooperate on proposed initiatives.

In 2000, just at the moment when the North was once again engaged in saber-rattling, an amazing breakthrough occurred: The two sides agreed to hold a summit in P'yongyang, the North Korean capital. President Kim, who had vigorously pushed his sunshine policy and had already achieved some improvements in communications between the two halves of the peninsula, was greeted with cheering crowds and feted at state dinners in the North. His counterpart, the reclusive Kim Jong-Il, seemed ready to make substantial concessions, including opening the border for family visits, halting the nonstop broadcasting of anti-South propaganda, and seeking a solution for long-term peace and reunification. The impetus behind the dramatic about-face appears to be North Korea's dire economic situation and its loss of solid diplomatic and economic partners now that the Communist bloc of nations no longer exists. Reunification will, of course, take many years to realize, and some estimate that it would cost close to a trillion dollars. Still, it is a goal that virtually every Korean wants. South Korea made the first installment toward this goal in 2002 when it agreed to give the North $25 million to rebuild rail and road links between the two countries—links that have been severed for more than 50 years.

However, for every step forward in the move toward reconciliation, there are usually several steps backward.

FURTHER INVESTIGATION

Despite years of animosity between Japan and South Korea, the two nations share much in common. To learn more about their close historical and cultural ties, see www.nytimes.com/2002/03/11/world/japan-rediscovers-its-korean-past.html

The most serious recent problems were the detonation of a nuclear bomb by the North in 2006 and the firing of a number of long-range missiles in 2008 and 2009. The South (and the rest of the world) condemned the explosion and the firings and immediately sought assurances from the United States that it would continue its nuclear umbrella protection policy. The United States quickly gave that assurance, but it also repeated its desire that the South take more of an active role in military leadership. The current plan is for the South to resume wartime command of its own armed forces by 2012 (currently, the United States has the right to command South Korean troops in the event of war).

South Korean government leaders have to face a very active, vocal, and even violent populace when they initiate controversial policies. Among the more vocal groups for democracy and human rights are the various Christian congregations and their Westernized clergy. Other vocal groups include the college students who hold rallies annually to elect student protest leaders and to plan anti-government demonstrations. In addition to the military-bases question, student protesters are angry at the South Korean government's willingness to open more Korean markets to U.S. products. The students want the United States to apologize for its alleged assistance to the South Korean government in violently suppressing an antigovernment demonstration in 1981 in Kwangju, a southern city that is a frequent locus of antigovernment as well as labor-related demonstrations and strikes. Protesters were particularly angered by then-president Noh Tae-woo's silencing of part of the opposition by convincing two opposition parties to merge with his own to form a large Democratic Liberal Party, not unlike that of the Liberal Democratic Party that governed Japan almost continuously for more than 40 years.

Ironically, demands for changes have increased at precisely the moment that the government has been instituting changes designed to strengthen both the economy and civil rights. Under Noh's administration, for example, trade and diplomatic initiatives were launched with Eastern/Central European nations and with China and the former Soviet Union. Under Kim Young-sam's administration, 41,000 prisoners, including some political prisoners, were granted amnesty, and the powerful chaebol business conglomerates were brought under a tighter rein. Similarly, relaxation of the tight controls on labor-union activity gave workers more leverage in negotiating with management. Unfortunately, union activity, exploding after decades of suppression, has produced crippling industrial strikes—as many as 2,400 a year—and the police have been called out to restore order. In fact, since 1980, riot police have fired an average of more than 500 tear-gas shells a day, at a cost to the South Korean government of tens of millions of dollars.

Domestic unrest notwithstanding, the South Korean people seemed to have put aside the era of dictatorial government and are moving proactively to solidify their role on the world stage. South Korea recently established

Snapshot: SOUTH KOREA

Summarized below is a quick look at the country with regard to its development, freedom, health/welfare, and achievements.

mass uprisings of the people. Democratic reforms have been enacted under Presidents Roh Tae-woo, Kim Young-Sam, Kim Dae Jung, and Roh Moo-hyun, and are likely to continue under Lee Myung-Bak, although Lee has been criticized for trying to control the media.

Development

The South Korean economy was so strong in the 1980s and early 1990s that many people thought Korea was going to be the next Japan of Asia. The standard of living was increasing for everyone until a major slowdown in 1997–1998. The resulting difficulties forced companies to abandon plans for wage increases or to decrease work hours. The Kaesong Industrial Park, one hour's travel from Seoul in North Korea, opened in 2005; by 2012 it could have up to 1,000 South Korean companies and 700,000 workers.

Freedom

Suppression of political dissent, manipulation of the electoral process, and restrictions on labor union activity have been features of almost every South Korean government from 1948 until the late 1980s. Martial law has been frequently invoked, and governments have been overthrown by

Health/Welfare

Korean men usually marry at about age 27, women at about 24. In 1960, Korean women, on average, gave birth to 6 children; in 2009, the average numbers of births per woman was 1.21. The average South Korean baby born today can expect to live to almost 80 years old. More than 90 percent of all infants are immunized.

Achievements

In 1992, Korean students placed first in international math and science tests. South Korea achieved self-sufficiency in agricultural fertilizers in the 1970s and continues to show growth in the production of grains and vegetables. The formerly weak industrial sector is now a strong component of the economy. In 2006, South Korean diplomat Ban Ki-moon was chosen as the secretary-general of the United Nations. South Korea will host the 2018 Winter Olympics.

---

unofficial diplomatic ties with Taiwan in order to facilitate freer trade. It then signed an industrial pact with China to merge South Korea's technological know-how with China's inexpensive labor force. Free-trade agreements are underway with Australia, China, and even Japan. Relations with Japan have improved somewhat since Japan "sort of" apologized for atrocities committed by its troops in Korea. But the refusal of Japan's Ministry of Education to fully acknowledge its aggressive role against Korea by rewriting its history textbooks for school students continues to anger South Koreans. Tensions over Dokdo (Takeshima) Island, which is claimed by both South Korea and Japan, intensified in 2006. Significant amounts of natural gas are thought to be under the ocean near the island (actually a small group of mostly uninhabited islands), so both countries want exclusive access.

■ KOREAN CULTURE

Koreans occupy a beautiful peninsula, and they have created a vibrant culture to match it. Mountains make

up about 70 percent of the Korean peninsula, and some 3,500 mountaintops poke out of the surrounding Yellow Sea and Sea of Japan (called the East Sea by Koreans), making Korea also a land of small islands. Monsoon rains bring water to the rice paddies and to the barley, potatoes, and soybean fields; Siberian winds bring snow to the mountains. Families are central to Korean life, and many rural families consist of 2 or even 3 generations. Households of city dwellers, who often live in modern high-rise apartment buildings, are usually smaller. Although Western foods such as hamburgers and fries are available everywhere in Seoul, Koreans are famous for the hot and spicy foods of their own making: *kimchi,* made from pickled cabbage, garlic, onions, oysters, and other ingredients, and *bulgogi,* a dish in which thin strips of meat are soaked in a mixture of sesame oil, garlic, onions, and black pepper, dipped in soy sauce and eaten using chopsticks. Some Korean cultural forms, such as *taekwondo,* have been exported all over the world.

Statistics

Geography

Area in Square Miles (Kilometers): 38,502 (99,720) (about the size of Indiana)

Capital (Population): Seoul (9.8 million)

Environmental Concerns: air and water pollution; overfishing

Geographical Features: mostly hills and mountains; wide coastal plains in west and south

Climate: temperate, with rainfall heaviest in summer

People
Population

Total: 48,754,657

Annual Growth Rate: 0.23%

Rural/Urban Population (Ratio): 17/83

Major Language: Korean; English

Ethnic Makeup: homogeneous Korean (except approximately 20,000 Chinese)

Religions: 26% Christian; 23% Buddhist; 49% none; 2% other or unknown

Health

Life Expectancy at Birth: 76 years (male); 82.5 years (female)

Infant Mortality: 4/1,000 live births

Physicians Available: 1/508 people

HIV/AIDS Rate in Adults: Less than 0.1%

Education

Adult Literacy Rate: 98%

Compulsory (Ages): 6–12; free

Communication

Telephones: 19.3 million (2009)

Mobile Phones: 48 million (2009)

Internet Users: 39.4 million (2010)

Internet Penetration (% of Pop.): 81%

Transportation

Roadways in Miles (Kilometers): 64,019 (103,029)

Railroads in Miles (Kilometers): 2,101 (3,381)

Usable Airfields: 116

Government

Type: republic

Independence Date: August 15, 1945 (from Japan)

Head of State/Government: President Lee Myung-bak; Prime Minister Kim Hwang-sik

Political Parties: Democratic Party; Democratic Labor Party; Pro-Park Alliance; Grand National Party; Liberty Forward Party; New Progressive Party; Renewal Korea Party

Suffrage: universal at 19

Military

Military Expenditures (% of GDP): 2.7%

Current Disputes: Demarcation Line and maritime boundary disputes with North Korea; Liancourt Rocks claim dispute with Japan

Economy

Currency ($ U.S. Equivalent): 1,154 won = $1

Per Capita Income/GDP: $30,000/$1.5 trillion

GDP Growth Rate: 6.1%

Inflation Rate: 3%

Unemployment Rate: 3.7%

Labor Force by Occupation: 69% services; 24% industry; 7% agriculture

Natural Resources: coal; tungsten; graphite; molybdenum; lead; hydropower

Agriculture: rice; root crops; barley; vegetables; fruit; livestock; milk; eggs; fish

Industry: electronics; telecommunications; automobile production; chemicals; shipbuilding; steel

Exports: $464 billion (primary partners China, United States, Japan)

Imports: $422 billion (primary partners China, Japan, United States)

Suggested Websites

www.korea.net/

www.english.tour2korea.com/

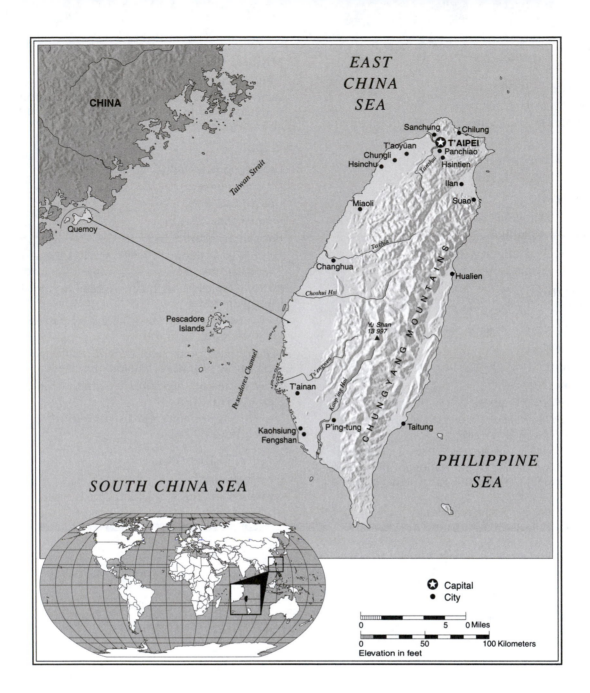

EAST CHINA SEA

CHINA

Taiwan Strait

Quemoy

Pescadore Islands

Pescadores Channel

Sanchung · Chilung
T'aoyüan ⭐ T'AIPEI
Chungli · Panchiao
Hsinchu Hsintien
Ilan ·
Miaoli Suao ·

Tanshui

Tachia

Changhua
Hualien ·
Choshui Hsi

Yu Shan
13 997 ▲

CHUNGYANG MOUNTAINS

Ts'engwen

Kaop'ing Hsi

T'ainan

P'ing-tung Taitung

Kaohsiung
Fengshan

PHILIPPINE SEA

SOUTH CHINA SEA

⭐ Capital
· City

| 0 | | 5 | 0 Miles |

| 0 | 50 | 100 Kilometers |
Elevation in feet

Taiwan

It has been called "beautiful island," "treasure island," and "terraced bay island," but to the people who have settled there, Taiwan (formerly known as Formosa) has come to mean "refugee island."

Typical of the earliest refugees of the island were the Hakka peoples of China, who, tired of persecution on the mainland, fled to Taiwan (and to Borneo) before A.D. 1000. In the seventeenth century, tens of thousands of Ming Chinese soldiers, defeated at the hands of the expanding Manchu Army, sought sanctuary in Taiwan. In 1949, a third major wave of immigration to Taiwan brought thousands of Chinese Nationalists, retreating in the face of the victorious Red Chinese armies. Hosting all these newcomers were the original inhabitants of the islands, various Malay-Polynesian-speaking tribes. Their descendants live today in mountain villages throughout the island, where some have been since 3000 B.C. These original inhabitants, generally grouped into about 10 tribes, often feel neglected or abused by modern Taiwanese society, and about 2,000 of them, dressed in traditional clothing, protested in Taipei in 2004 when the vice-president made statements about them they felt were untrue. They often resent the taking over of their island by the Chinese people.

Since 1544, other outsiders have shown interest in Taiwan, too: Portugal, Spain, the Netherlands, Britain, and France have all either settled colonies or engaged in trade along the coasts. But the non-Chinese power that has had the most influence is Japan. Japan treated parts of Taiwan as its own for 400 years before it officially acquired the entire island in 1895, at the end of the Sino-Japanese War. From then until 1945, the Japanese ruled Taiwan with the intent of fully integrating it into Japanese culture. The Japanese language was taught in schools, college students were sent to Japan for their education, and the Japanese style of government was implemented. Many Taiwanese resented the harsh discipline imposed, but they also recognized that the Japanese were building a modern, productive society. Indeed, the basic infrastructure of contemporary Taiwan—roads, railways, schools, and so on—was constructed during the Japanese colonial era (1895–1945). Japan still lays claim to the Senkaku Islands, a chain of uninhabited islands near Taiwan that the Taiwanese say belong to Taiwan.

In 1949, after Communist leader Mao Zedong defeated his rivals on the mainland, Taiwan became the island of refuge for the anti-Communist leader Chiang Kai-shek and his 3 million Kuomintang (KMT, or Nationalist Party) followers, many of whom had been prosperous and well-educated businesspeople and intellectuals in China. These Mandarin-speaking mainland Chinese, called Mainlanders, now constitute about 14 percent of Taiwan's people.

During the 1950s, Mao Zedong, the leader of the People's Republic of China, planned an invasion of Taiwan. However, Taiwan's leaders succeeded in obtaining military support from the United States to prevent the attack. They also convinced the United States to provide substantial amounts of foreign aid to Taiwan (the U.S. government saw the funds as a way to contain communism) as well as to grant it diplomatic recognition as the only legitimate government for all of China. For many years Taiwan

TIMELINE

A.D. 1544
Portuguese sailors are the first Europeans to visit Taiwan

1700s
Taiwan becomes part of the Chinese Empire

1895
The Sino-Japanese War ends; China cedes Taiwan to Japan

1945
Taiwan achieves independence from Japan

1947–49
Nationalists, under Chiang Kai-shek, retreat to Taiwan

1950s
A de facto separation of Taiwan from China; Chinese aggression is deterred with U.S. assistance

1971
China replaces Taiwan in the United Nations

1975
Chiang Kai-shek dies and is succeeded by his son, Chiang Ching-Kuo

1980s
The first two-party elections in Taiwan's history are held; 38 years of martial law end

1990s
Relations with China improve; the United States sells F-16 jets to Taiwan; China conducts military exercises to intimidate Taiwanese voters

2000s
Trade and communication with China continue to expand

The opposition Democratic Progressive Party wins the presidency

Pro-independence president Chen Shu-bian is re-elected in 2004, but anti-independence candidates gain enough strength to deny his party a majority in the legislature.

Taiwan is denied membership in the United Nations for the 15th time

The Kuomintang assumes control of the government once again

(© Getty Images RF)

The Taipei 101 building towering over Taipei. Taiwan has one of the highest population densities in the world.

maintained offices in its government for every province in China, even though no one could ever go there.

China was denied membership in the United Nations for more than 20 years because Taiwan held the "China seat." World opinion on the "two Chinas" issue began to change in the early 1970s. Many countries believed that a nation as large and powerful as the People's Republic of China should not be kept out of the United Nations, nor out of the mainstream of world trade, in favor of the much smaller Taiwan. In 1971, the United Nations withdrew its China seat from Taiwan and gave it to the P.R.C. Taiwan has consistently reapplied for

membership, arguing that there is nothing wrong with there being either two China seats or one China seat and one Taiwan seat; because of the vehement objection of China, requests have been denied each time for some 15 tries, most recently in 2007.

The United States and many other countries wished to establish diplomatic relations with China but could not get China to cooperate as long as they recognized the sovereignty of Taiwan. In 1979, desiring access to China's huge market, the United States, preceded by many other nations, switched its diplomatic recognition from Taiwan to China. Foreign-trade offices in Taiwan remained unchanged but embassies were renamed; the U.S. Embassy, for example, was called the American Institute in Taiwan. As far as official diplomacy with the United States was concerned, Taiwan became a non-nation, but that did not stop the two countries from engaging in very profitable trade, including a controversial U.S. agreement in 1992 to sell $4 billion to $6 billion worth of F-16 fighter jets to Taiwan. In 2011, Taiwan postponed the purchase of U.S.-made Black Hawk helicopters and Patriot missile systems. They were part of a US$6.4 billion military weapons package, a deal that has been strongly criticized by China. The postponement was not based on China's disapproval, but rather on budgetary shortfalls and manufacturing delays.

Taiwan has not been able to maintain diplomatic ties, yet continues to trade, with nations that recognize the mainland Chinese authorities as a legitimate government. In 1992, for instance, when South Korea established ties with mainland China, Taiwan immediately broke off formal relations with South Korea and suspended direct airline flights. However, Taiwan continued to permit trade in many commodities. Recognizing a potentially strong market in Vietnam, Taiwan also established air links with Vietnam in 1992, links that had been broken since the end of the Vietnam War. In 1993, a Taiwanese company collaborated with the Vietnamese government to construct a $242 million highway in Ho Chi Minh City (formerly Saigon). Only 23 countries formally recognize Taiwan today versus 171 for China, but Taiwan nevertheless maintains close economic ties with more than 140 countries. Of special interest to Taiwan are those countries with large populations of "overseas Chinese," or people of Chinese background who permanently live abroad. This community is estimated to number in excess of 33 million people, of whom 27 million live in the countries of Southeast Asia and 3.8 million live in the United States. Given that many of these overseas Chinese have become quite successful in business, Taiwan, which established a "Go South Policy" toward Southeast Asia, sees them as a source of potential investment as well as a way to bypass the diplomatic restrictions under which it suffers due to the United Nations'

FURTHER INVESTIGATION

The 2012 elections hold the possibility of Taiwan having its first female president. To learn more, see: www.taipeitimes.com/News/taiwan/archives/2011/11/19/2003518688

rejection of Taiwan as the singular, legitimate government of China. On this point, the rivalry between the mainland and Taiwan remains quite intense, but unlike the past, when Taiwanese were forbidden to openly discuss independence from China, the debate is now quite open.

■ AN ECONOMIC POWERHOUSE

Diplomatic maneuvering has not affected Taiwan's stunning postwar economic growth. Like Japan, Taiwan has been described as an economic miracle. In the past two decades, Taiwan has enjoyed more years of double-digit economic growth than any other nation. With electronics leading the pack of exports, a substantial portion of Taiwan's gross domestic product comes from manufacturing. Taiwan has been open to foreign investment and, of course, to foreign trade. However, for many years, Taiwan insisted on a policy of no contact and no communication with mainland China. Private enterprises eventually were allowed to trade with China—as long as the products were transshipped through a third country, usually Hong Kong. In 1993, government-owned enterprises such as steel and fertilizer plants were allowed to trade with China, on the same condition. In 2000, the Legislature lifted the ban on direct trade, transportation, and communications with China from some of the Taiwanese islands nearest to the mainland. Little by little, the door toward China is opening. Taiwanese business leaders want direct shipping, aviation, and communication links with all of China.

By 2010, Taiwanese annual trade with China exceeded US$100 billion dollars. In fact, the trade ties had become so strong by 2007 that former President Shui-ban Chen warned his compatriots not to become too dependent on the mainland economically if they wanted to maintain their de facto independence. As it happened, however, the 2008 presidential election brought to power the Kuomintang which immediately launched an effort to increase trade with China. The head of the Kuomintang visited China in 2009 and signed a number of business contracts increasing trade both ways. In banking, air transportation, and even cultural exchanges such as museum exhibits, the amount of cross-strait activity has increased substantially, and this despite strong opposition to Chinese investment in Taiwan by the opposition party. After Ma Ying-jeou of the Kuomintang was elected president, thousands marched in Taipei against the new, more open policies toward China. The liberalization of trade between China (especially its southern and coastal provinces), Taiwan, and Hong Kong has made the region, now known as Greater China, an economic dynamo. Economists predict that Greater China will someday pass the United States to become the largest economy in the world.

As one of the newly industrializing countries of Asia, Taiwan certainly no longer fits the label "underdeveloped." Taiwan holds large stocks of foreign reserves and carries a trade surplus with the United States (in

Taiwan's favor) far greater than Japan's, when counted on a per capita basis. The Taipei stock market has been so successful—sometimes outperforming both Japan and the United States—that a number of workers reportedly have quit their jobs to play the market, thereby exacerbating Taiwan's already serious labor shortage. (This shortage has led to an influx of foreign workers, both legal and illegal). The world economic slowdown hurt Taiwan's export-oriented economy in 2008 and 2009 with three quarters of contraction in the gross domestic product. But, no one can deny the economic powerhouse that Taiwan has become.

Successful Taiwanese companies have begun to invest heavily in other countries where land and labor are plentiful and less expensive. In 1993, the Philippines accepted a $60 million loan from Taiwan to build an industrial park and commercial port at Subic Bay, the former U.S. naval base; and Thailand, Australia, and the United States have also seen inflows of Taiwanese investment monies. By the early 1990s, some 200 Taiwanese companies had invested $1.3 billion in Malaysia alone, as Taiwan supplanted Japan as the largest outside investor in Malaysia. Taiwanese investment in mainland China has also increased.

Taiwan's economic success is attributable in part to its educated population, many of whom constituted the cultural and economic elite of China before the Communist revolution. Despite resentment of the mainland immigrants by native-born Taiwanese, everyone, including the lower classes of Taiwan, has benefited from this infusion of talent and capital. Yet the Taiwanese people are beginning to pay a price for their sudden affluence. Taipei, the capital, is awash not only in money but also in air pollution and traffic congestion. The traffic in Taipei is rated near the worst in the world. Concrete high-rises have displaced the lush greenery of the mountains. Many residents spend their earnings on luxury foreign cars and on cigarettes and alcohol, the consumption rate of which has been increasing by about 10 percent a year. Many Chinese traditions—for instance, the roadside restaurant serving noodle soup—are giving way to 7-Elevens selling Coca-Cola and ice cream.

Some Taiwanese despair of ever turning back from the growing materialism; they wish for the revival of traditional Chinese (that is, mostly Confucian) ethics. They doubt that it will happen. Still, the government, which has been dominated almost continuously since 1949 by the conservative Mandarin migrants from the mainland, sees to it that Confucian ethics are vigorously taught in school. And there remains in Taiwan more of traditional China than in China itself, because, unlike the Chinese Communists, the Taiwanese authorities have had no reason to attempt an eradication of the values of Buddhism, Taoism, or Confucianism. Nor has grinding poverty—often the most serious threat to the cultural arts—negatively affected literature and the fine arts, as it has in China. Parents, with incense sticks burning before small religious altars, still emphasize respect for authority, the benefits of

harmonious cooperative effort, and the inestimable value of education. Traditional festivals dot each year's calendar, among the most spectacular of which is Taiwan's National Day parade. Marching bands, traditional dancers, and a huge dragon carried by more than 50 young men please the crowds lining the streets of Taipei. Temples are filled with worshipers praying for health and good luck.

But the Taiwanese will need more than luck if they are to escape the consequences of their intensely rapid drive for material comfort. Some people contend that the island of refuge is being destroyed by success. Violent crime, for instance, once hardly known in Taiwan, is now commonplace. Six thousand violent crimes, including rapes, robberies, kidnappings, and murder, were reported in 1989—a 22 percent increase over the previous year, and the upward trend continued until the mid-2000s, when, in response to public opinion polls showing most Taiwanese thought their country's morals were in steep decline, the government began taking extra steps to reduce crime, especially from gangs. The result was that violent crime declined somewhat. But problems remain. Extortion against wealthy companies and abductions of the children of successful families have created a wave of fear among the rich. Travelers at Chiang Kai Shek international airport find they are targets of pickpockets, who rob about 10 victims per day, on average. In 2010, the number of violent crimes was still 5,277—not much of a decline from the peak in 1989.

Like other Asian countries, Taiwan was affected by the Asian financial crisis of the late 1990s. Labor shortages forced some companies to operate at 60 percent of capacity, and low-interest loans were hard to obtain because the government feared that too many people would simply invest in get-rich stocks instead of in new enterprises. In 2001, Taiwan recorded its first-ever year of negative growth, and, as we have seen, the global economic stress of 2008 and 2009 further damaged Taiwan's growth. But the economy rebounded quickly, and by 2010, the annual growth rate was just shy of 11 percent, far ahead of Japan and many other Asia Pacific countries.

■ POLITICAL LIBERALIZATION

In recent years, the Taiwanese people have had much to be grateful for in the political sphere. Until 1986, the government, dominated by the influence of the Chiangs, had permitted only one political party, the Nationalists, and had kept Taiwan under martial law for nearly four decades. A marked political liberalization began near the time of Chiang Ching-Kuo's death, in 1987. The first opposition party, the Democratic Progressive Party, was formed; martial law (officially, the "Emergency Decree") was lifted; and the first two-party elections were held, in 1986. In 1988, for the first time, a native-born Taiwanese, Lee Teng-hui, was elected to the presidency. He was reelected in 1996 in the first truly democratic, direct presidential election ever held in Taiwan. Although Lee never promoted the independence of Taiwan, his high-visibility campaign raised the ire of China, which attempted to intimidate the Taiwanese electorate into voting for a more pro-China candidate by conducting military exercises and firing missiles just 20 miles off the coast of Taiwan. As expected, the intimidation backfired, and Lee soundly defeated his opponents.

Under the Nationalists, it was against the law for any group or person to advocate publicly the independence of Taiwan—that is, to advocate international acceptance of Taiwan as a sovereign state, separate and apart from China. When the opposition Democratic Progressive Party (DPP) resolved in 1990 that Taiwan should become an independent country, the ruling Nationalist government immediately outlawed the DPP platform.

But times are changing. In 2000, after 50 years of Nationalist Party rule in Taiwan, the DPP was able to gain control of the presidency. Chen Shui-bian, a native-born Taiwanese, was elected. Both Chen and his running mate, Annette Lu, who were barely re-elected in a hotly contested 2004 election after an assassination attempt on their lives, had spent time in prison for activities that had angered the ruling Nationalists. Although his party had openly sought independence, Chen toned down that rhetoric and instead invited the mainland to begin talks for reconciliation. As usual, China had threatened armed intervention if Taiwan declared independence; but after the vote for Chen, the mainland seemed to moderate its position.

The bilateral talks that were initiated in 1998 continue to hold promise. Some believe that such talks will eventually result in Taiwan being annexed by China, just as in the cases of Hong Kong and Macau (although from a strictly legalistic viewpoint, Taiwan has just as much right to annex China). Others believe that dialogue will eventually diminish animosity, allowing Taiwan to move toward independence without China's opposition. China's position on Taiwan ("It belongs to China and we are going to get it back") seems clearer to the world than does Taiwan's position ("We sort of want independence, but we won't say so, and we would sure like to have close business ties with China").

Under the Chen administration, Taiwan increased the rhetoric around independence. In a video conference aired in Japan, Chen came close to declaring independence, stating that, in reality, separate countries already exist on both sides of the 100-mile-wide Taiwan Strait. His administration also upped the pressure on China's dismal human-rights record, issuing a critical report on rights abuses in the mainland and warning that relations between the two sides would not improve until human rights are protected on the mainland. He also proposed a national referendum that would have demanded of China that it redeploy the 800 missiles it has aimed at Taiwan (although he later softened the proposed wording), and also back-pedaled on a proposed new constitution that would have taken effect in 2008 and probably would have included language suggesting the near independence of Taiwan.

In the midst of this pressure, China, interestingly, dropped its objection to allow Taiwan to participate as an observer at a World Health Organization meeting. No doubt this was due to the world outrage over the way China had handled the SARS epidemic in 2003.

Still, the mainland seems to have the upper hand on most matters. For instance, the World Trade Organization succumbed to Chinese pressure not to allow Taiwan into the organization until China itself had been admitted, even though Taiwan had qualified years earlier. Thus, China became the 143rd member of the WTO, and Taiwan (technically "Chinese Taipei") became the 144th. Moreover, China continues to block Taiwan's membership in many other international organizations.

Opinion on the independence issue is clearly divided. Even some members of the anti-independence Nationalist Party have bolted and formed a new party (the New KMT Alliance, or the New Party) to promote closer ties with China. As opposition parties proliferate, the independence issue could become a more urgent topic of political debate. In the meantime, contacts with the P.R.C. increase daily; Taiwanese students are now being admitted to China's universities, and Taiwanese residents by the thousands now visit relatives on the mainland. In 2006, Taiwan and China agreed to allow direct charter passenger flights during major holidays and some direct cargo flights throughout the year. Nearly 200 direct flights now cross the Taiwan Strait. Under the Ma administration, it is likely that the independence language will soften toward China, but it is also likely that Taiwan will continue to act as a de facto independent country (many Taiwanese accept former President Chen's view that Taiwan is, and always has been, independent). Taiwan will likely continue to court opportunities in the Philippines, Thailand, Indonesia, and South Korea. Despite a 2002 invitation from China for high-level visits from Taiwanese leaders, China continues to hold firm to its vow to invade Taiwan if it should ever declare independence. Under these circumstances, many—probably most—Taiwanese will likely remain content to let the rhetoric of reunification continue while enjoying the reality of de facto independence.

In 2009, the wife and some family members of former president Chen pleaded guilty to money laundering, and Chen, who had done two hunger strikes to protest the charges, continued to have corruption allegations made against him personally. Chen is currently serving a 20-year jail sentence for a series of corruption changes. Many believe the charges were politically motivated and promoted by the KMT, especially when his successor, Ma Ying-jeou of the KMT, declassified government documents that aided in Chen's convictions on embezzlement. But the plot thickened when, in 2011, former President Lee Teng-hui (President 1988–2000) was indicted for embezzlement. He had once been a KMT member but recently announced he would support the Democratic Progressive Party's candidate for President.

Snapshot: TAIWAN

Summarized below is a quick look at the country with regard to its development, freedom, health/welfare, and achievements.

Taiwan now seems to be on a path toward greater democratization. In 1991, 5,574 prisoners, including many political prisoners, were released in a general amnesty. Slowly, the government is withdrawing from direct involvement in the private business sector.

Development

Taiwan has vigorously promoted export-oriented production, particularly of electronic equipment. In the 1980s, manufacturing became a leading sector of the economy, employing more than one third of the workforce. Virtually all Taiwanese households own color televisions, and other signs of affluence are abundant. Taiwan became the 144th member of the WTO, right after China which insisted it be granted membership first.

Freedom

For nearly 4 decades, Taiwan was under martial law. Opposition parties were not tolerated, and individual liberties were limited. A liberalization of this pattern began in 1986.

Health/Welfare

Taiwan has one of the highest population densities in the world. Education is free and compulsory to age 15, and the country boasts more than 100 institutions of higher learning. Social programs, however, are less developed than those in Singapore, Japan, and some other Pacific Rim countries.

Achievements

From a largely agrarian economic base, Taiwan has been able to transform its economy into an export-based dynamo with international influence. Today, only about 5 percent of the population work in agriculture, and Taiwan ranks among the top 20 exporters in the world.

Statistics

Geography

Area in Square Miles (Kilometers): 22,357 (35,980) (about the size of Maryland and Delaware combined)

Capital (Population): Taipei (2.62 million)

Environmental Concerns: water and air pollution; poaching; contamination of drinking water; radioactive waste; trade in endangered species

Geographical Features: mostly rugged mountains in east; flat to gently rolling plains in west

Climate: tropical; marine; persistent cloudiness

People

Population

Total: 23,071,779

Annual Growth Rate: 0.2%

Rural/Urban Population Ratio: 25/75

Major Languages: Mandarin Chinese; Taiwanese; Hakka dialects

Ethnic Makeup: 84% Taiwanese; 14% Mainlander Chinese; 2% aborigine

Religions: 93% mixture of Buddhism, Confucianism, and Taoism; 4.5% Christian; 2.5% others

Health

Life Expectancy at Birth: 75.5 years (male); 81 years (female)

Infant Mortality: 5/1,000 live births

Physicians Available: 1/424 people

HIV/AIDS Rate in Adults: na

Education

Adult Literacy Rate: 96%

Compulsory (Ages): 6–15; free

Communication

Telephones: 14.6 million main lines (2009)

Mobile Phones: 27 million (2009)

Internet Users: 16 million (2011)

Internet Penetration (% of Pop.): 70%

Transportation

Roadways in Miles (Kilometers): 25,771 (41,475)

Railroads in Miles (Kilometers): 982 (1,580)

Usable Airfields: 41

Motor Vehicles in Use: 5,300,000

Government

Type: multiparty democratic regime

Head of State/Government: President Ma Ying-jeou; Premier Wu Den-yih

Political Parties: Democratic Progressive Party; Nationalist Party (Kuomintang); Non-Partisan Solidarity Union; People First Party

Suffrage: universal at 20

Military

Military Expenditures (% of GDP): 2.2%

Current Disputes: disputes with various countries over islands

Economy

Currency ($ U.S. Equivalent): 31.64 New Taiwan dollars = $1

Per Capita Income/GDP: $35,700/$821.8 billion

GDP Growth Rate: 10.8%

Inflation Rate: 3.1%

Unemployment Rate: 5.2%

Labor Force by Occupation: 59% services; 36% industry; 5% agriculture

Population Below Poverty Line: 1.2%

Natural Resources: coal; natural gas; limestone; marble; asbestos

Agriculture: rice; vegetables; fruit; tea; flowers; pigs; poultry; fish

Industry: electronics; communications and information technology products; petroleum refining; armaments; chemicals; textiles; iron and steel; machinery; cement; food processing; vehicles; consumer products; pharmaceuticals

Exports: $274 billion (primary partners China, Hong Kong, United States)

Imports: $251 billion (primary partners Japan, China, United States)

Suggested Websites

www.gio.gov.tw/

www.eng.taiwan.net.tw

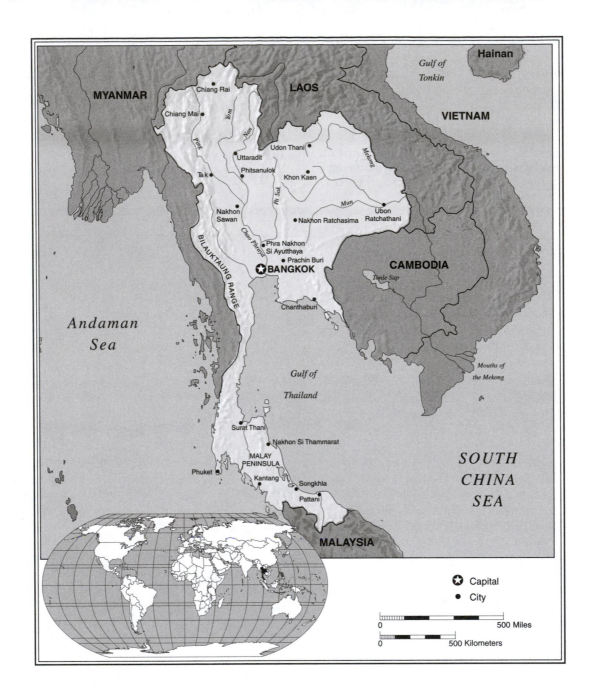

Gulf of
Tonkin

Hainan

MYANMAR

LAOS

VIETNAM

Chiang Rai

Chiang Mai

Udon Thani

Uttaradit

Phitsanulok

Khon Kaen

Tak

Nakhon
Sawan

Phra Nakhon
Si Ayutthaya

Prachin Buri

BANGKOK

Nakhon Ratchasima

Ubon
Ratchathani

CAMBODIA

Tonle Sap

Chanthaburi

Andaman
Sea

Gulf of

Thailand

Mouths of
the Mekong

BILAUKTAUNG RANGE

Surat Thani

Nakhon Si Thammarat

MALAY
PENINSULA

Phuket

Kantang

Songkhla

Pattani

SOUTH
CHINA
SEA

MALAYSIA

★ Capital
● City

0   500 Miles

0   500 Kilometers

Thailand

(Kingdom of Thailand)

One afternoon in late 2004, more than 50 Thai warplanes roared over the southern Thailand provinces of Yala, Pattani, and Narathiwat, where violence by Islamic extremists had taken the lives of over 500 people in several months of unrest. Instead of bombs, however, the air force dropped 120 million paper cranes, folded for the people of southern Thailand by their countrymen in the north. Like symbolic white doves of peace in Western cultures, cranes are Asian symbols of peace and reconciliation, and the airdrop was intended to show that, despite religious and other differences, Thai people care for one another. The prime minister, who promised to restore a mosque damaged in the violence, reported that the airdrop had an "enormous positive psychological effect," and that once people knew what was happening, crowds started lining the roads to catch the paper birds. Such is the kind of leadership that has made Thailand unique among Southeast Asian countries.

The roots of Thai culture extend into the distant past. People were living in Thailand at least as early as the Bronze Age. By the time Thai people from China (some scholars think from as far away as Mongolia) had established the first Thai dynasty in the Chao Phya Valley, in A.D. 1238, some communities, invariably with a Buddhist temple or monastery at their centers, had been thriving in the area for 600 years. Early Thai culture was greatly influenced by Buddhist monks and traders from India and Sri Lanka (Ceylon).

By the seventeenth century, Thailand's ancient capital, Ayutthaya, boasted a larger population than that of London. Ayutthaya was known around the world for its wealth and its architecture, particularly its religious edifices. Attempts by European nations to obtain a share of the wealth were so inordinate that in 1688, the king expelled all foreigners from the country. Later, warfare with Cambodia, Laos, and Malaya yielded tremendous gains in power and territory for Thailand, but it was periodically afflicted by Burma (present-day Myanmar), which briefly conquered Thailand in the 1760s (as it had done in the 1560s). The Burmese were finally defeated in 1780, but the destruction of the capital required the construction of a new city, near what, today, is Bangkok.

Generally speaking, the Thai people have been blessed over the centuries with benevolent kings, many of whom have been open to new ideas from Europe and North America. Gathering around them advisers from many nations, they improved transportation systems, education, and farming while maintaining the central place of Buddhism in Thai society. Occasionally royal support for religion overtook other societal needs, at the expense of the power of the government.

The gravest threat to Thailand came during the era of European colonial expansion, but Thailand—whose name means "Free Land"—was never completely conquered by European powers. Today, the country occupies a land area about the size of France.

◼ MODERN POLITICS

Since 1932, when a constitutional monarchy replaced the absolute monarchy, Thailand (formerly known as Siam) has weathered 18 attempted or successful military or political coups d'etat, most recently in 2006. The

TIMELINE

A.D. 1200s
The formal beginning of Thailand as a nation
1782
King Rama I ascends the throne, beginning a nine-generation dynasty
1932
Coup; constitutional monarchy
1939
The country's name is changed from Siam to Thailand
1942
Thailand joins Japan and declares war on the United States and Britain
1946
Thailand resumes its historical pro-Western stance
1960s–1970s
Communist insurgency threatens Thailand's stability
1973
Student protests usher in democratic reforms
1990s
Currency decisions in Thailand precipitate the Southeast Asian financial crisis; mass demonstrations force a return to civilian rule
2000s
Thailand rebounds from the economic crisis of the late 1990s
Bangkok looks for ways to reduce its terrible pollution and traffic congestion
The 2004 Indian Ocean earthquake and tsunami took the lives of up to 11,000 people, many of them European tourists vacationing on the powdery resort beaches of Phuket Island
A bloodless military coup in 2006 deposes Prime Minister Thaksin
Battles between the "red shirts" and the "yellow shirts" in 2008 and 2009 shut down airports and tarnish Thailand's image
Red Shirts staged massive demonstrations in 2010 calling for the dissolution of parliament
Thailand and Cambodia restore diplomatic relations after Thaksin, in exile in Cambodia, steps down as advisor to Cambodian government. Thailand and Cambodia exchange fire over disputed territory along the border.

Buddhism is an integral part of the Thai culture. Six hundred years ago, Buddhist monks traveled from India and Ceylon (present-day Sri Lanka) and built temples and monasteries throughout Thailand. These Buddhist monks are reading and praying outside a temple.

Constitution has been revoked and replaced numerous times; governments have fallen under votes of no-confidence; students have mounted violent demonstrations against the government; and the military has, at various times, imposed martial law or otherwise curtailed civil liberties.

Clearly, Thai politics are far from stable, and in recent years, they have become even more volatile. Nevertheless, over the years there have been threads of stability that have allowed Thailand to develop far ahead of its neighbors. For instance, its people were spared the

? DID YOU KNOW?

Thailand's extensive rice paddies depend on monsoon rains each year. But in 2011, monster amounts of rainfall and back-to-back typhoons swamped 4 million acres of land—the worst flooding in 50 years. Some 200 famous temples in the former royal capital of Ayutthaya were damaged, and residents of Bangkok were sandbagging everything they could as the filthy waters seeped into more parts of the city. Some 500 people drowned, including many children. Every year, throughout all of Asia, some 240,000 children drown in flood waters, often because they do not know how to swim.

direct ravages of the Vietnam War, which raged nearby for 20 years. Despite all the political upheavals, the same royal family has been in control of the Thai throne for nine generations, although its power has been substantially delimited for some 60 years. Furthermore, before the first Constitution was enacted in 1932, the country had been ruled continuously, for more than 700 years, by often brilliant and progressive kings. At the height of Western imperialism, when France, Britain, the Netherlands, and Portugal were in control of every country on or near Thailand's borders, Thailand remained free of Western domination, although it was forced—sometimes at gunpoint—to relinquish sizable chunks of its holdings in Cambodia and Laos to France, and holdings in Malaya to Britain. The reasons for this singular state of independence were the diplomatic skill of Thai leaders, Thai willingness to Westernize the government, and the desire of Britain and France to let Thailand remain interposed as a neutral buffer zone between their respective armies in Burma and Indochina. Moreover, Thailand has always curried favor with countries more powerful than itself. In 1833, for example, it was the first Asian country to establish diplomatic relations with the United States.

The current king, Phumiphon Adunyadet, born in the United States and educated in Switzerland, is highly respected as head of state. The king is also the nominal

(Courtesy of Lisa Clyde Nielsen)

Bangkok is one of the largest cities in the world. The city is interlaced with canals, and the population crowds along river banks. With the enormous influx of people who are lured by industrialization and economic opportunity, the environment has been strained to the limit.

head of the armed forces, and his support is critical to any Thai government. Despite Thailand's structures of democratic government, any administration that has not also received the approval of the military elites, many of whom hold seats in the Senate, has not prevailed for long. The military has been a rightist force in Thai politics, resisting reforms from the left that might have produced a stronger labor union movement, more freedom of expression (many television and radio stations in Thailand are controlled directly by the military), and less economic distance between the social classes. Military involvement in government increased substantially during the 1960s and 1970s, when a Communist insurgency threatened the government from within and the Vietnam War destabilized the external environment.

Before the February 1991 coup, there had been signs that the military was slowly withdrawing from

FURTHER INVESTIGATION

To learn more about the long-standing border dispute in Preah Vihear between Thailand and Cambodia which has caused the two militaries to engage in artillery fire exchanges and other acts: www.dur.ac.uk/resources/ibru/publications/.../ bsb1-4_john.pdf

direct meddling in the government. This may have been because the necessity for a strong military appeared to have lessened with the end of the Cold War. In late 1989, for example, the Thai government signed a peace agreement with the Communist Party of Malaya, which had been harassing villagers along the Thai border for more than 40 years. Despite these political/military improvements, Commander Suchinda Kraprayoon led an army coup against the legally elected government in 1991 and, notwithstanding promises to the contrary, promptly had himself named prime minister. Immediately, Thai citizens, tired of the constant instability in government occasioned by military meddling, began staging mass demonstrations against Suchinda. The protesters were largely middle-class office workers who used their cellular telephones to communicate from one protest site to another. The demonstrations were the largest in 20 years, and the military responded with violence; nearly 50 people were killed and more than 600 injured. The public outcry was such that Suchinda was forced to appear on television being lectured by the king; he subsequently resigned. An interim premier dismissed several top military commanders and removed military personnel from the many government departments over which they had come to preside. Elections followed in 1992, and Thailand returned to civilian rule, with the military's influence greatly diminished.

Then, in late 2006, while populist Prime Minister Thaksin Shinawatra was delivering a speech at the United Nations in New York, General Sondhi Boonyaratglin and his army colleagues took over the government. They suspended the Constitution, declared martial law, and installed retired army commander Surayu Chuilanont as interim premier. The coup leaders quickly met with the King, received his blessing, and announced—using a typically Asian explanation—that Thaksin was deposed because he had "destroyed harmony in society."

In earlier elections, Thaksin's party had won all the contested seats, but the Supreme Court later nullified the vote as undemocratic. Though popular in the country-side, Thaksin had alienated the middle class in the cities through questionable business deals involving Singapore and by being less than respectful of the King. The coup was entirely bloodless, and with the King obviously sup-portive of the move, the nation, it was hoped, would quickly return to normal. But the growing divide between the middle and upper classes (the "yellow shirts" as they were called because they wore yellow t-shirts at their pro-tests) and the rest of the population (the "red shirts"), erupted into mass demonstrations again in late 2008. Thousands of protestors stormed Government House—the prime minister's office—and attacked his car as he was leaving a cabinet meeting. They dragged his aide from another car and beat him. Ignoring a state of emer-gency, they stormed the nation's airports, forcing two of them to be shut down. They also forced the cancellation of a pan-Asian summit in April 2009 when 600 protes-tors entered the conference center and surrounded the hotels where the presidents of China, Japan, and South Korea were staying. The leaders had to be evacuated by helicopter and boat. The bad international attention hurt tourism, damaged the stock market, and paralyzed the government. Protestors also swarmed around a court building. The protests turned violent, with scores injured and seven killed in bomb blasts and violent clashes with police. The protestors wanted the prime minister, a brother-in-law to the deposed and self-exiled Thaksin, to step down.

The courts responded by dissolving the prime min-ister's People Power Party and banning dozens of party officials from politics for five years. The conservative Democrat Party head Abhisit Wetchachiwa was named as prime minister. In the 2009 elections, the power of the "red shirts" seemed to be waning as 255 of the 480 seats in parliament went to the conservative or "yellow shirts" party. But tensions remain just below the surface, and insinuations that the king himself may have been behind Thaksin's ouster continue to circulate.

Unlike some democratic governments that have one dominant political party and one or two smaller opposi-tion parties, party politics in Thailand is characterized by diversity. Indeed, so many parties compete for power that no single party is able to govern without forming coalitions with others. Parties are often founded on the strength of a single charismatic leader rather than on a distinct political philosophy, a circumstance that makes the entire political

setting rather volatile. The Communist Party remains banned. Campaigns to elect members of parliament often turn violent. In past elections, 10 candidates were killed when their homes were bombed or sprayed with rifle fire, and nearly 50 gunmen-for-hire were arrested or killed by police, who were attempting to protect the candidates of the 11 political parties vying for office.

In 1999, corruption charges by the New Aspira-tion Party against the ruling Democrat Party failed to topple the government. Many citizens seemed to credit the government with bringing Thailand out of eco-nomic recession (but the International Monetary Fund loan of $17.2 billion probably helped more). Still, cor-ruption was everywhere evident in 2000, when massive fraud was uncovered in Senate elections. It was the first time that senators were being directly elected instead of appointed, and many of them resorted to wholesale vote-buying and ballot-tampering. Out of 200 senators, 78 were disqualified as a result of election fraud, and the elections had to be held again. In 2010, the government seized assets worth US$1.4 billion belonging to former Prime Minister Thaskin, and, in a continuation of the pro-Thaskin movement, thousands of "red shirt" protes-tors once again took to the streets, causing the govern-ment to declare a state of emergency for the fourth time since 2008. Protestors blocked roads, hotels, and shop-ping malls in the business district and demanded that the government hold immediate elections. The protests continued for three months; nearly 90 people were killed and 1,400 injured. The protests clearly had a "haves ver-sus have-nots" quality, with the "red shirts" coming from mostly poor, rural areas. They intentionally obstructed the businesses of those they saw as wealthy, causing busi-ness owners to complain that the violence had cost them US$12 million a day. The government finally quashed the uprising in May 2010, but the social inequality behind the protests remained.

In the 2011 elections, after considerable political maneuvering and much delay, the Election Commission certified the election of Yingluck Shinawatra as the new Prime Minister. She became the first female to hold the post. Former Prime Minsiter Somchai Wongsawat was banned from politics for five years by the Constitutional Court. The new Prime Minister is the sister of former Prime Minister Thaksin.

■ FOREIGN RELATIONS

Thailand is a member of the United Nations, the Asso-ciation of Southeast Asian Nations, and many other regional and international organizations. Throughout most of its modern history, Thailand has maintained a pro-Western political position. During World War I, Thailand joined with the Allies; and during the Vietnam War, it allowed the United States to stage air attacks on North Vietnam from within its borders, and served as a major rest and relaxation center for American soldiers. During World War II, Thailand briefly allied itself with Japan but made decided efforts after the war to rees-tablish its former Western ties. When the United States launched the war against Iraq, Thailand responded by sending 450 medical engineers to the war zone.

Thailand's international positions have seemingly been motivated more by practical need than by ideology. During the colonial era, Thailand linked itself with Britain because it needed to offset the influence of France; during World War II, it joined with Japan in an apparent effort to prevent its country from being devastated by Japanese troops; during the Vietnam War, it supported the United States because the United States seemed to offer Thailand its only hope of not being directly engaged in military conflict in the region.

Economically, Thailand occasionally distances itself from the West. Thailand is seen by the world business community as the third-best place for investment (after China and India), and investment funds from around the world, especially from Asia itself, have poured in, although the pace has slowed due to the political unrest since 2006. At the moment, Thailand is cultivating a closer relationship with Japan. In the late 1980s, disputes with the United States over import tariffs and international copyright matters cooled the countries' warm relationship (the United States accused Thailand of allowing the manufacture of counterfeit brand-name watches, clothes, computer software, and many other items, including medicines). Moreover, Thailand found in Japan a more ready, willing, and cooperative economic partner than the United States. Many Thais also find Japanese culture to be interesting, and sign up for Japanese etiquette classes to improve their abilities in the business world.

During the cold war and especially during the Vietnam War era, the Thai military strenuously resisted the growth of Communist ideology inside Thailand, and the Thai government refused to engage in normal diplomatic relations with the Communist regimes on its borders. Because of military pressure, elected officials refrained from advocating improved relations with the Communist governments. However, in 1988, Prime Minister Prem Tinslanond, a former general in the army who had been in control of the government for eight years, stepped down from office, and opposition to normalization of relations seemed to mellow. The subsequent prime minister, Chatichai Choonhavan, who was ousted in the 1991 military coup, invited Cambodian leader Hun Sen to visit Thailand; he also made overtures to Vietnam and Laos. Chatichai's goal was to open the way for trade in the region by helping to settle the agonizing Cambodian conflict. He also hoped to bring stability to the region so that the huge refugee camps in Thailand, the largest in the world, could be dismantled and the refugees repatriated. Managing regional relations will continue to be difficult: Thailand fought a brief border war with Communist Laos in 1988. The influx of refugees from the civil wars in adjacent Cambodia and Myanmar continues to strain relations. Currently some 100,000 Karen refugees live precariously in 20 camps in Thailand along the border with Myanmar. The Karens, many of whom practice Christianity and are the second-largest ethnic group in Myanmar, have fought the various governments in their home country for years in an attempt to create an independent Karen state. Despite the patrol efforts of Thai troops, Myanmar soldiers frequently cross into Thailand at night to raid, rape, and kill the Karens. Thailand has tolerated the massive influx of war refugees from Myanmar, but its patience seems to have been wearing thin in recent years.

This became clear in 2009 when the Thai military started arresting Muslim immigrants from Myanmar, whipping them, and sending them out to sea in boats without engines or sails. Some 1,000 people were thus treated, and the United Nations called for a full investigation. It was believed that the military did not want the Muslims to join their Islamic brothers in southern Thailand and add to the religious unrest there.

THE ECONOMY

Part of the thrust behind Thailand's diplomatic initiatives is the changing needs of its economy. For decades, Thailand saw itself as an agricultural country; indeed, 42% of the labor force works in agriculture today, with rice as the primary commodity. Rice is Thailand's single most important export and a major source of government revenue. Every morning, Thai families sit on the floor of their homes around bowls of hot and spicy *tom yam goong* soup and a large bowl of rice; holidays and festivals are scheduled to coincide with the various stages of planting and harvesting rice; and, in rural areas, students are dismissed at harvest time so that all members of a family can help in the fields. So central is rice to the diet and the economy of the country that the Thai verb equivalent of "to eat," translated literally, means "to eat rice." Thailand has the fifth-largest amount of land under rice cultivation and is the largest exporter of rice in the world.

Unfortunately, Thailand's dependence on rice subjects its economy to the cyclical fluctuations of weather (sometimes the monsoons bring too little moisture) and market demand. Thus, in recent years, the government has invested millions of dollars in economic diversification. Not only have farmers been encouraged to grow a wider variety of crops, but tin, lumber, and offshore oil and gas production have also been promoted. Thailand is the world's largest rubber-producing country, but with prices at 30-year lows, that industry is struggling to survive. Foreign investment in export-oriented manufacturing has been warmly welcomed. Japan in particular benefits from trading with Thailand in food and other commodities, and it sees Thailand as one of the more promising places to relocate smokestack industries. For its part, Thailand seems to prefer Japanese investment over that from the United States, because the Japanese seem more willing to engage in joint ventures and to show patience while enterprises become profitable. Indeed, economic ties with Japan are very strong. For instance, in recent years Japan has been the largest single investor in Thailand and has accounted for more than 40 percent of foreign direct investment (Taiwan, Hong Kong, and the

United States each have accounted for about 10 percent). About 20 percent of Thai imports come from Japan, while over 11 percent of its exports go to Japan.

Thailand's shift to an export-oriented economy paid off until 1997, when pressures on its currency, the baht, required the government to allow it to float instead of having it pegged to the U.S. dollar. That action triggered the Southeast Asian financial crisis. Until that time, Thailand's gross domestic product growth rate had averaged about 10 percent a year—one of the highest in the world, and as high, or higher than, all the newly industrializing countries of Asia (Hong Kong, South Korea, Singapore, Taiwan, and China). Furthermore, unlike the Philippines and Indonesia, Thailand was able to achieve this incredible growth without very high inflation. The 1997–1998 financial crisis hit Thailand very hard, but the country recovered to a growth rate in excess of 4 percent only to be hit by the global meltdown of 2008–2009, slowing growth to just over 3 percent. Nevertheless, by 2010, the growth rate had rebounded to 7.8 percent.

■ SOCIAL PROBLEMS

Industrialization in Thailand, as everywhere, draws people to the cities. Bangkok is one of the largest cities in the world. Numerous problems, particularly air pollution, traffic congestion, and overcrowding, complicate life for Bangkok residents. An international airport that opened near Bangkok in 1987 was so overcrowded just four years later that a new one had to be planned, and new harbors had to be constructed south of the city to alleviate congestion in the main port. Demographic projections indicate that there will be a decline in population growth in the future as the birth rate drops and the average Thai household shrinks from the six people it was in 1970 to only three people by 2015. This will alter the social structure of urban families, especially as increased life expectancy adds older people to the population and forces the country to provide more services for the elderly.

Today, however, many Thai people still make their living on farms, where they grow rice, rubber, and corn, or tend chickens and cattle, including the ever-present water buffalo. Thus, it is in the countryside (or "up-country," as everywhere but Bangkok is called in Thailand) that the traditional Thai culture may be found. There, one still finds villages of typically fewer than 1,000 inhabitants, with houses built on wooden stilts alongside a canal or around a Buddhist monastery. One also finds, however, unsanitary conditions, higher rates of illiteracy, and lack of access to potable water. Of increasing concern is deforestation, as Thailand's growing population continues to use wood as its primary fuel for cooking and heat. The provision of social services does not meet demand even in the cities, but rural residents are particularly deprived.

Relative tolerance has mitigated ethnic conflict among most of Thailand's numerous minority groups. The Chinese, for instance, who are often disliked in other Asian countries because of their dominance of the business sectors, are able to live with little or no discrimination

Coup Lite

In September 2006, a coup led by Army General Sondhi overthrew the elected government. Yet, the event had little of the trauma typically associated with coups d'etat. There was no resistance, no one was killed or injured, and except for tanks around government buildings, the military presence was minimal. Top members of the former government were literally "invited" to come to the coup leaders' offices to be interviewed. Coup leaders announced that schools and businesses were to be closed for a day, so most students and workers treated the day as a holiday and then went back to school and work the next day. Radio stations in some areas were temporarily silenced, but in Bangkok the stations played jazz music along with patriotic tunes. Photos of the King were shown on television, and the coup leaders went on the air to apologize to the citizens and ask their "forgiveness for the inconvenience" of the coup. Such is the power of the middle class, which supported the coup, and of the King (the world's longest reigning monarch), who quickly put his imprimatur on the coup and thus—temporarily—calmed the public yet again.

in Thailand; indeed, they constitute the backbone of Thailand's new industrial thrust.

In addition to the Chinese, Thailand is home to many other ethnic groups: the Lisu, the Mien, the Hmong, the Akha, the Karen, and the Lahu, to name a few. As many as 550,000 semi-nomadic tribal peoples live in Thailand's northern mountains, where they have little in the way of modern conveniences, and live more or less as they were living hundreds of years ago. In fact, many of them have little or no emotional link with the nation of Thailand, preferring to crisscross the borders between Tibet, Myanmar, Laos, and China as they make their subsistence living without electricity or running water. Ancestors of the 100,000 members of the Lahu tribe, for example, entered Thailand from Tibet about 200 years ago (large numbers also live in China, Laos, Vietnam, and Myanmar). They live in the midst of a lush jungle, where they build their bamboo huts on stilts and wear homespun clothing. With no phones, no television, and no cars, the Lahu use wooden plows pulled by oxen to prepare their fields for rice planting.

Culturally, Thai people are known for their willingness to tolerate (although not necessarily to assimilate) diverse lifestyles and opinions. Buddhist monks, who shave their heads and make a vow of celibacy, do not find it incongruous to beg for rice in districts of Bangkok known for prostitution and wild nightlife. And worshippers seldom object when a noisy, congested highway is built alongside the serenity of an ancient Buddhist temple. However, the mammoth scale of the proposed $3.2 billion, four-level road-and-railway system in the city and its likely effect on cultural and religious sites prompted the Thai cabinet to order the construction underground; but

Snapshot: THAILAND

Summarized below is a quick look at the country with regard to its development, freedom, health/welfare, and achievements.

Development

Many Thais are small-plot or tenant farmers, but the government has energetically promoted economic diversification. Despite high taxes, Thailand has a reputation as a good place for foreign investment. Nearly 1,800 foreign companies have invested in Thailand. Electronics and other high-tech industries from Japan, the United States, and other countries have been very successful in Thailand.

Freedom

Since 1932, when a constitutional monarchy replaced absolutism, Thailand has endured eighteen military coups and countercoups, most recently in September 2006. Combined with the threat of Communist and Muslim insurgencies, these have resulted in numerous declarations of martial law, press censorship, and suspensions of civil liberties. Thai censors banned the movie *Anna and the King* in 1999, just as they had done with *The King and I* in 1956. It is believed by many that people close to the king were behind the 2006 coup that removed a democratically elected prime minister.

Health/Welfare

About 35% of college-age Thais actually attend universities, which is three times that of neighboring Vietnam. Thailand has devoted substantial sums to the care of refugees from Cambodia and Vietnam. The rate of nonimmigrant population growth has dropped substantially since World War II. HIV/AIDS is a significant problem in Thailand, as is bird flu, which has infected thousands of chickens, ducks, Bengal tigers, and some humans.

Achievements

Thailand is the only Southeast Asian nation never to have been colonized by a Western power. It was also able to remain detached from direct involvement in the Vietnam War. Unique among Asian cultures, Thailand has a large number of women in business and other professions. Thai dancing is world-famous for its intricacy. In 1996, boxer Somluck Khamsing became the first Thai to win an Olympic gold medal.

---

the cabinet had to recant when the Hong Kong firm designing the project announced that it was technically impossible to build it underground.

Despite their generally tolerant attitudes, Thais have been sore pressed to handle Muslim unrest in the provinces near Malaysia, where residents feel neglected and persecuted by the government. In 2004, violence by Muslims provoked a strong reaction by Thai police, resulting in the shooting deaths of some who were rioting and the deaths by suffocation by many more detainees as they were being transported in overcrowded military vehicles. Militants have attacked schools and police stations and raided military depots. With Muslim preachers using increasingly volatile speech during worship services and in schools, it is not likely that the unrest—already close to a full-blown separatist movement—will be calmed any time soon. Nine bombs set by Muslim agitators caused officials to cancel New Year's festivities in Bangkok in January 2007, and despite the appointment of a Muslim as head of the military, the Muslim fundamentalists in the three southern provinces of Pattani, Yala, and Narathiwat continued to engage in terrorist acts against the government.

The Thai government seems to be making headway on at least two serious social problems, HIV/AIDS and smoking. Thailand has been known for years as a hotbed for the transmission of the virus that causes AIDS. Education programs and tighter regulation of the sex industry have produced dramatic results, with the number of new HIV infections on the decline. In 2002, the government passed a strict antismoking law. Smoking is now outlawed in virtually all indoor spaces, including Buddhist temples. Not only can smokers be fined if they light up indoors, but so can the establishment that allows it. According to the World Health Organization, Thailand now joins Hong Kong and Singapore as the Asian countries with the strongest antismoking laws. Thailand is known as a laissez-faire culture where "never mind" is the solution to vexing problems; thus the implementation of these strict policies demonstrates how seriously the Thai government regards the health of its citizens.

Statistics

Geography

Area in Square Miles (Kilometers): 198,117 (513,120) (about twice the size of Wyoming)

Capital (Population): Bangkok (6.9 million)

Environmental Concerns: air and water pollution; poaching; deforestation; soil erosion

Geographical Features: central plain; Khorat Plateau in the east; mountains elsewhere

Climate: tropical monsoon

People

Population

Total: 66,720,153

Annual Growth Rate: 0.57%

Rural/Urban Population Ratio: 66/34

Major Languages: Thai; English; various dialects

Ethnic Makeup: 75% Thai; 14% Chinese; 11% Malay and others

Religions: 94.6% Buddhist; 4.6% Muslim; .8% others

Health

Life Expectancy at Birth: 71 years (male); 76 years (female)

Infant Mortality: 16/1,000 live births

Physicians Available: 1/3,333 people

HIV/AIDS Rate in Adults: 1.3%

Education

Adult Literacy Rate: 93%

Compulsory (Ages): 6–15

Communication

Telephones: 7 million main lines (2009)

Mobile Phones: 83 million (2009)

Internet Users: 17.5 million (2010)

Internet Penetration (% of Pop.): 26%

Transportation

Roadways in Miles (Kilometers): 111,880 (180,053)

Railroads in Miles (Kilometers): 2,528 (4,071)

Usable Airfields: 104

Government

Type: constitutional monarchy

Independence Date: founding date 1238; never colonized

Head of State/Government: King Phumiphon Adunyadet; Prime Minister Yinglak Chinnawat

Political Parties: Thai Nation Development Party; Democrat Party; Motherland Party; For Thais Party; Thai Pride; Royalist People's Party; Thai Unity Party

Suffrage: universal and compulsory at 18

Military

Military Expenditures (% of GDP): 1.8%

Current Disputes: avian flu; Malaysian border closure; border disputes with Laos and Cambodia; disputes regarding refugees from Burma

Economy

Currency ($ U.S. Equivalent): 32 baht = $1

Per Capita Income/GDP: $8,700/$586.9 billion

GDP Growth Rate: 7.8%

Inflation Rate: 3.3%

Unemployment Rate: 1.1%

Labor Force by Occupation: 42% agriculture; 38% services; 20% industry

Population Below Poverty Line: 10%

Natural Resources: tin; rubber; natural gas; tungsten; tantalum; timber; lead; fish; gypsum; lignite; fluorite; arable land

Agriculture: rice; cassava; rubber; corn; sugarcane; coconuts; soybeans

Industry: tourism; textiles and garments; agricultural processing; beverages; tobacco; cement; electric appliances and components; electronics; furniture; plastics; automobiles; tungsten; tin

Exports: $193.5 billion (primary partners China, Japan, United States)

Imports: $161 billion (primary partners Japan, China, United States, Malaysia)

Suggested Websites

www.tourismthailand.org/

www.bangkok.usembassy.gov/

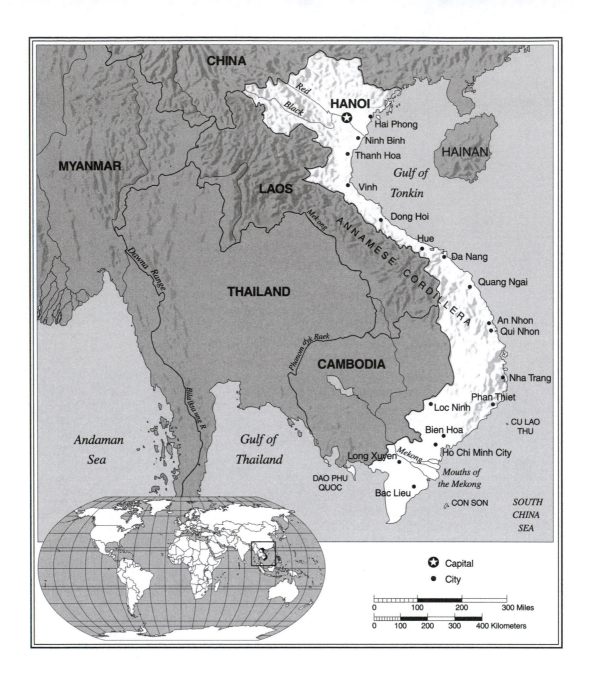

CHINA

Red

Black

HANOI ✪

Hai Phong

Ninh Binh

Thanh Hoa

HAINAN

Gulf of Tonkin

MYANMAR

LAOS

Vinh

Dong Hoi

Hue

Mekong

A N N A M E S E   C O R D I L L E R A

Đa Nang

Dawna Range

THAILAND

Quang Ngai

An Nhon
Qui Nhon

Phnom dyk Raek

CAMBODIA

Nha Trang

Phan Thiet

Loc Ninh

CU LAO THU

Bia hua uik. R

Andaman Sea

Gulf of Thailand

Bien Hoa

Long Xuyen

Mekong

Ho Chi Minh City

DAO PHU QUOC

Mouths of the Mekong

Bac Lieu

CON SON

SOUTH CHINA SEA

✪ Capital

● City

0   100   200   300 Miles

0 100 200 300 400 Kilometers

Vietnam

(Socialist Republic of Vietnam)

Foreign powers have tried to control Vietnam for 2,000 years. Most of that time it has been the Chinese who have had their eye on control—specifically of the food and timber resources of the Red River Valley in northern Vietnam.

Most of the northern Vietnamese were originally ethnic Chinese themselves; but over the years, they forged a separate identity for themselves and came to resent Chinese rule. Vietnam was conquered by China as early as 214 B.C. and again in 111 B.C., when the Han Chinese emperor Wu Ti established firm control. For about 1,000 years (until A.D. 939, and sporadically thereafter by the Mongols and other Chinese), the Chinese so thoroughly dominated the region that the Vietnamese people spoke and wrote in Chinese, built their homes like those of the Chinese, and organized their society according to Confucian values. In fact, Vietnam (*viet* means "people" and *nam* is Chinese for "south") is distinct among Southeast Asian nations because it is the only one whose early culture—in the north, at least—was influenced more by China than by India.

The Chinese did not, however, directly control all of what constitutes modern Vietnam. Until the late 1400s, the southern half of the country was a separate kingdom known as Champa. It was inhabited by the Chams, who originally came from Indonesia. For a time Champa was annexed by the north. However, between the northern region called Tonkin and the southern Chams-dominated region was a narrow strip of land occupied by Annamese peoples (a mixture of Chinese, Indonesian, and Indian ethnic groups), who eventually overthrew the Cham rulers and came to dominate the entire southern half of the country. In the 1500s, the northern Tonkin region and the southern Annamese region were ruled separately by two Vietnamese family dynasties. In the 1700s, military generals took power, unifying the two regions and attempting to annex or control parts of Cambodia and Laos as well.

In 1787, Nguyen-Anh, a general with imperial ambitions, signed a military-aid treaty with France. The French had already established Roman Catholic missions in the south, were providing mercenary soldiers for the Vietnamese generals, and were interested in opening up trade along the Red River. The Vietnamese eventually came to resent the increasingly active French involvement in their internal affairs and took steps to curtail French influence. The French, however, impressed by the resources of the Red River Valley in the north and the Mekong River Delta in the south, were in no mood to pull out. Vietnam's geography contains rich tropical rain forests in the south, valuable mineral deposits in the north, and oil deposits offshore.

War broke out in 1858, and by 1863 the French had won control of many parts of the country, particularly in the south around the city of Saigon (now known as Ho Chi Minh City). Between 1884 and 1893, France solidified its gains in Southeast Asia by taking the northern city of Hanoi and the surrounding Tonkin region and by putting Cambodia, Laos, and Vietnam under one administrative unit, which it named *Indochina*.

Ruling Indochina was not easy for the French. For one thing, the region included hundreds of different ethnic groups, many of whom had

TIMELINE

214 B.C.
China begins 1,000 years of control or influence over the northern part of Vietnam

A.D. 1500s
Northern and southern Vietnam are ruled separately by two Vietnamese families

1700s
Military generals overthrow the ruling families and unite the country

1787
General Nguyen-Anh signs a military-aid treaty with France

1863
After 5 years of war, France acquires its first holdings in Vietnam

1887–1893
France establishes Indochina including Vietnam, Cambodia, and Laos

1930
Ho Chi Minh founds the Indochinese Communist Party

1940s
The Japanese control Vietnam

post-1945
France attempts to regain control

1950s
The United States begins to aid France to contain the spread of communism

1954
Geneva agreements end 8 years of warfare with the French; Vietnam is divided into North and South

1961
South Vietnam's regime is overthrown by a military coup

1965
The United States begins bombing North Vietnam

1973
The United States withdraws its troops and signs a cease-fire

1975
North Vietnamese troops capture Saigon and reunite the country; a U.S. embargo begins

1979
Vietnamese troops capture Cambodia; China invades Vietnam

1980s
Communist Vietnam begins "doi moi" economic liberalization policy; naval clashes with China over the Spratly Islands

(continued)

1990s
The U.S. economic embargo of Vietnam is lifted; the United States establishes full diplomatic relations

2000s
Relations with China improve with the resolution of a border dispute
A bilateral trade agreement is signed with the United States
Vietnam and the United States sign an agreement allowing for direct air flights from the two countries for the first time in three decades
Vietnam resumes fully normalized trade with the United States in 2004 and joins the World Trade Organization in 2007
Vietnam assumes a non-permanent seat on the UN Security Council

been traditional enemies long before the French arrived. Within the borders of Vietnam proper lived Thais, Laotians, Khmers, northern and southern Vietnamese, and mountain peoples whom the French called Montagnards. Most of the people could not read or write—and those who could wrote in Chinese, because the Vietnamese language did not have a writing system until the French created it. Most people were Buddhists and Taoists, but many also followed animist beliefs.

In addition to the social complexity, the French had to contend with a rugged and inhospitable land filled with high mountains and plateaus as well as lowland swamps kept damp by yearly monsoon rains. The French were eager to obtain the abundant rice, rubber, tea, coffee, and minerals of Vietnam, but they found that transporting these commodities to the coast for shipping was extremely difficult.

■ VIETNAMESE RESISTANCE

France's biggest problem, however, was local resistance. Anti-French sentiment began to solidify in the 1920s when Vietnamese university students started returning from France where they had learned about Western democracy; by the 1930s, Vietnamese youths were beginning to engage in open resistance. Prominent among these was Nguyen ai Quoc, who founded the Indochinese Communist Party in 1930 as a way of encouraging

? DID YOU KNOW?

The Ho Chi Minh Trail was famous during the Vietnam War as the communist supply route. Crisscrossing Vietnam and spilling over into Laos and Cambodia as well, the route allowed North Vietnam to supply its troops fighting in the south. In 2000, the government began an ambitious plan to modernize the route, turning parts of it into a multi-lane highway. Eventually (it will not be completed until 2020), it will run from the Chinese border to the Mekong Delta, a distance of almost 2,000 miles. Once bombed heavily by U.S. and South Vietnamese forces, the completed portions are already a tourist attraction, allowing people to safely travel though some of the most beautiful parts of Vietnam. It is estimated that the cost of the project will be well over US$3 billion.

the Vietnamese people to overthrow the French. He is better known to the world today as *Ho Chi Minh*, meaning "He Who Shines."

Probably none of the resisters would have succeeded in evicting the French had it not been for Nazi Germany's overrunning of France in 1940 and Japan's subsequent military occupation of Vietnam. These events convinced many Vietnamese that French power was no longer a threat to independence; the French remained nominally in control of Vietnam, but everyone knew that the Japanese had the real power. In 1941, Ho Chi Minh, having been trained in China by Maoist leaders, organized the League for the Independence of Vietnam, or Viet Minh. Upon the defeat of Japan in 1945, the Viet Minh assumed that they would take control of the government. France, however, insisted on reestablishing a French government. Within a year, the French and the Viet Minh were engaged in intense warfare, which lasted for eight years.

The Viet Minh initially fought the French with weapons supplied by the United States when that country was helping local peoples to resist the Japanese. Communist China later became the main supplier of assistance to the Viet Minh. This development convinced U.S. leaders that Vietnam under the Viet Minh would very likely become another Communist state. To prevent this occurrence, U.S. president Harry S. Truman decided to back France's efforts to recontrol Indochina (although the United States had originally opposed France's desire to regain its colonial holdings). In 1950, the United States gave $10 million in military aid to the French—an act that began a long, costly, and painful U.S. involvement in Vietnam.

In 1954, the French lost a major battle in the north of Vietnam, at Dien Bien Phu, after which they agreed to a settlement with the Viet Minh. The country was to be temporarily divided at the 17th Parallel (a latitude above which the Communist Viet Minh held sway and below which non-Communist Vietnamese had the upper hand), and country-wide elections were to be held in 1956. The elections were never held, however; and under Ho Chi Minh, Hanoi became the capital of North Vietnam, while Ngo Dinh Diem became president of South Vietnam, with its capital in Saigon.

■ THE UNITED STATES GOES TO WAR

Ho Chi Minh viewed the United States as yet another foreign power trying to control the Vietnamese people through its backing of the government in the South. The United States, concerned about the continuing attacks on the south by northern Communists and by southern Communist sympathizers known as Viet Cong, increased funding and sent military advisers to help prop up the increasingly fragile southern government. Unlike the French who had wanted to control Vietnam for its resources, the United States wanted to control the spread of communism. In 1955, The U.S. had helped airlift a million Vietnamese Catholics who were fleeing communism in the north, and U.S. President Eisenhower had

FURTHER INVESTIGATION

To learn more about the imbalance in the ratio of boys to girls in Vietnam, go to www .asiaone.com/News/Latest+News/Asia/Story/ A1Story20110712-288691.html

been urged, at that early date, to either send in troops or to even drop two small nuclear bombs (!) on the Communist north. He did neither, but the fear of a "domino effect" in which it was believed that one country after another would fall to communism, propelled U.S. President John F. Kennedy, in 1963, to send 12,000 military "advisers" to Vietnam. In 1964, an American destroyer ship was attacked in the Gulf of Tonkin by North Vietnam (some people believe, the attack was a hoax, intentionally fabricated to justify direct U.S. military involvement in Vietnam). The U.S. Congress responded by giving then-president Lyndon Johnson a free hand in ordering U.S. military action against the north; before this time, U.S. troops had not been involved in direct combat.

By 1969, some 542,000 American soldiers and nearly 66,000 soldiers from 40 other countries were in battle against North Vietnamese and Viet Cong troops. Despite unprecedented levels of bombings and the use of sophisticated electronic weaponry and powerful chemicals such as Agent Orange (used to strip trees of foliage so the enemy could be spotted more easily), U.S. and South Vietnamese forces continued to lose ground to the Communists. The Communists used guerrilla tactics and built their successes on latent antiforeign sentiment among the masses as well as extensive Soviet military aid. At the height of the war, as many as 300 U.S. soldiers were being killed every week.

Watching the war on the evening television news, many Americans began to feel that the war was a mistake. Anti-Vietnam rallies became a daily occurrence on American university campuses, and many people began finding ways to protest U.S. involvement: dodging the military draft by fleeing to Canada, burning down ROTC buildings, and publicly challenging the U.S. government to withdraw. President Richard Nixon had once declared that he was not going to be the first president to lose a war; but after his expansion of the bombing into Cambodia to destroy Communist supply lines and after significant battlefield losses, domestic resistance became so great that an American withdrawal seemed inevitable. The U.S. attempt to "Vietnamize" the war by training South Vietnamese troops and supplying them with advanced weapons did little to change South Vietnam's sense of having been sold out by the Americans.

Secretary of State Henry Kissinger negotiated a cease-fire settlement with the North in 1973; but most people believed that as soon as the Americans left, the North would resume fighting and would probably take control of the entire country. This indeed happened, and in April 1975, under imminent attack by victorious North Vietnamese soldiers, the last Americans lifted off in helicopters from the roof of the U.S. Embassy in Saigon. The South Vietnamese government subsequently surrendered, and South Vietnam ceased to exist. The country had, at last, been reunited, but in a way that no one would have predicted.

The war wreaked devastation on Vietnam. It has been estimated that nearly 2 million people were killed during just the American phase of the war; another 2.5 million were killed during the French era. In addition, 4.5 million people were wounded, and nearly 9 million lost their homes. U.S. casualties included more than 58,000 soldiers killed and 300,000 wounded. In the end, the Vietnamese got what they had wanted all along: the right to run their own country without outside interference.

■ A CULTURE, NOT JUST A BATTLEFIELD

Because of the Vietnam War (which the Vietnamese call the American War), many people think of Vietnam as if it were just a battlefield. But Vietnam is much more than that. It is a rich culture made up of peoples representing diverse aspects of Asian life. In good times, Vietnam's dinner tables are supplied with dozens of varieties of fish and shrimp, as well as the ever-present bowls of rice and beef soup, or *pho* from the French word "feu" for fire. *Pho* is a mixture of beef soup, rice noodles, meat spiced with cloves, Asian basil, white onions, and star anise. It is enjoyed at both breakfast and dinner. Sugarcane and bananas are also favorites. Because about 70 percent of the people live in the countryside, the population as a whole possesses a living library of practical know-how about farming, livestock raising, fishing, and home manufacture. About 13 percent of appropriate age students go to universities, and the government vigorously promotes higher education. Most children attend elementary school, and 94 percent of the adult population can read and write.

Literacy was not always so high; much of the credit is due the Communist government, which, for political-education reasons, has promoted schooling throughout the country. Another government initiative—unifying the two halves of the country—has not been as easy, for, upon the division of the country in 1954, the North followed a socialist route of economic development, while in the South, capitalism became the norm.

Religious belief in Vietnam is an eclectic affair and reflects the history of the nation; on top of a Confucian and Taoist foundation, created during the centuries of Chinese rule, rests Buddhism (a modern version of which is called Hoa Hao and claims 1 million believers); French Catholicism, which claims about 7 percent of the population; and a syncretist faith called Cao Dai, which claims about 2 million followers. Cao Dai models itself after Catholicism in terms of hierarchy and religious architecture, but it differs in that it accepts many Gods—Jesus, Buddha, Mohammed, Lao-Tse, and others—as part of its pantheon.

Many Vietnamese pray to their ancestors and ask for blessings at small shrines located inside their homes.

Animism, the worship of spirits believed to live in nature, is also practiced by many of the Montagnards. These mountain peoples, most of whom are Protestants, actively resisted the Communists during and after the Vietnam War. The government has now officially labeled them a separatist movement and has banned their demonstrations for religious freedom and private land rights. Many of them fled to Cambodia in 2001, and, as of 2011, many were resettled in Canada and the United States.

Religious tension remains high in Vietnam. In 2004, over a thousand, mostly Protestant, ethnic Christians rioted over the Easter weekend in Vietnam's Central Highland city of Daklak. What started out as a peaceful demonstration against religious repression and land confiscation turned violent, and government troops ended up arresting scores of protestors and sealing off the entire region. Despite these problems, religious freedom has improved enough in recent years that the United States has removed Vietnam from its list of human rights abusers. Signs of improvement in other areas of human rights are also appearing. For instance, Pham Hong Son, a physician who was imprisoned for four years for translating items from a U.S. government website and posting them on the Internet in Vietnamese, was released in 2006.

Vietnamese citizens are permitted to believe in God and to participate in religious rituals, but Western religions are regarded warily, and those who practice them are often harassed, especially if, along with Christianity, they begin demanding democratic reforms. Periodically, the government launches a wave of arrests against those advocating democratic reforms. When one Vietnamese actor, Don Duong, appeared in Hollywood movies about the Vietnam War that the government did not like, he was labeled a "national traitor" and told to prepare for deportation. These, and many other examples confirm that the Communist government intends to remain in control and to prevent a "gradual evolution" toward Western-style democracy—a formidable task, given that half the country already was just such a place before the North won the war.

■ THE ECONOMY

When the Communists won the war in 1975 and brought the capitalist South under its jurisdiction, the United States imposed an economic embargo on Vietnam, which most other nations honored and which remained in effect for 19 years, until President Bill Clinton ended it in 1994. As a consequence of war damage as well as the embargo and the continuing military involvement of Vietnam in the Cambodian War and against the Chinese along their mutual borders, the first decade after the end of the Vietnam War saw the entire nation fall into a severe economic slump. Whereas Vietnam had once been an exporter of rice, in the 1980s it had to import rice from abroad. Inflation raged at 250 percent a year, and the government was hard-pressed to cover its debts. Many South Vietnamese were, of course, opposed to Communist rule and

attempted to flee on boats—some 100,000 of them—but, contrary to popular opinion, most refugees left Vietnam because they could not get enough to eat, not because they were being persecuted.

Beginning in the mid-1980s, the Vietnamese government began to liberalize the economy. Under a renovation and restructuring plan called *doi moi* (similar in meaning to Soviet *perestroika*), the government began to introduce elements of free enterprise into the economy. Moreover, despite the Communist victory, the South remained largely capitalist. For a while after the U.S. trade embargo was lifted outside investors poured money into the Vietnamese economy. Australia, Japan, and France invested $3 billion in a single year; and in 1994, more than 540 firms from Singapore, Hong Kong, Japan, and France were doing business with Vietnam. Furthermore, the World Bank, the Asian Development Bank, and the International Monetary Fund were providing development loans.

Unfortunately, the financial crises in nearby Malaysia, Indonesia, and other countries caused Vietnam to devalue its currency in 1998. Combined with El Niño–caused drought, forest fires (some 900 in 1998), and floods (1999 floods killed more than 700 people), the currency devaluation slowed economic growth and gave government leaders an excuse to trumpet the failures of capitalism. But the worst shock was the withdrawal of many of the companies that had wanted to invest in Vietnam. They withdrew, reluctantly, because of crippling government red tape, unfair business policies, and blatant corruption (a situation often heard in China today as well).

Much to the worry of government traditionalists, the Vietnamese people seem fascinated with foreign products. They want to move ahead and put the decades of warfare behind them. Western travelers in Vietnam are treated warmly, and the Vietnamese government has cooperated with the U.S. government's demands for more information about missing U.S. soldiers—the remains of five more of whom were located as recently as 2004, thirty years after the war's end. In 1994, after a 40-year absence, the United States opened up a diplomatic mission in Hanoi as a first step toward full diplomatic recognition. So eager are the Vietnamese to reestablish economic ties with the West that the Communist authorities have opened up the possibility of allowing the U.S. Navy to lease its former port at Cam Ranh Bay, one of the finest deep water ports in Asia. In 2010, the government announced it would allow foreign ships to dock at the port after Russia finishes upgrading the repair facilities in 2014. Diplomatic bridge-building between the United States and Vietnam increased in the 1990s, when a desire to end the agony of the Cambodian conflict created opportunities for the two sides to talk together. Telecommunications were established in 1992, and in the same year, the United States gave US$1 million in aid to assist handicapped Vietnamese war Veterans. In 2004, Vietnam was the only Asian country included in the U.S. government's US$15 billion AIDS program. Two decades after the end of the Vietnam War, the United States established

(© Tim Hall/Getty/Photodisc RF)
Outdoor market in Hanoi, Vietnam.

full diplomatic relations with Vietnam. Then, in 2001, after difficult and lengthy negotiations, the two countries signed a trade pact. With tariffs reduced or eliminated, exports to the United States doubled, and cotton, fertilizer, seafood, shoes, and other products began being shipped across the Pacific. The agreement was difficult for many of the hard-line Communist leaders; but after a 1997 farmers' riot and other evidences of citizen unrest, they recognized that they had to do something to salvage the struggling economy. The trade agreement was the first step toward Vietnam joining the World Trade Organization in 2007.

Despite this gradual warming of relations, however, anti-Western sentiment remains strong in some parts of the population, particularly the military. As recently as 1996, police were still tearing down or covering up signs advertising Western products, and anti-open-door policy editorials were still appearing in official newspapers. The situation was not helped by the visit, in 2000, of U.S. Senator John McCain, a former prisoner of war in Vietnam, during which he publicly declared that the wrong side had won the war. Still, the Vietnamese gave a formal state welcome to U.S. Defense Secretary William Cohen, who discussed military and other topics, especially the missing-in-action issue, with Vietnamese leaders. After joining the WTO, Vietnam began privatizing banks and other businesses as Vietnam at last moved decidedly away from the socialist model of economic control. In 2009, the first products from Vietnam's new US$3 billion Dung Quat oil refinery were shipped. The refinery could eventually produce enough to handle about one-third of Vietnam's fuel needs. Crude oil is the country's largest export item, but refinery capacity is still low.

As money flows into government coffers, corruption seems to increase. For instance, in 2006, a number of transport ministry officials were accused of using World Bank money (some US$7 million of it) to gamble on European soccer matches.

■ HEARTS AND MINDS

As one might expect, resistance to the current Vietnamese government comes largely from the South Vietnamese, who, under both French and American tutelage, adopted Western values of capitalism and consumerism. Many South Vietnamese had feared that after the North's victory, South Vietnamese soldiers would be mercilessly killed by the victors; thousands were, in fact, killed, but many—up to 400,000—former government leaders and military officers were instead sent to "reeducation camps," where, combined with hard labor, they were taught the values of socialist thinking. Several hundred such internees still remain incarcerated. Many of the well-known leaders of the South fled the country when the Communists arrived and have long since made new lives for themselves in the United States, Canada, Australia, and other Western countries. Those who have remained—for example, Vietnamese members of the Roman Catholic Church—have occasionally resisted the Communists openly, but their protests have been silenced. Hanoi continues to insist on policies that remove the rights to which the South Vietnamese had become accustomed. For instance, the regime has halted publication and dissemination of books that it judges to have "harmful contents." There is not much that average Vietnamese can do to change these policies except passive obstruction, which many are doing even though it damages the efficiency of the economy.

Over the years, Vietnam has improved relations with some of its neighbors. In 1999, it settled a long-standing, 740-mile border dispute with China; and in 2000, it accepted the last repatriated Vietnamese from refugee camps in Hong Kong. Also in 2000, the government

Where are the Girls?

Vietnam, like China, is undergoing an upheaval in the ratio of boys to girls. In 2010, for every 111 boys born, there were only 100 girls born. This is likely the result of traditional beliefs that favor having boys to care for parents in their old age (girls, once married, join their husband's extended family and so are not available for their own parents). Usually, there are about five more boys per hundred, but due to such modern medical devices as ultrasound machines, parents are choosing to have sex-selected abortions (technically illegal, but heavily practiced). In the countryside, some villages have nearly 30 more boys per hundred, a situation that raises serious questions about the ability of men to find women to marry when they are adults. Some studies suggest that the presence of large numbers of unmarried men in a society is a precursor of social and political unrest. Unless things change, some people predict Vietnam's ratio could rise as high as 120 boys for every 100 girls.

Snapshot: VIETNAM

Summarized below is a quick look at the country with regard to its development, freedom, health/welfare, and achievements.

Development

In 2010, due to a rapidly growing economy, the government achieved its goal of being removed from the IMF's list of undeveloped countries. In 2007, Vietnam joined the World Trade Organization and promised to privatize over 70 major corporations, including several banks. US$8 billion (8% of the GDP) of the economy comes in each year from remittances sent from overseas Vietnamese, especially those in the United States. Vietnam now exports rice, cement, fertilizer, steel, and coal. Vietnam is the largest producer of cashew nuts in the world.

Freedom

Vietnam is nominally governed by an elected National Assembly that meets once every five years. Real power, however, resides in the Communist Party and with those military leaders who helped defeat the U.S. and South Vietnamese armies. Civil rights, such as the right of free speech, are curtailed. Private-property rights are limited. In 1995, Vietnam adopted its first civil code providing property and inheritance rights for citizens. The U.S. State Department has dropped Vietnam from its list of human rights abusers.

Health/Welfare

Health care, along with a social security system, is nationalized, but aspects of a free market system have been seeping in. Nutrition for the population has improved significantly and infant mortality rates are down. Polio was eradicated in 1996. HIV/AIDS is a growing problem. Fertility rates are now lower than replacement level. The World Health Organization has been involved in disease-abatement programs since reunification of the country in 1975.

Achievements

Vietnam provides free and compulsory schooling for all children. The curricular content has been changed in an attempt to eliminate Western influences. New Economic Zones have been created in rural areas to try to lure people away from the major cities of Hanoi, Hue, and Ho Chi Minh City (formerly Saigon). Vietnam has one of the highest school enrollment rates in the world.

---

announced that it would build a multi-lane highway along the route of the famous Ho Chi Minh Trail. The highway will benefit Laos and Cambodia, as well as Vietnam.

One country with which Vietnam still has a very shaky relationship is China. In mid-2011, hundreds of Vietnamese protested at the Chinese Embassy against China's live-fire military drills in South China Sea waters claimed by Vietnam. They demanded that China stop violating Vietnam's sovereign territory and allow Vietnam to engage in oil exploration without further harassment from China. In response to China's activities in the area, Vietnam's navy conducted live-fire drills of its own. China has provoked similar protests in the Philippines and other countries claiming parts of the area.

On health matters, the government has been concerned with the devastation caused by the bird flu. Over 600,000 chickens in Vietnam (and thousands more in Hong Kong, China, and others countries) had to be destroyed to try to stop the disease from spreading. The illness also affects humans, and at the height of the outbreak in 2009, over 2,000 Vietnamese were infected and over 40 had died.

Statistics

Geography

Area in Square Miles (Kilometers): 127,881 (331,210) (slightly larger than New Mexico)

Capital (Population): Hanoi (2,668,000)

Environmental Concerns: deforestation; soil degradation; overfishing; water and air pollution; groundwater contamination

Geographical Features: low, flat delta in south and north; central highlands; hilly and mountainous in far north and northwest

Climate: tropical in south; monsoonal in north

People

Population

Total: 90,549,390

Annual Growth Rate: 1.08%

Rural/Urban Population Ratio: 70/30

Major Languages: Vietnamese; English; French; Chinese; Khmer; mountain area languages

Ethnic Makeup: 86% Vietnamese; 9% Tay, Thai, Muong and other mountain tribes; 5% other

Religions: Buddhist 9%; Catholic 7%; Hoa Hao 1.5%; Cao Dai 1%; Protestant 0.5%; Muslim 0.1%; none 80.9%

Health

Life Expectancy at Birth: 70 years (male); 75 years (female)

Infant Mortality: 21/1,000 live births

Physicians Available: 1/1,000 people

HIV/AIDS Rate in Adults: 0.4%

Education

Adult Literacy Rate: 94%

Compulsory (Ages): 6–11; free

Communication

Telephones: 17.4 million main lines (2009)
Mobile Phones: 98.2 million (2009)
Internet Users: 28.6 million (2011)
Internet Penetration (% of Pop.): 32%

Transportation

Roadways in Miles (Kilometers): 106,498 (171,392)
Railroads in Miles (Kilometers): 1,458 (2,347)
Usable Airfields: 44

Government

Type: Communist state
Independence Date: September 2, 1945 (from France)
Head of State/Government: President Truong Tan Sang; Prime Minister Nguyen Tan Dung
Political Parties: Communist Party of Vietnam; other parties proscribed
Suffrage: universal at 18

Military

Military Expenditures (% of GDP): 2.5%
Current Disputes: boundary/border disputes with Cambodia and other countries

Economy

Currency ($ U.S. Equivalent): 19,148.9 dong = $1
Per Capita Income/GDP: $3,100/$103.6 billion
GDP Growth Rate: 6.8%
Inflation Rate: 9%
Unemployment Rate: 4.4%
Labor Force by Occupation: 54% agriculture; 26% services; 20% industry
Population Below Poverty Line: 10.6%
Natural Resources: phosphates; coal; manganese; bauxite; chromate; oil and gas deposits; forests; hydropower
Agriculture: rice; corn; potatoes; rubber; soybeans; coffee; tea; cotton; pepper; cashews; sugarcane; peanuts; bananas; poultry; fish; seafood
Industry: food processing; textiles; machine building; mining; cement; chemical fertilizer; glass; tires; petroleum; fishing; shoes; steel; coal; paper
Exports: $72 billion (primary partners United States, Japan, China)
Imports: $80 billion (primary partners China, South Korea, Japan, Taiwan)

Suggested Websites

www.lcweb2.loc.gov/frd/cs/ntoc.html
www.vietnamtourism.com/

Articles From the World Press

Topic Guide

All the articles that relate to each topic are listed below the bold-faced term.

1 Deep Danger: Competing Claims in the South China Sea

Marvin C. Ott

"The South China Sea is a growing focus of concern in Washington, at the headquarters of the US Pacific Command in Honolulu, and in a number of Southeast Asian capitals."

Learning Objectives

After reading this article, you will more clearly understand the following:

- Mischief Reef
- South China Sea claims
- International waterways
- ASEAN Regional Forum (ARF)
- Freedom of navigation
- Political ambiguity

The waters of the South China Sea are dotted with hundreds of atolls, reefs, and small islands—only one of which has sufficient fresh water to qualify, under traditional international law, as capable of supporting human habitation. Nonetheless, these land features and the 1.35 million square miles of water that surround them are the subject of competing territorial claims by China and Taiwan (whose claims appear to encompass the entire South China Sea and all of its land features) and by five Southeast Asian countries (Malaysia, Brunei, Vietnam, the Philippines, and Indonesia, though Indonesia's claim is limited to waters at the sea's extreme southern tip). Major island groups in dispute include the Paracels, which are occupied by China, and the Spratlys, where multiple claimants have placed outposts.

Despite the sea's evident potential for generating conflict, it remained an obscure afterthought in international politics until the mid-1990s. Even China's military occupation of the Paracels in 1974, which involved a naval engagement with South Vietnamese forces, was barely noticed by the international press. Claimants to the rest of the sea were not in position to enforce claims or exercise effective authority over such a logistically daunting area. Moreover, the conflicts that inflamed Southeast Asia during the cold war were land-based, and the dominant naval power in the region, the United States, had neither claims of its own nor interest in championing the claims of others.

All this began to change in 1995 when the Philippines discovered that China had erected a fortified military outpost on the Spratly Islands' remote but aptly named Mischief Reef. The facility was striking in large part because of its location—120 nautical miles from the Philippines (Palawan) but over 600 nautical miles from China

(Hainan). The Philippines protested to China. Manila also attempted unsuccessfully to enlist US military support, but did succeed in persuading the Association of Southeast Asian Nations (ASEAN) to express strong collective concern to China.

Beijing responded with sustained outreach to Southeast Asia, and sought to strengthen ties in the region and burnish its image as a "good neighbor." A centerpiece of this was a Declaration on the Conduct of Parties in the South China Sea, signed by ASEAN and China in 2002, in which all parties pledged good behavior pending the resolution of conflicting claims. China also began to tout its willingness to engage in the "joint development" of presumed petroleum and mineral resources in the South China Sea while setting aside conflicting claims. China's efforts at reassurance did not, however, involve abandoning its claims to the South China Sea or its facility on Mischief Reef—where construction and upgrades proceeded apace.

By the beginning of 2010 the geopolitics of the South China Sea had settled into three themes. First, the incipient conflict between China's claims and those of Indonesia, Malaysia, the Philippines, and Brunei had been put on the back burner as the governments concerned proceeded with other business. Second, Vietnam had emerged as the exception to this rule, as Chinese naval and maritime patrol vessels seized Vietnamese fishing boats and pressured Western oil companies exploring under Vietnamese licenses to cease activities. Despite the largely successful delineation of the land border between China and Vietnam, their sea boundary in the Gulf of Tonkin remained in serious dispute. Third, a few disturbing incidents occurred in which Chinese patrol craft challenged and harassed US naval surveillance ships operating in international waters within China's exclusive economic zone (EEZ). The most prominent incident, in March 2009, involved the *Impeccable,* which Chinese ships and aircraft tried to force from the area in which it was operating, about 75 miles from Hainan Island. In sum, the South China Sea remained relatively quiescent, yet some ominous storm clouds were visible on the horizon.

■ CALL AND RESPONSE

In July 2010, US Secretary of State Hillary Clinton spoke at a meeting of the ASEAN Regional Forum (ARF) in Hanoi. (The ARF convenes 27 nations at a ministerial-level conclave to discuss security issues in Asia. Because it is organized around the 10 nations of ASEAN, the ARF's regional focus tends to be Southeast Asian.) At the apparent urging of host Vietnam, Clinton directed her remarks to issues involving the South China Sea. After noting that several Southeast Asian nations lay claim to at least some portion of the South China Sea or its land features, she urged the attendees to endorse two traditional principles of international diplomacy: multilateral negotiations for multilateral disputes and the international status of established commercial sea-lanes.

> **By asserting sovereignty over sea-lanes, China has taken a position that no major country in the world can support.**

Regarding the sea-lanes through the South China Sea—which by some measures are the world's busiest—Clinton said: "The United States has a national interest in freedom of navigation, open access to Asia's maritime commons, and respect for international law." That statement reiterated the long-standing US policy that sea-lanes in the South China Sea were not subject to control or ownership (sovereignty) by any country and that their status was a vital interest of the United States, which regularly traverses them with commercial and naval vessels. Twelve of the twenty-seven representatives at the ARF, including a majority from ASEAN, spoke in support of the US position.

Had the Chinese foreign minister, Yang Jiechi, chosen not to respond, Clinton's statement and the meeting itself would have received only perfunctory attention outside a small circle of specialists and regional officials. But the foreign minister did react—angrily. He accused the United States of meddling in matters that did not concern it and seemed particularly incensed that other countries, by supporting the United States, had engaged in orchestrated opposition to China.

Lest there be any doubt, a Foreign Ministry spokesperson in Beijing subsequently stated, "We resolutely oppose any country which has no connection to the South China Sea getting involved in the dispute, and we oppose the internationalization, multilateralization, or expansion of the issue." Meanwhile (following the Hanoi meeting), a spokesman for the Ministry of Defense declared on the record that "China has indisputable sovereignty" over the South China Sea.

The effect of these developments has been to highlight the importance of the South China Sea as an arena of international tension and potential conflict, but also as a test and indicator of China's strategic intentions toward Southeast Asia. Thus, the South China Sea is a growing focus of concern in Washington, at the headquarters of the US Pacific Command in Honolulu, and in a number of Southeast Asian capitals.

For most of the past two decades, a remarkable amount of uncertainty has surrounded China's strategic intent to its south. Academic students of China and government officials, notably in Southeast Asia, have been—and remain—unsure and divided in their views. China has sent conflicting signals, whether inadvertently or by design, that have contributed substantially to the confusion. But since the ARF meetings, close observers have detected a discernible coalescence of opinion—and concern.

■ PEACE AND CHARM

For Southeast Asian governments, geography and population dictate that China will be a major, if not overwhelming, issue in their foreign relations. China shares a long land boundary with the region; through more than two millennia of history, imperial China saw the Nanyang (South Seas), through the lens of Confucian civilization, as subordinate and tributary. From this perspective the European and American colonization of most of Southeast Asia interrupted a long-established and natural relationship. But colonization also attracted large numbers of Chinese settlers to the region. This has left contemporary Southeast Asia with a problematic legacy—large, economically prominent, ethnic Chinese populations.

Southeast Asian governments in the decades since the People's Republic was established have seen China move through the full spectrum of capabilities and behaviors. In the 1950s and even later, China promoted the idea of communist revolutionary power, which championed Marxist insurgencies and urban movements that were intended to overthrow first the colonial and then the postcolonial regimes in the region. Ultimately communist movements came to power in the former states of French Indochina but not beyond.

From the mid-1960s to the mid-1970s China was consumed by Mao Zedong's campaign of extreme domestic radicalization—the Great Proletarian Cultural Revolution. During this time China virtually ceased to be a factor in Southeast Asia, and elsewhere overseas. But with Mao's death, and with Deng Xiaoping's ascent to the position of paramount leader and his embrace of Western-style economic reform, China's overall trajectory and its presence in Southeast Asia took a sharp and welcome turn. Beijing became the champion of increased economic ties as well as regional growth and stability.

The message to Southeast Asia, capsulized as "China's peaceful rise," was that of positive-sum, mutually advantageous relationships. The contrast with earlier periods—all within the professional memory of senior Southeast Asian officials—could hardly have been more dramatic.

In terms of strategic outlook, China's contemporary leaders evoke the classic realists of nineteenth-century Europe—vitally concerned with prerogatives of sovereignty and the sanctity of borders, animated by calculations of power and influence. From the standpoint of the Chinese regime, Southeast Asia is properly understood as a natural and rightful Chinese sphere of influence, a region where China's interests are paramount. When these are properly acknowledged, Beijing is prepared to adopt policies that benefit Southeast Asia as well as China—a dominion of Confucian harmony and benevolence.

China's presentation of itself to Southeast Asia as a benign neighbor, sometimes characterized as a "charm offensive," reached full flower beginning in the mid-1990s. Diplomatic efforts produced a series of tangible achievements including an ASEAN-China Free Trade Agreement; framework agreements for security cooperation between China and each ASEAN member; the aforementioned Declaration on the Conduct of Parties in the South China Sea; and an elaborate "dialogue" process of regular, structured interaction on diplomatic, economic, and defense issues.

All of this was underpinned by trade and investment ties, which have grown to the point that China has replaced Japan and America as ASEAN's largest trading partner. Perhaps even more significantly, China has invested heavily in infrastructure (rail, roads, river transport, pipelines, and electrical grids), an undertaking designed to link Southeast Asia and southern China as a single economic unit. At the same time, China is building a "cascade" of massive hydroelectric dams on the upper Mekong River in southern China. These dams will not only produce electric power but will also give China the ability to control the flow of the Mekong River system, with untold consequences for downstream states.

China portrays all of these developments as natural and benign consequences of its "peaceful rise," and as substantial, tangible benefits for Southeast Asia. But one need not be paranoid to see these same developments as consistent with—or precursors to—a Chinese strategy for dominance over Southeast Asia. In the region's capitals, after years of giving credence to China's portrayal of itself in soft-power terms, unease and doubt have grown perceptibly regarding China's growing hard-power capabilities and apparent strategic intent. These doubts are provoked—not exclusively, but in substantial part—by China's statements, actions, and military buildup with regard to the South China Sea.

■ FOLLOW THE DOTTED LINE

Since its founding the People's Republic of China has published maps adopting a maritime boundary ("the nine-dotted line") first promulgated by the Republic of China in 1936 and encompassing the entire South China Sea. While some other boundary claims by Beijing have sparked immediate controversy (for example, regarding India, Tibet, and the Soviet Union), the expansive notion of China's maritime boundary has usually generated little attention.

This changed only briefly with the Mischief Reef episode—which was followed by China's efforts to assuage Southeast Asian concerns and effectively remove the South China Sea from the diplomatic front burner. It was in Beijing's interest to soft-pedal the issue: An aggressive claim to the entire South China Sea would have pitted China against ASEAN, and China in any case lacked the military capacity to enforce its claim. Deng had often reminded his countrymen of a traditional Chinese aphorism that roughly translates as "bide your time and conceal your capabilities until you are ready to act." For Beijing, clarity was a danger and ambiguity was an asset when it came to the South China Sea.

In the years following, a dense conceptual fog enveloped the Chinese position. Some of this was a natural byproduct of the fact that different Chinese voices (academic, diplomatic, military, journalistic) addressed the issue without clear guidance from on high. But much of it was calculated, and the result was uncertainty and disagreement in the small community of outside observers and officials who followed the issue. The prevailing view was that China was claiming something less than full sovereignty—largely because Beijing refrained from using that word. According to this view, the dotted line denoted something other than a legal international boundary, but just what it did denote was murky.

Ample grounds for confusion existed. At various times Chinese officials have cited as a basis for China's claim different and mutually inconsistent rationales, including historic presence, an archipelagic principle, an EEZ principle, and a continental-shelf principle. China rebuts Japan's claims to outcroppings in the East China Sea, noting that they are not habitable as international law requires—but China has cited the same kind of land features to justify its own claims to the South China Sea. China's 1958 "Declaration on China's Territorial Sea" refers to "high seas" in the South China Sea—which contradicts the notion of a territorial sea.

> The Chinese foreign minister stared across the table at his ASEAN counterparts and observed that some countries are "small" and China is "big."

In addition, legislation adopted by China in 1992 that put the dotted line into law referred to "historic Chinese waters"—a category that has no standing under international law. Beijing has drawn archipelagic baselines around the Paracels (which it claims) but not around the Spratlys (which it also claims). And China has ratified the UN Law of the Sea, but with reservations that make ratification almost meaningless.

Moreover, China has, by declaring a "coastal economic exclusion zone," given the concept of an EEZ an interpretation unrecognized in international law. In an effort to rebut a joint Malaysian-Vietnamese submission to the United Nations, China in 2009 submitted a map that included its dotted-line boundary but contained no justification for it. (Indonesia responded with a formal request to the United Nations that Beijing clarify its claim; China has remained silent.) Indeed, the dotted line has never been precisely demarcated, and large sections of it (for example, near the Natuna Islands) remain entirely opaque.

The fog dissipates when we examine the proposition that China's dotted line is intended to be exactly what Chinese officials have said it is—a demarcation of China's maritime border. Inside the line is Chinese sovereign territory.

Consider, first, that the dotted line that appears on all Chinese-produced maps extends around Taiwan—and no doubt whatsoever exists that Beijing views Taiwan as sovereign Chinese territory. China in 1974 deployed naval forces to seize from Vietnam the Paracels, an archipelago that is not characterized by China as separate and distinct from the South China Sea. And the People's Liberation Army (PLA) has built an impressive military outpost on a reef located over 600 nautical miles from China.

Consider, too, that China's 1992 territorial law affirmed the dotted line and mandated that Chinese armed forces defend the country's maritime territory. China's rapid buildup in military capabilities has focused on projection of naval and air power beyond China's shores. The Chinese navy, in the meantime, has stopped Vietnamese fishermen from operating well within Vietnam's EEZ, while Beijing has warned international oil companies away from Vietnamese offshore leaseholds.

What is more, while China agreed to sign the Declaration on the Conduct of Parties in the South China Sea, it refused to make the agreement legally binding or to refrain from building new structures. Two PLA senior colonels in a public symposium hosted by the US Pacific Command, when asked if the American Seventh Fleet had a right to traverse the South China Sea without China's permission, answered "no." And in a recent display of technological prowess, a Chinese submersible descended to the deepest portion of the South China Sea and planted a Chinese flag there. In various discussions Chinese officials have referred to the South China Sea as a "core interest"—a term previously reserved for Taiwan and Tibet.

Against this backdrop, the ARF meetings in Hanoi provided a clarifying moment—perhaps no more so than when the Chinese foreign minister stared across the table at his ASEAN counterparts and pointedly observed that some countries are "small" and China is "big."

■ SOUTHEASTERN STRICTURES

America's willingness to stake out a position in support of a maritime commons, not a territorial sea, and in favor of multilateral diplomacy, as opposed to China's determination to deal with the Southeast Asian countries one at a time, was welcome in many regional capitals. It provided a vital and long-overdue signal that the ASEAN governments did not have to cope with China alone and enjoyed the support of a powerful friend. In this sense, Clinton's initiative has provided ASEAN a dose of courage and self-confidence in its relationship with China.

That said, US policy makers must maintain a healthy awareness of what Southeast Asian governments are in fact able and willing to do. To employ an overused metaphor, at least some ASEAN members may be prepared to

hold America's coat if Washington duels Beijing. But, for a number of compelling reasons, they cannot be expected to enter the arena themselves in any but carefully circumscribed ways.

First, it has long been a truism that Southeast Asian governments fear being forced to choose between China and the United States. The regional consensus is that the US-China relationship is vitally important to all concerned. When leaders in the region are asked what kind of relationship best protects Southeast Asian interests, the answer is a variation on the Goldilocks principle—"not too hot and not too cold." A cooperative relationship, but not a deeply collaborative one, would be just right. Just as they fear China-US conflict, the ASEAN countries also fear its opposite—a great power condominium deciding regional issues with little input from Southeast Asia.

Second, China's influence over and strategic reach into Southeast Asia are deep, powerful, and growing. This is particularly evident in the economic sphere. Between 2009 and 2010, aggregate trade increased roughly 50 percent year-on-year. Not coincidentally, the China-ASEAN Free Trade Area entered into force at the beginning of 2010.

Third, despite significant investments in military modernization, no Southeast Asian country is prepared to confront China militarily. The only country that has done so in recent decades is Vietnam, in response to China's 1979 invasion across its northern boundary. Vietnamese forces acquitted themselves well in that encounter, but Hanoi is under no illusion that such success could be replicated today. The only naval and air forces that can credibly face off against China in the South China Sea are American—and if it came to that, US commanders could expect little or no operational support from ASEAN, with the possible and limited exception of Vietnam.

Fourth, ASEAN is not the feckless cave of winds that some Westerners describe—but it is also not a unified, purposeful actor regarding the South China Sea. Several ASEAN governments, including those of Laos, Cambodia, and Myanmar, are highly responsive to Chinese interests and have no dog in the fight over the South China Sea. The best that Washington can expect—and only if it is assiduously nurtured—is cautious diplomatic support along the lines of what one saw at the ARF meeting in Hanoi.

■ BEYOND SERIOUS

The ramifications of a serious Chinese claim to the entire South China Sea are profound. By asserting sovereignty over the sea-lanes, China has taken a position that no major country in the world can support—not the Europeans, not Japan, not India, not Australia, not the United States, and not the principal ASEAN states. Obviously, when a rapidly rising global power takes such a step, the implications are beyond serious.

In addition, the South China Sea, like Taiwan, has the clear potential to spark armed conflict between the United States and China. This is a specter that keeps military planners at the US Pacific Command awake at night. The danger is made greater by China's evident assumption that the United States is on the decline (along with its defense expenditures) while China is on the rise (including its defense expenditures).

Meanwhile, though it is generally underappreciated, a remarkable and unique security architecture has emerged in Southeast Asia. It is, in Victor Cha's apt phrase, a "complex patchwork" of multilateral dialogue mechanisms and bilateral security commitments involving the United States. It has effectively kept the peace in the region over the past 35 years and holds promise for continuing to do so for at least the medium term. A major confrontation in the South China Sea has the potential to harm that architecture beyond repair.

Recent events and statements have clearly framed the current strategic landscape in the South China Sea. On one hand we have seen several gestures by China that might be broadly characterized as conciliatory. General Chen Bingde, the PLA chief of staff, paid a weeklong visit to Washington in May 2011. In a major address to a US military audience, he stated that "China never intends to [militarily] challenge the [United States]," while noting the continued superiority of American armed forces. Meanwhile, Chinese diplomats have been at pains to suggest that previous references to a Chinese "core interest" in the South China Sea may have been misunderstood. At the annual Shangri-la Dialogue in Singapore in June, China's defense minister declared that China did not "seek hegemony" in the region.

However, at virtually the same time, both Vietnam and the Philippines have registered public complaints over what they view as China's hegemonic behavior. Vietnam in May 2011 complained that Chinese patrol boats confronted a Vietnamese oil exploration vessel operating off the coast of southern Vietnam and deliberately cut the ship's cables—the second such incident in two weeks. This produced an anti-China protest demonstration in Hanoi. Manila in June accused the Chinese navy of firing on Filipino fishermen, intimidating an oil exploration ship from the Philippines, and placing markers (posts and a buoy) in areas of the Spratlys claimed by the Philippines.

> **For most of the past two decades, a remarkable amount of uncertainty has surrounded China's strategic intent to its south.**

What is most interesting and significant is the Chinese reaction to these and similar events. A Foreign Ministry spokesman demanded that both countries stop infringing on China's sovereign territory. The authoritative *China Daily* carried an opinion piece by a prominent Chinese academic (Gong Jianhua) claiming that Vietnam and the Philippines had taken advantage of China's restraint by trying to convert what was a bilateral dispute into a multilateral one.

"In the beginning," the piece said, "the South China Sea dispute was not referred to any international or regional organization. But after the formation of ASEAN, Vietnam, the Philippines, and some other countries used it as a regional platform to coordinate their positions to 'speak in one voice' and gain strategic advantage against China. . . . [And] now the United States has jumped into the dispute." The author went on to assert that China was at a disadvantage, "with only a small number of disputed islands under its control." Also, "without a formidable navy . . . China is in an unfavorable position. To become an influential power, China has to transform from a 'continental power' to a 'maritime power.' And the South China Sea dispute is a real test for it to achieve that goal."

◼ NO ILLUSIONS

Then-US Secretary of Defense Robert Gates, at his valedictory appearance before the Shangri-la Dialogue, offered an alternative view on the South China Sea. "The US position on maritime security remains clear," he said. "We have a national interest in freedom of navigation, in unimpeded economic development and commerce, and in respect for international law . . . [including] . . . equal and open access to international waterways."

Gates described America's continuing and growing security presence in East and Southeast Asia: "Taken together, all of these developments demonstrate the commitment of the United States to sustaining a robust military presence in Asia, one that underwrites stability by supporting and reassuring allies while deterring, and if necessary defeating, potential adversaries."

In sum, the South China Sea is a strategic arena of growing significance and not inconsiderable danger. Viewed globally, an era in American strategy is ending as US forces begin their withdrawal from Iraq and Afghanistan. The next strategic era will surely have Asia at its center—the rapid growth of economic and military capability in that region makes it inevitable.

China constitutes the geographic and economic core of Asia, and China's rising power and ambition will drive events and compel a response from other countries. The United States has long enjoyed dominance in the maritime domain. But Beijing's growing naval and air capabilities seem clearly intended to challenge that dominance in the South China Sea and in the sea-lanes on Asia's rim—and thereby challenge a vital American interest in freedom of the seas. The senior leadership of the US Pacific Command has no illusions regarding the dimension of the emerging challenge.

Challenge Questions

The following questions will increase understanding of the contents of this article:

1. What is the status quo of territorial claims in the South China Sea, and what were the results of the Philippines finding a Chinese military outpost in the Spratly Islands?

2. Why is China a major part of foreign relations for Southeast Asian governments?

3. Describe some of the inconsistencies in China's reasons for claiming the South China Sea. How does China benefit from the resulting confusion?

4. Provide some examples of Chinese resistance to regional and international interference in its sovereignty claim.

5. What is America's position on the South China Sea, and why don't Southeast Asian countries take sides?

6. Why isn't ASEAN more effective against China's actions in the South China Sea?

2 Does Economic Integration Augur Peace in East Asia?

Scott L. Kastner

"At the end of the day, it is not clear that economic ties will act as a significant constraint on state behavior when high-stakes issues are on the table."

Learning Objectives

After reading this article, you will more clearly understand the following:

- Economic integration
- Taiwan Strait
- China and Taiwan disputes
- Taiwan independence
- Pragmatism vs ideology

East Asia faced an uncertain future as it emerged from the cold war two decades ago. There were certainly reasons for pessimism. The United States' relationship with China, which had blossomed in the decade after normalization, took a sharp turn for the worse in the aftermath of the government's Tiananmen Square crackdown. Relations between Washington and Tokyo were strained as the two countries wrangled over trade and other economic issues, and some wondered whether America would maintain a robust security presence in East Asia as the cold war wound down. Deep-seated historical animosities persisted in the region, aggravating numerous territorial and maritime disputes.

Moreover, several of the factors identified by liberal international relations theory as helping to stabilize interstate relations were weak or absent in East Asia. Important states such as China remained authoritarian. Regional institution-building was in its early stages. And, though many states in the region had embraced economic development strategies premised on integration into global markets, intraregional trade was limited compared to Europe or North America.

Twenty years later, considerable uncertainty still surrounds prospects for stability in East Asia. China's rapid economic development has enabled the country to pursue a vigorous military modernization program, which in turn has fueled concerns about a regional security dilemma. Countries in the region have been unable to resolve the North Korean nuclear issue. Maritime disputes in the South and East China seas persist. And key relationships, including those between Japan and China

and the United States and China, continue to suffer from high levels of mistrust.

But East Asia has also changed a great deal since the early 1990s. Among the most striking trends has been a shift toward regionalism, as evidenced by increasing economic integration and cooperation among countries along the Western Pacific. Nations in East Asia have become even more dependent on trade than they were at the end of the cold war, and intraregional trade has expanded as a share of East Asian states' total trade—a trend driven largely by the development of regional production networks. Meanwhile, interest in preferential trade agreements has proliferated: According to the Asian Development Bank, 93 free trade agreements involving Asian countries were in effect in 2011, up from 25 in 2000.

In short, perhaps more than any other region in the world, East Asia has come to be characterized by both extensive economic integration and persistent and serious political frictions. Will increasing regional economic integration help to mitigate the chances that these frictions could escalate to armed conflict?

■ TRADE PARTNERS

The People's Republic of China (PRC) lies at the center of changing economic trends in Asia. China's economy has averaged nearly 10 percent annual growth for the past three decades, and it has achieved this remarkable record in part by becoming deeply integrated into global markets. The PRC is now the world's largest exporting country, and China attracts more foreign direct investment than any other developing nation. In recent years China's outbound foreign direct investment and foreign aid have also increased rapidly.

Furthermore, China today is the primary trading partner of many neighboring countries, and several nations in the region that have had contentious political relations with the PRC have become deeply intertwined with China economically.

For example, economic relations between China and Japan have continued to develop despite a cool political relationship between the two countries. China's military modernization has generated considerable anxiety in Japan, while Japan's perceived unwillingness to show sufficient contrition for historical injustices—exemplified by visits to the Yasukuni military shrine by Japanese leaders—has sparked anger in China. Tensions between the two sides have flared in recent years over issues such as maritime rights in the East China Sea and ownership of the Senkaku/Diaoyu islands; the two countries became involved in a diplomatic standoff last year after Japan detained the captain of a Chinese fishing ship for ramming a Japanese patrol vessel.

Nevertheless, China today is Japan's top trading partner, and exports to China accounted for nearly 20 percent of all Japanese exports in 2010. Japan is China's second most important trading partner behind the United States, and Japan has also been a key source of foreign direct investment in the PRC.

Political relations between South Korea and China have generally been stable since the two countries established formal diplomatic relations in 1992; they have certainly been less frosty than Sino-Japanese relations. Still, the relationship can be testy at times, and China's historically close ties with North Korea are sometimes a source of tension between Seoul and Beijing. Relations became strained last year, for instance, when China remained reluctant to criticize Pyongyang for its role in the sinking of the South Korean vessel Cheonan, even after international investigators blamed North Korea for the incident. At the same time, Beijing was strongly critical of subsequent US–South Korean joint naval exercises in the Yellow Sea. Yet despite these occasional tensions, China has become, by far, South Korea's largest trading partner. Today South Korean exports to China are more than double South Korean exports to Seoul's second largest trading partner, the United States.

Elsewhere in Asia the list goes on. China today is Vietnam's largest trading partner, even though the two countries fought a war as recently as 1979, and despite continued frictions between Beijing and Hanoi in the South China Sea. In the spring of this year, relations became especially acrimonious after Vietnam accused Chinese boats of cutting the exploration cables of Vietnamese survey ships; in response, Hanoi undertook live-fire naval exercises and clarified rules for conscription in the event of war. Similarly, lingering border disputes with India and continued unease in New Delhi about Beijing's relations with Pakistan have not prevented a surge in Sino-Indian trade; China is now India's second largest trading partner. And decades of persistent tensions across the Taiwan Strait did not prevent crossstrait economic relations from taking off beginning in the late 1980s. Today China is Taiwan's principal trading partner and the primary destination for the island's outbound foreign direct investment.

In short, as China has grown more prosperous, countries in the region have increasingly come to depend on its booming economy—even if those countries have not always enjoyed close political relations with Beijing.

At the same time, China has embraced East Asia's move toward regional free trade agreements. Perhaps most notable in this regard, China and the Association of Southeast Asian Nations agreed to create a free trade area that went into effect in 2010; the new free trade region is the world's largest in terms of population. China and Taiwan last year signed a sweeping Economic Cooperation Framework Agreement (ECFA), which will serve as a roadmap to an eventual free trade agreement linking the two sides of the Taiwan Strait.

The PRC has also signed several bilateral free trade agreements with countries such as New Zealand, Thailand, and Singapore. And China, Japan, and South Korea are discussing the possibility of a trilateral free trade agreement linking the major Northeast Asian economies; the three nations announced in May that they hope to complete a joint study of such an agreement by next year.

■ A REGION TRANSFORMED?

Is deepening economic integration transforming East Asia into a less conflict-prone region? There is some reason for optimism in this regard. Indeed, a vast amount of scholarship has explored the general relationship between trade and conflict, and contributors to this literature have pointed to several ways that international economic integration can reduce the danger of war.

Most obviously, economic integration makes armed conflict more costly. Wars can be highly disruptive to cross-border economic transactions, and such disruptions have the potential to impart substantial adjustment costs on national economies. For instance, it is not hard to imagine the negative impact that a serious military confrontation in the Taiwan Strait could have on the economies of both sides. The risks of shipping to and from trade-dependent Taiwan would escalate dramatically, especially if the PRC were to pursue a blockade against the island. A serious increase in cross-strait tensions—even short of war—could have a deeply chilling effect on Taiwanese businesses operating in China, perhaps leading many to leave. And if the United States were to become involved in a cross-strait conflict, US-China economic relations in all likelihood would not emerge unscathed. The high economic costs of military conflict, in turn, may lead governments to be more cautious in risking such conflicts in pursuit of international goals.

Economic integration can also facilitate a more fundamental transformation in the underlying goals pursued by states. For example, international economic ties can spark coalitional change at the domestic level, as groups with a vested interest in foreign economic ties grow larger and hence become more influential. These groups, in turn, are less likely to value foreign policy goals—such as those motivating territorial disputes—that can propel countries into military conflict. If economic integration is extensive, the actors benefiting from international economic ties will have the ability to influence state goals and perhaps effect a change in the makeup of the governing coalition itself.

Furthermore, as countries become more intertwined economically, citizens and officials from their respective societies come into greater contact with each other. According to Japanese statistics, for instance, more than 100,000 Japanese citizens have resided in China in recent years, and over half of a million Chinese nationals live in Japan. Not surprisingly, this phenomenon is even more pronounced across the Taiwan Strait: Many estimate that a million Taiwanese—primarily businesspeople and their families—currently live in mainland China. Recent years have seen over 4 million annual Taiwanese visits to the mainland, and mainland visits to Taiwan have surged since restrictions were relaxed under Taiwan's current president, Ma Ying-jeou.

> **Deepening economic ties have been an important factor driving the recent improvement in cross-strait political relations.**

Recent economic cooperation between Taiwan and the PRC has also necessitated extensive contact between officials on the two sides of the strait. Indeed, the recently concluded ECFA agreement institutionalizes such contact by establishing a Cross-Strait Economic Cooperation Committee, composed of representatives from both sides, to handle negotiations and disputes related to the agreement. This increased societal and official interaction, in turn, can help to facilitate deeper understanding and trust, and over the longer term could potentially help countries in East Asia begin to overcome long-standing historical animosities that often exacerbate regional political tensions.

Finally, economic integration can reduce the likelihood of the sorts of dangerous miscalculations that help give rise to armed conflict. Disputes between states can turn into wars when state leaders drastically misjudge the resolve and the power of their adversaries. For example, the United States and China found themselves at war on the Korean Peninsula in the 1950s after Washington seriously misjudged Beijing's willingness and ability to intervene should US forces enter North Korea. More recently, Saddam Hussein appears to have dramatically underestimated US resolve to invade Iraq in the run-up to the 2003 war.

Economic integration can make this type of miscalculation less likely in part because of the increased official contact and communication it fosters: State leaders may simply come to better understand each other's red lines that must not be crossed. But perhaps more importantly, economic integration makes it easier for states to communicate credibly with each other on issues of war and peace. Verbal threats to use military force become more costly because they can discourage international investors; state leaders will be more judicious in making such threats, and other leaders will be less likely to view the threats as empty bluster.

Economic ties also make it possible to impose economic sanctions when disputes arise. Sanctions, because they are costly, can likewise add more credibility to threats, and can also act as a substitute for armed conflict as a tool of coercion. Many have pointed out, for instance, that the PRC could cause tremendous harm to Taiwan without firing a shot.

■ THE LIMITS OF LIBERALISM

Nevertheless, the idea that economic ties make armed conflict less likely remains controversial and disputed. Some of this skepticism is grounded in the empirical record. Proponents of the realist school of international relations often point out, for example, that extensive trade flows in Europe in the early twentieth century did not prevent the outbreak of the First World War. At the end of the day, it is not clear that economic ties will act as a significant constraint on state behavior when high-stakes issues are on the table. Indeed, economic ties can even exacerbate disagreements between countries and help give rise to new disputes.

In today's world, for instance, a war between the United States and China would be almost unimaginably costly, in part because of the burgeoning economic relationship between the two countries. But the high costs of war could paradoxically encourage lower-level provocations that aim to reshape the status quo for the same reasons that nuclear weapons give rise to what nuclear theorists have called a "stability/instability paradox."

The idea is straightforward. Nuclear weapons dramatically raise the potential costs of a war between two nuclear powers. These increased costs are stabilizing since they make nuclear powers much more reluctant to fight wars against each other. But this very reluctance gives nuclear powers cover to pursue lower-level provocations, as they can be confident that other powers will want to avoid escalation.

Deepening economic integration in East Asia could have an analogous effect. For instance, it is not inconceivable that some countries in the region will act more, rather than less, assertively regarding their territorial and maritime claims to the extent they believe the increasing costs of war make other countries reluctant to escalate.

In addition, international economic ties are often themselves a significant source of disagreements between countries. Persistent disputes over the value of China's currency, for example, will not lead to militarized conflict in East Asia—but they could act as a significant barrier to more cordial relations. Furthermore, while the beneficiaries of deepening cross-border economic ties can become powerful new proponents of stability, international economic relations also generate dislocations that can help mobilize competing voices. Surging Chinese imports into Southeast Asia have been a source of concern in the region because of their impact on local industries. As a case in point, thousands of workers took to the streets in Indonesia to protest the 2010 implementation of the ASEAN-China Free Trade Agreement.

In sum, though there are a number of ways in which deepening economic integration in East Asia could help make the region less susceptible to armed conflict, such an outcome should not be taken as an article of faith. Ultimately, to be more confident that growing economic linkages in the region are having a pacific effect, we would want evidence that the specific causal processes that could link trade to peace are actually occurring in East Asia. The Taiwan Strait, long characterized by both intense political conflict and a burgeoning economic relationship, offers a good place to take a closer look.

■ A LESS DANGEROUS STRAIT

For nearly two decades the Taiwan Strait has epitomized the "cold politics, hot economics" relationships that have characterized post–cold war East Asia. Cross-strait rivalry dates to the end of the Chinese civil war, when Chiang Kai-shek's Nationalist regime retreated to Taiwan. At the time, the newly established PRC lacked the capacity to "liberate" the island. But establishing PRC sovereignty over Taiwan remained an important long-term goal.

By the early 1990s, Taiwan (known as the Republic of China, or ROC) was well along a path of democratization that would help transform the nature of the cross-strait rivalry. In an increasingly free Taiwan, the nature of the island's relationship with China became the subject of open debate, while the ROC's first native Taiwanese president, Lee Teng-hui (1988–2000), pushed for an expanded international role for Taiwan. Taipei began actively seeking re-entry into the United Nations, and in 1995 Lee traveled to the United States to attend a reunion at Cornell University, his alma mater. These events, and others, convinced PRC leaders that Lee's ultimate goal was Taiwanese independence. In the aftermath of Lee's trip to the United States, Beijing undertook a series of military exercises near Taiwan—culminating in missile tests in the vicinity of Taiwanese ports shortly before the ROC's first presidential election in 1996.

Tensions flared again in 1999 after Lee characterized cross-strait relations as "special state-to-state," and tensions persisted during much of the presidency of Chen Shui-bian (2000–2008), a member of the officially pro-Taiwanese independence Democratic Progressive Party (DPP). Chen's 2004 reelection—after campaigning on a strongly nationalist message—in particular led to a deep sense of pessimism in Beijing about long-term trends in cross-strait relations. Chinese officials repeatedly emphasized a willingness to "pay any costs" to prevent Taiwanese independence. In 2005 the PRC National People's Congress went so far as to pass an anti-secession law that mandated the use of "non-peaceful means" in the event of "Taiwan's secession from China" or "major incidents entailing Taiwan's secession."

Even so, while the cross-strait political relationship was generally adversarial during the Lee and Chen administrations, cross-strait economic relations grew rapidly. During the 1990s China became the primary destination of Taiwan's outbound foreign direct investment, and by late in the decade Taiwan's high-technology industries were increasingly moving production facilities to the mainland. In 2002 China became Taiwan's largest export market, and in 2005 the PRC became Taiwan's largest trading partner overall. These developments occurred even though many in Taiwan worried about over-reliance on the market of an adversary, and despite efforts by the Taiwanese government to manage the flow of investment to the PRC.

> **East Asia has come to be characterized by both extensive economic integration and persistent political frictions.**

Taiwan's current president, Ma Ying-jeou, has aimed both to improve the cross-strait political environment and to increase cross-strait economic cooperation. Since he came into office in 2008, relations between the PRC and Taiwan have stabilized considerably. Representatives from the two sides restarted a dialogue that had been frozen since the late 1990s, and negotiators have succeeded in reaching numerous agreements on issues such as bilateral tourism, direct transportation linkages, and financial cooperation. The process culminated in 2010 when Beijing and Taipei signed the landmark ECFA agreement.

Deepening economic ties between China and Taiwan clearly have been an important factor driving the recent improvement in cross-strait political relations. Part of the reason Ma has prioritized improving cross-strait relations is that he believes a stable relationship with China is crucial to Taiwan's economic future. Ma also believes that increasing economic integration will reinforce stability in cross-strait political relations. In order to facilitate recent economic cooperation with China, Ma has been pragmatic on sovereignty issues. For example, he has consistently emphasized that he will pursue neither independence nor unification while president.

Ma's pragmatism in this regard generally resonates with Taiwanese public opinion. Surveys consistently show that a substantial majority of Taiwanese voters support maintaining the status quo in cross-strait relations. Studies aiming to better understand this finding have revealed that, although a sizable majority of voters would support a formally independent Taiwan if peace with the PRC could be maintained, an equally large majority would oppose independence if it were to provoke a PRC attack.

This pragmatism doubtless arises in part because Taiwanese voters recognize that a war with China would be incredibly costly for Taiwan; deepening cross-strait economic ties almost certainly help to reinforce this belief. Pragmatism on the part of Taiwan's citizens, in turn, makes it hard for strongly proindependence leaders to be elected in Taiwan, as voters fear that such leaders might provoke conflict with China.

Although it is hard to predict how events will unfold if the DPP comes back to power, and in particular how the PRC would react to a victory by DPP candidate Tsai Ing-wen in next year's presidential election, it is a good bet that Tsai would seek ways to extend the current détente in cross-strait relations. There is reason to believe, in short, that recent stability in cross-strait relations has emerged in part as a consequence of deepening cross-strait economic integration.

■ SOVEREIGN SENTIMENTS

Nevertheless, there is also reason to be at least somewhat skeptical about how deeply entrenched the causal processes linking increased economic integration to a reduced likelihood of military conflict actually are in the Taiwan Strait. For example, while deepening cross-strait economic ties are undoubtedly raising the costs of military conflict for both sides, other trends seem to be at least partially offsetting these costs for Beijing. Importantly, the military balance of power in the Taiwan Strait has been shifting sharply in the PRC's favor. The ability and resolve of the United States to intervene in the event of a China-Taiwan conflict are also increasingly uncertain, at least looking forward to the longer term.

Moreover, while the possibility of economic coercion does give Beijing a way to punish Taiwan without firing a shot, economic sanctions could also backfire for China. Economic sanctions would most seriously hurt individuals and businesses in Taiwan that already have a direct stake in the cross-strait relationship. Punishing these sorts of actors could damage Beijing's long-term goals in Taiwan by alienating a constituency that tends to support stable cross-strait relations to begin with. In other words, it is not clear that the PRC would view economic coercion against Taiwan as a useful or desirable substitute for military conflict, nor is it clear that Beijing would necessarily impose economic sanctions as a signal of resolve before initiating military conflict against Taiwan.

Perhaps most importantly, although economic ties may be contributing to a sense of pragmatism among Taiwan's voters, there is little evidence to suggest a deeper shift in the preferences of most Taiwanese on cross-strait sovereignty issues. Surveys have found that the percentage of Taiwan's citizens who identify themselves solely as "Chinese" has declined to near negligible levels over the course of the past decade. On the other hand, the percentage of respondents identifying themselves solely as "Taiwanese" has increased dramatically since the early 1990s, and now represents a majority (having surpassed in recent years the percentage who identify themselves as "both Taiwanese and Chinese").

Support in Taiwan for political unification with the PRC, even over the longer term, remains miniscule. Meanwhile, the PRC has given no indication that increasing cross-strait economic ties are leading to a softening of its position on sovereignty issues. The core sovereignty dispute between Beijing and Taipei is likely to remain untransformed for the foreseeable future.

■ NO GUARANTEES

Relations across the Taiwan Strait have improved greatly in recent years, and China-Taiwan economic integration appears to have been an important factor contributing to this new stability. But even the cursory discussion presented here suggests an exceedingly complex relationship between cross-strait economic ties and the possibility of a China-Taiwan military confrontation. Economic ties can influence the likelihood of armed conflict through several different causal pathways, and there are grounds for both optimism and pessimism concerning how deeply entrenched these different pathways have become in the Taiwan Strait.

The case, in short, offers both a cautionary and a hopeful tale for East Asia as a whole. On the one hand, increased economic integration and cooperation do not guarantee that the many disagreements in the region will be managed peacefully. Just as the cross-strait sovereignty dispute persists despite burgeoning economic ties, so maritime disputes in the South and East China seas, for example, likely will continue despite growing economic interdependence between China and its neighbors. It remains as important as ever to manage these disputes carefully.

On the other hand, it is hard to argue that expanded cross-strait economic linkages have not contributed to the current moderation in China-Taiwan tensions. If economic ties can have a stabilizing effect, even if only on the margins, in a conflict as intractable as this, there is cause for cautious optimism about the stabilizing potential of increased economic integration in the region as a whole.

Challenge Questions

The following questions will increase understanding of the contents of this article:

1. What is the primary reason for Taiwan not actively pursuing *de jure* independence?
2. What is the "stability/instability paradox," and how does it relate to other aspects of international relations?
3. According to the article, which individual has helped stabilize relations with Taiwan and the PRC, and what has he done to encourage this?
4. What does the phrase "cold politics, hot economics" mean, and how does the relationship between Taiwan and China exhibit it?
5. Identify one country that has not been successful in creating an economic partnership with a group that they ideologically disagree with. Describe the reason for this difference.

3 Two-speed Australia

Robert Milliken

The dynamism shifts to the frontier states.

Learning Objectives

After reading this article, you will more clearly understand the following:

- Hung parliament
- Emerging economies in Asia
- Mining in resource-rich states
- Trade partners
- Budget goals and taxation
- Economic booms
- City planning

Australians will start 2011 with a quite different frame of mind from the one that greeted 2010. Back then, it seemed that nothing could shake a sense of stability and confidence that came from surviving the global downturn without recession. That came unstuck when political turbulence uncharacteristically rocked the lucky country: an election left it with a hung federal parliament and the first minority government in 70 years. For Julia Gillard, the Labor prime minister, that means 2011 will be riddled with problems simply keeping in power a government that depends on support from independents and Greens. Policy caution will be the name of the game.

But in one sense, at least, boldness will have its day. Australia's march towards a "two-speed" economy will only gain momentum. Dynamism is shifting from the older, more populous states, New South Wales (NSW) and Victoria, to the wide open spaces of the frontier states, Queensland and Western Australia (WA). Relatively thin with people they may be (they account for about a third of Australia's 22m population); but their red dirt and turquoise oceans hold the iron ore, coal, gas and other riches that are helping to fuel the growth of China, India and other emerging economies in Asia.

> **Australia's terms of trade will rise by 17% to their highest level on record.**

These resource-rich states barely noticed the downturn. Even so, the federal government stepped in with a stimulus package that helped to keep the economy afloat by throwing money at building projects. But the coming year will mark the return of growth in private-sector spending as the economy's big driving force. Businesses will spend about A$123 billion ($122 billion) over the financial year to mid-2011, a quarter more than in the previous year. Mining alone will account for almost half the spending.

The world's biggest players will be among the top spenders. Chevron heads a consortium building Australia's biggest-ever resource project: Gorgon, a A$43 billion liquefied-natural-gas venture off the WA coast. Workers will start pouring in to Barrow Island, Gorgon's base, in 2011 to help meet its target of delivering the first gas to Asia three years later. BHP Billiton and Rio Tinto will churn out even more iron ore in WA's Pilbara region to satisfy China, Australia's biggest trading partner.

◼ SHARING OUT THE SWAG

As iron ore and coal prices soar, the Treasury reckons Australia's terms of trade will rise by 17% to their highest level on record. Meanwhile, taxes from the miners' higher profits, in turn, will help the government to meet its pledge of returning the budget to surplus by 2013.

Ric Battellino, the deputy governor of Australia's central bank, notes that mining booms stretching back to the 1850s gold rush typically lasted about 15 years: until the riches ran out or the world stopped wanting them. Although the growth of China and India at the moment seems never-ending, there is no guarantee that this boom will be any different. So Australia will have to start confronting how to manage the two-speed economy while the luck lasts. Debates will heat up about the danger that the mining states' successes will come at the expense of manufacturing and other industries elsewhere. And there will be louder calls to bank the taxman's proceeds in a sovereign fund, to be spent on roads, railways, ports and hospitals.

The slower-speed states will start fighting back against the drift of capital and investment to the north. NSW, home to almost half Australia's finance and insurance industries, will push Sydney as an Asia-Pacific finance hub. The city recently opened a centre to arbitrate international business disputes, competing with Singapore and Hong Kong.

And if Sydney's own planning disputes can be solved, work will start in 2011 on towers for the proposed finance centre at Barangaroo, a site on Sydney Harbour named after an aboriginal woman who lived there when the British settled in 1788. People will keep flooding into Sydney, making it one of the world's least affordable cities in which to buy a home. The NSW Labor government's failure to build transport and other infrastructure to match the growth of the city will play strongly in a state election due in March. If opinion polls are right, the government there will crash to defeat after 16 years in power.

With its dependence on coal (both its biggest export and also the source of about 80% of its electricity), Australia has the misfortune in terms of its public image of being one of the world's high-carbon economies. In 2011 Australia's changed political landscape will offer a remedy for the bungled approaches by the Labor and conservative Liberal parties in taking action against climate change. The election in 2010 gave the Greens enough support to hold the balance of power in the Senate, the upper house, from mid-2011. Fixing a price on carbon will then move a big step closer.

Challenge Questions

The following questions will increase understanding of the contents of this article:

1. What are the effects of Australia having a minority government?
2. Why is power shifting to the frontier states?
3. Who is Australia's biggest trading partner?
4. What is Australia's budget goal, and how does it plan to meet it?
5. How can gold rushes of the past prepare Australia for the future?
6. Explain what is meant by the phrase, "Two-speed Australia."

4 Silencing Cambodia's Honest Brokers

Elizabeth Becker

Learning Objectives

After reading this article, you will more clearly understand the following:

- Paris peace accords
- Cambodian civil society groups
- International aid
- Global pressure
- Protecting constitutional rights
- Foreign investors

This year is the 20th anniversary of the Paris peace accords that ended the Cambodian war and any further threat from the murderous Khmer Rouge. It required all the major powers—the United States, leading European countries, the former Soviet Union and China—as well as most Asian nations to come up with an accord, a rare achievement. In a speech last week, Gareth Evans said that during his eight years as the Australian foreign minister "nothing has given me more pleasure and pride than the Paris peace agreement concluded in 1991."

I reported from Paris on the negotiations, which took several years of convoluted diplomacy since few countries or political parties had clean hands in the rise and fall of the Khmer Rouge. When the deal was finally signed in October of 1991 there were self-congratulations all around, champagne and a huge sigh of relief that Cambodia could move on to peace and democracy.

It didn't turn out that way. Cambodia today is essentially ruled by a single political party with little room for an opposition, has a weak and corrupt judiciary, and the country's most effective union leaders have been murdered.

That wasn't the scenario envisioned in Paris. Now, just as 20th anniversary commemorations are approaching, one of the few groups still enjoying the freedoms created under the peace accords are about to be silenced. The government of Cambodia is poised to enact a law that will effectively hamstring the country's lively civil society and NGOs, among the last independent voices in Cambodia.

In Paris, the framework for Cambodia's democracy was a much debated element of the peace accords. That debate led to Cambodia's Constitution and its guarantee of freedom of association and speech. The proposed law on civil society would deprive these independent Cambodian groups of those rights and undermine much of their work representing the country's most vulnerable citizens—advocating for their rights and dispensing aid, largely paid for with foreign donations. Most recently, these civil society groups exposed the government's eviction of the poor from valuable land in Phnom Penh. As a result, the World Bank is suspending all new loans to Cambodia until those made homeless receive proper housing.

Under the new law, these independent citizen groups would have to register with the government and win approval to operate under vague criteria; if the government disapproves of a group's behavior it can dissolve it using equally vague criteria. There would be no right of appeal.

The normally fractious Cambodian civil groups have joined together against the new law and asked the government for serious amendments to protect basic constitutional rights. They were rejected and only superficial changes were made. With little time left, one of their NGO leaders made an emergency trip to to meet with international organizations, foreign embassies and the U.S. government, asking them to speak out loudly against the measure before it passes in the coming weeks.

"If this law is passed we will be silenced. Foreign donors will give us less money. The people who will suffer are the poor," said Borithy Lun, the head of the Cooperation Committee for Cambodia. He led a meeting at the offices of Oxfam America, where I am a member of the board of directors. The law would diminish the ability of international NGOs, like Oxfam, to help the poor in Cambodia as well, since it requires all foreign nonprofit organizations to work directly with official agencies, essentially becoming an arm of the government.

All of this will have a direct impact on Cambodia's impressive economic gains. Foreign businesses have come to rely on Cambodia's civil society groups to act as honest brokers, pointing out the pitfalls in an economy marked by corruption and weak law enforcement. Foreign governments and institutions have already warned the Cambodian government that if the proposed civil society law is passed, they will rethink the $1 billion in aid given to Cambodia every year, which is roughly half of the country's budget. Secretary of State Hillary Rodham Clinton has spoken up repeatedly in favor of strong, independent civil societies and Cambodia has made no secret of its desire to continue improving relations with the United States.

As the commemorations of the Paris peace accords begin, with more champagne and seminars, instead of looking backward to past glory, it might be better to focus on today and reinforce the accords. Countries that are rightfully proud of their role in bringing peace to Cambodia are in a good position to require preserving the independence of civil society when Cambodia comes asking for their votes at the United Nations this fall.

The Cambodian government has two big objectives: It wants to win one of the nonpermanent seats on the United Nations Security Council, and to get the United Nations to help resolve the Thai-Cambodia border dispute centered on the temple of Preah Vihear. Cambodia has dispatched senior diplomats to countries large and small to win their votes and has initiated border talks with the government of the new Thai prime minister, Yingluck Shinawatra. The price for greater influence and prestige in the world should be reinforcing democracy, not diminishing it.

Challenge Questions

The following questions will increase understanding of the contents of this article:

1. What was significant about the 1991 Paris Peace Accord?
2. What effect would the proposed law have on Cambodians' constitutional rights? Give an example.
3. In what ways can external governments and international groups influence Cambodia's policies?
4. Why are Cambodia's civil society groups important to foreign investors?
5. According to the author, what are two major goals of the Cambodian government? In light of these goals, what advice would you give to Cambodian officials?

5 For Chinese Students, Smoking Isn't All Bad

Daryl Loo

The government tobacco maker sponsors schools, earning goodwill.

"Tobacco helps you become talented," children are told.

Learning Objectives

After reading this article, you will more clearly understand the following:

- Tobacco monopoly
- Smoking in China
- Anti-tobacco efforts
- Funding for primary education
- Internal government lobbying
- Tobacco marketing strategies
- Roles of philanthropy

In dozens of rural villages in China's western provinces, one of the first things primary school kids learn is what helps make their education possible: tobacco. The schools are sponsored by local units of China's state-owned cigarette monopoly, China National Tobacco. "On the gates of these schools you'll see slogans that say 'Genius comes from hard work—tobacco helps you become talented,'" says Xu Guihua, secretary general of the Chinese Association on Tobacco Control, a privately funded lobbying group. "They are pinning their hopes on young people taking up smoking."

Anti-tobacco groups say efforts in China to reduce sales, including a ban on smoking in public places introduced in May, have been hampered by light penalties, a lack of education about the dangers of smoking, and the fact that the regulator, the State Tobacco Monopoly Administration, also runs the world's biggest cigarette maker.

While Chinese law bans tobacco advertising on radio, television, and in newspapers, they "do not have clear restrictions on sales and sponsorship activities," according to a report published in January by Yang Gonghuan, a former deputy director of China's Center for Disease Control & Prevention, and Tsinghua University professor Hu Angang. Regional units of the monopoly funded construction of more than 100 primary schools throughout China, such as the Sichuan Tobacco Hope Primary School, the official Xinhua News Agency reported in May. Some schools are named after local tobacco companies such as Hongta or top-selling

cigarette brands like Zhongnanhai, named after the compound next to the Forbidden City where China's top leaders live and work. The state tobacco company in September 2010 announced it was sponsoring an additional 42 primary school libraries in Xinjiang and 40 in Tibet, and in November made a ¥10 million donation to a women's development fund for a "Healthy Mothers' Express" campaign.

China National Tobacco lists charitable activities on its website. In a survey of more than 2,000 adults conducted in 2009 by the Association on Tobacco Control, 7 percent had a good impression of the tobacco industry due to its charity work, while 18 percent said they would pick a cigarette brand because of its good works. State Tobacco's press office didn't respond to interview requests or faxed questions about sponsorship.

China has more than 320 million smokers, a third of the world's total, and 53 percent of men there smoke. About 1 million Chinese die from tobacco-related illnesses every year. The tobacco industry grew at an average annual rate of 19 percent from 2006 to 2010, according to State Tobacco. Last year, earnings rose 17 percent, to ¥605 billion ($95 billion), including ¥499 billion paid in taxes.

China created the tobacco monopoly in the 1980s, when the industry supplied more than 10 percent of government revenue. Today, tobacco contributes 6.7 percent, according to Yang and Hu's report. "Especially in tobacco-growing provinces like Yunnan and Guizhou, the tobacco industry is a very important part of local government income," says Wang Shiyong, the World Bank's senior health specialist in Beijing. "There is a lot of internal government lobbying to make sure the health consequences of smoking are not addressed."

$95 BILLION—2010 earnings of China's state-owned tobacco monopoly, up 17 percent.

A government survey in 2010 found that two in five male doctors light up every day in China. Pfizer, whose Champix is the main prescription anti-smoking drug sold in China, funded a three-year program in 2008 to set up 60 smoke-free hospitals in Beijing, Shanghai, and Guangzhou. Smoking among the hospitals' leadership fell to 8.4 percent, from 19.1 percent, while overall rates for doctors fell to 6.8 percent from 10.7 percent, says Pfizer spokeswoman Neena Moorjani.

Still, the education drives have a long way to go. Only one in four adults in China believe exposure to tobacco smoke causes heart diseases and lung cancer, and the percentage among smokers is even lower—22 percent—according to the 2010 Global Adult Tobacco Survey for China.

"We've been trying to get the Ministry of Education to stop the tobacco companies from sponsoring these schools," says Xu, a former deputy director at the Chinese Center for Disease Control & Prevention. "But the ministry wants us to show them proof that this is causing harm."

From *Bloomberg BusinessWeek*, October 3–9, 2011. Copyright © Bloomberg 2011 by Bloomberg BusinessWeek. Reprinted by permission of Bloomberg LP via YGS Group.

Challenge Questions

The following questions will increase understanding of the contents of this article:

1. What is the relationship between China's tobacco monopoly and China's schools?

2. Discuss the positive and negative aspects of National Tobacco's school sponsorship.

3. Why are anti-tobacco efforts in China ineffective?

4. What incentives does the tobacco company have for being involved in education, healthy mother campaigns, and other charitable work?

5. How many of the world's smokers live in China?

6. How do the marketing campaigns of tobacco (or anti-tobacco) in China compare with strategies used in your own country?

6   The Recession's Real Winner

Fareed Zakaria

China Turns Crisis into Opportunity.

Learning Objectives

After reading this article, you will more clearly understand the following:

- Economic growth of China
- Global financial crisis
- Fiscal preparedness
- Chinese infrastructure
- Global leadership

One year ago, the leading governments of the world saved the global economy. Remember October 2008: Lehman Brothers had disappeared, AIG was teetering, every bank was watching its balance sheet collapse. Around the world, credit had frozen and trade was grinding to a halt. Then came a series of moves beginning in Washington—bank bailouts, rescue packages, fiscal stimuli, and, most crucially, monetary easing. It is not an exaggeration to say that these measures prevented a depression. But the crisis has still fueled a major slowdown that has affected every country in the world.

The great surprise of 2009 has been the resilience of the big emerging markets—India, China, Indonesia—whose economies have stayed vibrant. But one country has not just survived but thrived: China. The Chinese economy will grow at 8.5 percent this year, exports have rebounded to where they were in early 2008, foreign-exchange reserves have hit an all-time high of $2.3 trillion, and Beijing's stimulus package has launched the next great phase of infrastructure building in the country. Much of this has been driven by remarkably effective government policies. Charles Kaye, CEO of the global private-equity firm Warburg Pincus, lived in Hong Kong for years. After his last trip to China a few months ago he said to me, "All other governments have responded to this crisis defensively, protecting their weak spots. China has used it to move aggressively forward." It is fair to say that the winner of the global economic crisis is Beijing.

> **'All other governments have responded to this crisis defensively. China has used it to move aggressively forward.'**

Almost every country in the Western world entered the crisis ill prepared. Governments were spending too much money and running high deficits, so when they had to spend massively to stabilize the economy, deficits zoomed into the stratosphere. Three years ago, European countries were required to have a budget deficit of less than 3 percent of GDP to qualify for EU membership. Next year, many will have deficits of about 8 percent of GDP. The U.S. deficit will be higher, in percentage terms, than at any point since World War II.

China entered the crisis in an entirely different position. It was running a budget surplus and had been raising interest rates to tamp down excessive growth. Its banks had been reining in consumer spending and excessive credit. So when the crisis hit, the Chinese government could adopt textbook policies to jump-start growth. It could lower interest rates, raise government spending, ease up on credit, and encourage consumers to start spending. Having been disciplined during the fat years, Beijing could now ease up during the lean ones.

And look at the nature of China's stimulus. Most of U.S. government spending is directed at consumption—in the form of subsidies, wages, health benefits, etc. The bulk of China's stimulus is going toward investment for future growth: infrastructure and new technologies. Having built 21st-century infrastructure for its first-tier cities in the last decade, Beijing will now build similar facilities for the second tier.

China will spend $200 billion on railways in the next two years, much of it for high-speed rail. The Beijing-Shanghai line will cut travel times between those two cities from 10 hours to four. The United States, by contrast, has designated less than $20 billion, to be spread out over more than a dozen projects, thus guaranteeing their failure. It's not just rail, of course. China will add 44,000 miles of new roads and 100 new airports in the next decade. And then there is shipping, where China has become the global leader. Two out of the world's three largest ports are Shanghai and Hong Kong.

China is also well aware of its dependence on imported oil and is acting in surprisingly farsighted ways. It now spends more on solar, wind, and battery technology than the United States does. Research by the investment bank Lazard Freres shows that of the top 10 companies (by market capitalization) in these three fields, four are Chinese. (Only three are American.) It is also making a massive investment in higher education.

"For the last decade, as China's economy kept growing at unprecedented rates, most Western analysts kept discussing when it would crash," says Zachary Karabell, the author of a smart new book, *Superfusion,* on the Sino-U.S. economy. "Now with China surging ahead through this crisis, all they can discuss is, when will China stall? It's as if they see the facts, but they can't quite make sense of them." China's strange mixture of state intervention, markets, dictatorship, and efficiency is puzzling. But it's time to stop hoping for China's failure and start understanding and adapting to its success.

Challenge Questions

The following questions will increase understanding of the contents of this article:

1. What explains why China has thrived in a global economic crisis?

2. How does fiscal responsibility link to the stabilization (or destabilization) of a nation's economy?

3. What have been the advantages of China's financial preparedness?

4. What challenges does the United States face in responding to the financial crisis?

5. Compare China's response to the economic crisis with the United States' response.

6. Why might Westerners have dismissed China's success strategies before now?

7 Indonesia's Image and Reality

Donald E. Weatherbee

Learning Objectives

After reading this article, you will more clearly understand the following:

- Indonesia's economic success
- Political invitations
- Indonesian presidential elections
- ASEAN membership
- "Soft power"
- Regional challenges

In 2011, nearly midway through his second term, Indonesia's first directly elected president, Susilo Bambang Yudhoyono (popularly known as SBY), leads a nation that is globally recognized as a political and economic success story. Since the 1998 toppling of Suharto's authoritarian regime, the country, which at the turn of the millennium seemed on the verge of collapse, has overcome daunting political, economic, and social obstacles that elsewhere in the developing world have led to failed states.

Far from failing, Indonesia today has emerged as a signally important middle-power player on the international stage whose friendship and cooperation are sought by the world's greater powers. Yet, at the same time, domestic and regional problems continue to hinder the nation's progress.

■ MUSLIM AND ROBUST

Indonesia would enjoy global visibility in any case as the world's fourth most populous country with some 240 million citizens. And nearly 90 percent of Indonesians profess Islam, giving the nation the world's largest Muslim population. However, while the size and religion of the population are often underlined, it is not these factors alone that make Indonesia an important actor on the international scene. It is also that this population lives in a stable political democracy underpinned by a vibrant economy.

With a GDP over $700 billion, Indonesia has the largest economy in Southeast Asia. Bolstered by sound macroeconomic policies and strong domestic consumption, the economy achieved a real growth rate of 6.1 percent in 2010, and is headed to 6.5 percent in 2011 and a predicted 6.6 to 7 percent in 2012. With the exception of China and India, Indonesia's growth outpaces that of the other countries in the Group of 20, of which Indonesia is the only Southeast Asian member.

Indonesia's robust economic performance led Jakarta to hope for an invitation to the summitry of the BRIC grouping (Brazil, Russia, India, China) of emerging world economies. Rather than joining a possible BRIIC, Indonesia was disappointed when, in April 2011, South Africa participated in the third summit meeting of what became BRICS.

■ THE POLITICAL CHALLENGE

But can Indonesia stay its economic course? In the long run, greater investment in infrastructure and education will be crucial. More immediately, there is concern that support for strong macroeconomic policies will be sacrificed to political contingencies. This seemed to be the case when SBY refused to back reformist Finance Minister Sri Mulyani Indrawati in 2010 when she collided with the political and business interests of cabinet strongman Aburizal Bakrie. The real issue was her fight against the corruption that eats away at the legitimacy of state institutions, especially the judiciary and police.

Although SBY "talks the talk" regarding reforms, he often does not "walk the walk." In many respects, his presidential style is frustrating to action-oriented reformists. It is too reductionist simply to attribute his apparent indecisiveness, temporizing, and search for consensus to his Javanese acculturation. SBY operates in a parliamentary system of 560 legislators from 9 political parties. In the 2009 elections his Democrat Party with 20.9 percent of the vote garnered 148 seats, while he as a directly elected president crushed his opponent with nearly 61 percent of the vote. SBY governs through a dingy (as opposed to a rainbow) six-party coalition, which gives him a theoretical majority of 463 seats. In fact, the other parties in the coalition often behave as a not-so-loyal opposition.

SBY's second five-year term ends in 2014, and he is constitutionally barred from seeking a third. Some degree of "lame-duckness" already seems to be setting in. Thus, in what promises to be a relatively lengthy run-up to the 2014 election, SBY's decision-making will take place in an increasingly heated political atmosphere.

The ideal candidate choice in the presidential contest would be one that would consolidate and build on the nation's legacy of political stability, economic growth, development, and democratic and modernizing Islam. These are the components of Indonesia's "soft power" that give President Yudhoyono and his foreign minister Marty Natalegawa the confidence to lay claim openly to a greater global role, particularly through multilateral forums addressing nontraditional security issues.

SBY has even proposed Indonesia as a bridge between the West and the Muslim world. It is this Indonesia that US President Barack Obama embraced in November 2010 when he announced a new "comprehensive partnership" and hailed Indonesia as a global model of democracy and diversity. Still, within the country, concerns about effective governance remain.

■ REGIONAL RESISTANCE

Indonesia's regional reality, meanwhile, does not fit with its international image. Even as Jakarta seeks to spread its soft-power wings globally, its wings are being clipped in the Southeast Asian context.

In the 10-member grouping of the Association of Southeast Asian Nations (ASEAN), Indonesia sees itself as primus inter pares. It has, however, not been able to give effective momentum to the Indonesian-inspired goal of creating an ASEAN security community by 2015 "ensuring that the countries in the region live in peace with one another . . . in a just, democratic, and harmonious environment." Indeed, the realization of this vision seems as remote today as when originally laid out in 2003.

In 2011, Jakarta jumped the queue and assumed the chairmanship of ASEAN before its regular turn in 2013. The ostensible reason was that 2013 would be too bureaucratically burdensome, since Indonesia is hosting APEC (the Asia-Pacific Economic Cooperation summit) that year. Jakarta has sought to seize the opportunity to move a hesitant and faltering ASEAN toward greater unity of purpose and policy coherence. Yet Indonesia's desire to lead has met the resistance of ASEAN members who refuse to follow.

For example, every Indonesian effort, either bilaterally or through ASEAN, to engage Myanmar in a meaningful exchange on the need for democratic reform has been rebuffed. Indonesia's promise of a strong regional human rights mechanism has been thwarted. Escalating conflict between Thailand and Cambodia threatens to unravel ASEAN, and Jakarta's efforts at intermediation have been unavailing.

Indonesia, likewise, has been unable to orchestrate a unified response to China's increasingly aggressive claims and actions in the South China Sea. Absent the support of their ASEAN partners, frontline Vietnam and the Philippines have turned to the United States, which only heightens tensions and diminishes ASEAN's credibility. ASEAN's conceit that it is the lynchpin of East Asia's evolving security is undermined by the reality of a dysfunctional aspiring community with no political coherence or strategy to meet its internal and external challenges.

The disconnect between Indonesia's global ambitions and its inability to move its ASEAN partners will be magnified in coming years when Jakarta's chairmanship will be succeeded by Cambodia in 2012, followed by Brunei and Laos. And possibly Myanmar will interrupt the line in 2014, reclaiming a role it gave up under pressure in 2007. None of these successor states shares Indonesia's regional vision of peace, harmony, and democracy.

International political respect for Indonesia could be tarnished by the country's association with an ASEAN that continues to lose credibility. As Indonesia's international stature grows, its own national interest may call for a foreign policy that, while not necessarily post-ASEAN, does not make ASEAN its centerpiece.

Challenge Questions

The following questions will increase understanding of the contents of this article:

1. What factors helped Indonesia become an international player?
2. Why does Indonesia find resistance to its leadership at the regional level?
3. How would a unified ASEAN be more effective in the region?
4. Give an example of how "soft power" has given Indonesia greater confidence on the global stage.
5. How is the chairmanship for ASEAN determined?
6. What countries will lead ASEAN next?

8 Hoikuen or Yochien

Aimi Kono Chesky

Past, Present and Future of Japanese Early Childhood Education.

Learning Objectives

After reading this article, you will more clearly understand the following:

- Early education in Japan
- *Hoikuen, yochien,* and *nintei kodomo-en*
- Childcare vs schooling
- Historical transitions
- Early childhood curricula
- "Educare" ideology

For more than a century, Japanese early childhood education programs have been divided into two main institutions: *hoikuen* (child care) and *yocbien* (kindergarten). Hoikuen is operated under the auspices of the Ministry of Health, Labour and Welfare, whereas yochien falls under the Ministry of Education, Culture, Sports, Science and Technology. Infants as young as a few months old are eligible for hoikuen, while no children younger than 3 are accepted for yochien. While kindergarten education is not compulsory in Japan, a majority of children—over 95% of children between the ages of 3 and 6—attend some type of early childhood programs (WebCrews, Inc., 2009).

Recently, another type of early childhood program was developed with the main goal of offering a more comprehensive service to children and their families in an attempt to meet each family's unique needs (Rengo, n.d.). Clearly, an effort has been made to provide better services for young children, given the increase of working mothers. Still, Japanese early childhood education professionals continue to face a serious challenge in defining the best practices for young children and their families.

When defining best practices, it is important to understand the evolution of Japanese early childhood programs and re-examine the true purpose of Japanese early childhood education. Such information also would benefit early childhood educators in other parts of the world as they consider ways to improve their own practices. Thus, this article aims to delineate major historical transitions in Japanese early childhood programs, describe current trends and challenges, and provide implications for future directions.

HISTORIES

Hoikuen

The first form of hoikuen was launched in 1890, not long after the first yochien was established in 1876. In the early 1900s, many child care facilities were built for children whose fathers were at war and mothers at work. The years after World War I saw an increased need to provide services for working mothers and their infants. In response, a hoikuen for newborns and infants opened. In 1919, the first public hoikuen opened. Two years later, the city of Tokyo established a child care ordinance; children attending hoikuen were to be cared for and educated according to the yochien's (kindergarten's) standards. As a result of a major earthquake in 1923 that severely affected the Kanto region (which includes Tokyo), many new child care facilities were built in order to assist children who lost their families.

Most child care facilities had temporarily closed by the end of World War II, due to damage from air assaults. In 1947, two years after the end of the war, the "Child Welfare Law" was established. As a result, hoikuen was recognized as one of the welfare facilities for children ("History," n.d.).

As the Japanese economy improved during the 1950s, more women became recruited into the workforce, leading to increased demand for hoikuen. In 1963, the Ministry of Education and the Ministry of Welfare jointly expressed the distinction between yochien and hoikuen: "The purpose of yochien is to provide schooling for children, and that of hoikuen is to provide care for children who are not otherwise cared for. Therefore, these two entities clearly serve different functions" ("History," n.d.). In the mid-1960s, the Ministry of Welfare established guidelines for child care services. These were merely guidelines, however, and were not legally binding.

In 1990, modified guidelines for child care services were enforced, and the government made a major effort to meet children's individual needs. By 1999, the Ministry of Welfare had again revised guidelines for child care services, focusing on: 1) the unification of care and education, 2) care that respects the whole child, 3) the environment as an agent for child care, 4) broadening child care functions, and 5) the importance of caregivers' roles.

Yochien

The first yochien in Japan, which opened in 1876 in Tokyo, was a public entity The first private yochien opened four years later. In 1899, the Ministry of Education (now the Ministry of Education, Culture, Sports,

Science and Technology) established the requirements for yochien education and equipment. It was the first legal document written for yochien in Japan and specified:

> Ages of children: 3 through 6 years old The center's hours of operation: five hours or less Purpose of kindergarten: to supplement care at home Curricular areas: play, music, conversation, hands-on activities.

The Ministry of Education investigated theoretical orientations of Japanese yochien in the early 1920s. The results showed the Froebelian approach to be the most popular, followed by a combination of Froebelian and Montessori approaches. In 1926, "Kindergarten Enactment" was established, and yochien received a solid position as an educational setting. While the regulations differed little from the previous requirements ordered in 1899, a special note was made in regard to yochien's services to families with dual earners. It said that yochien may operate from early morning to late evening and may consider accepting young children who have not yet turned 3, if a family needed such accommodations ("History," n.d.).

The Pacific Theater of World War II broke out at the end of 1941, and a major air attack on Honshu (the mainland of Japan) severely hampered yochien education. Some resulting major changes to yochien included having air attack drills and setting up age-combined care. Learning styles also became more teacher-directed. In 1944, the government ordered all the kindergartens closed. As a result, yochien throughout Japan transformed into child care centers.

A year after the end of World War II, American education delegates visited Japan to observe the state of Japanese education. In the following years, the Ministry of Education established "child care education essentials—guidelines for educating young children." The curricula were divided into 12 areas: field trips; rhythm; recess; free play; music; story time; art; construction; nature; *gokko*-play (such types of play as hide-and-seek, tag, dramatic play, and puppet play); health; and special occasions. The main goal was to make everyday experiences fun for the children; therefore, their activities were planned around play ("History," n.d.).

After Japan's status as an independent country was reestablished in 1951, the country began efforts to formulate a coherent education system from kindergarten to upper secondary school (Oda & Mori, 2006). In 1952, kindergarten educators were urged to develop their own curriculum guidelines. By 1956, the kindergarten education guidelines had been revised a few times, yet it maintained the emphasis on everyday, life-oriented education. The 1956 guidelines divided the curricula into six areas (health, social, nature, language, music/rhythm, and art), leading many people to perceive yochien as a preparation for elementary school education (Oda & Mori, 2006).

As Japan's economy grew stronger and people believed they could make better choices for themselves and their children, some parents began to question the quality of yochien education. In the late 1980s, some early childhood teachers argued for revising the curriculum guidelines (Oda & Mori, 2006). The Ministry of Education conducted a nationwide survey focusing on yochien practices and instructional methods. The results showed that some kindergartens pursued inappropriate practices that did not follow the original intention of kindergarten education, which was to provide children with playful and enjoyable experiences. Thus, the ministry issued a strong warning against using a subject matter-oriented curriculum and skill-centered instruction (Oda & Mori, 2006).

The results of the study led to another revision of the kindergarten education guidelines in 1989. The goal of the revision was to move away from a subject-oriented practice toward one that values each child's characteristics and capabilities: "The basic ideal of kindergarten education is to understand the nature of children, and to educate them through their environment. For this purpose, teachers must build a trusting relationship with their pupils and create a good educational environment together" (Ministry of Education, 1989). The number of curricular areas was lowered from six to five: health, human relationships, environment, language, and expression.

The most recent revision of the guidelines took place in 2008; the revision stated that yochien is a place to raise and educate young children to assist their healthy growth and development through an appropriate environment, and that yochien builds a foundation for the education they will receive thereafter (The Ministry of Education, Culture, Sports, Science and Technology, 2008). The number and types of curricular areas remained the same.

■ CURRENT STATUS
Operational Differences

Hoikuen are still perceived as child care facilities rather than educational institutions. As reviewed, the primary function of hoikuen has been to accommodate children while they are apart from their parents. Presently, both public and private hoikuen are available, with no notable difference in their costs. While hoikuen are required to operate a minimum of 8 hours per day, there is no standard number of weeks per year they must operate (Kids Information Service, 2005).

Unlike hoikuen, yochien have been considered an educational institution from the onset, serving children ages 3 through 6. More private yochien exist than public ones, and private yochien typically cost about twice as much as public ones. Both public and private yochien typically operate four hours a day and are required to be in session for a minimum of 39 weeks per year (Kids Information Service, 2005). Unlike their equivalents in the United States, yochien tend to be independent entities and are not considered part of elementary school education. However, yochien sometimes co-exist with an elementary school, middle school, high school, and, in some instances, university, all under the umbrella of an incorporated educational institution. These schools often are built on the same site, a set-up referred to as "consistent education."

Differences in Classroom Setting

At hoikuen, there is no limit to the number of children being cared for in one classroom, and no regulation in regard to age grouping. However, the "child welfare minimum standards" require the adult-child ratio to be 1:3 with children under the age of 1, 1:6 with children between 1 and 3 years old, 1:20 with children between 3 and 4, and 1:30 with children older than 4 (Nozomi Yochien, n.d.).

On the other hand, no more than 35 children are allowed per yochien classroom. With the youngest group (3- and 4-year-olds), most places keep the number below 30. Yochien are typically divided into three grade levels: 3- and 4-year-olds, 4- and 5-year-olds, and 5- and 6-year-olds. Occasionally, a classroom will have more than one teacher, most often in the younger grades. There is no adult-child ratio requirement for yochien (Nozomi Yochien, n.d.).

■ RECENT TRENDS

The trend of a declining birth rate in Japan will likely continue unless a more effective support system for working mothers is established (Kosodate shien, 2008). The limited availability of programs for the very young has been a major hurdle for working mothers. Recently, the Japanese government introduced a program, translated as "Japan that supports children and the families," that includes a strategic plan to eliminate the waiting lists for admission to hoikuen, as well as to expand after-school activities. However, in 2007, federal financial support for families in Japan was less than 1% of gross domestic product (GDP), compared to 2% to 3% in most Western countries (Kosodate shien, 2008). Another concern is related to the efficiency of services. For example, a recent report suggests that child care facilities designed to serve sick children are rarely utilized, and that home visits by child care providers may be a better, more cost-effective alternative (Kosodate shien, 2008).

In recent years, a combined yochien and hoikuen program named "nintei kodomo-en" (accredited children's garden)—was launched. Nintei kodomo-en was first established in 2006, under the combined auspices of the Ministry of Education, Culture, Sports, Science and Technology and the Ministry of Health, Labour and Welfare. Initially, nintei kodomo-en was introduced as a protocol of a hoikuen-yochien united program and as a strategy for cutting down the number of children on hoikuen waiting lists (Benesse Corporation, 2008). In April 2009, 357 such programs were operating throughout Japan, an increase from 229 a year earlier (Ministry of Health, Labour and Welfare, 2009). Nintei kodomo-en is available to any family, regardless of the parents' working situation. Newborns through 6-year-olds are eligible. Another unique function of this program is to provide services rarely offered at either hoikuen or yochien, such as child-rearing counseling and family events.

It is too early to predict whether nintei kodomo-en will ultimately become a form of early childhood education program in Japan. One source reported that about 80% of families who utilize nintei kodomo-en are satisfied with the services they have received, yet dissemination of the program seems rather slow (Benesse Corporation, 2008). The number of programs more than doubled between 2007 and 2008, and then increased by over 50% between 2008 and 2009; yet, the ministries are not satisfied with the rate of increase and want the number to reach 2,000 programs (from 357 in 2009) in the very near future (Benesse Corporation, 2008). One critique of the program states that the ideal of such a program may have outpaced its planning and that many issues remain to be solved, including the need for a stronger partnership between the Ministry of Health, Labour, and Welfare and the Ministry of Education, Culture, Science, Sports and Technology. Many who manage nintei kodomo-en believe that the Japanese government has not provided enough financial support (Benesse Corporation, 2008).

■ FUTURE DIRECTIONS

Over a century ago, there were signs of unification of hoikuen and yochien when yochien tried to extend its operational hours and considered accepting very young children. Slowly but gradually, Japanese early childhood education has been moving toward establishing more comprehensive support and service systems to young children and their families. To accomplish this goal, the community must be involved. Oda and Mori (2006) state that Japanese early childhood educators are now challenged to view kindergarten as an early education center in a community; therefore, the role of kindergarten is to provide unique educational and caring services for the families of the children. In order to reach this status, the local government needs more authority in deciding on the types and locations of early childhood programs that are offered in a community. Second, updated and accurate information about the early childhood programs in the community should be readily available to the families with young children. Third, each community must continuously conduct needs assessments of families residing in the community. Finally, a plan for adding or alternating programs may be proposed and discussed further as a community.

Caldwell (as cited in Decker & Decker, 2005) proposed the term *educare* "as a way to enhance the field conceptually to embrace the many services provided" (p. 28). It is interesting that the Japanese term "education" itself signifies *educare*. "Education" in Japanese is written in two Kanji characters—the first character symbolizes *teaching*, while the second symbolizes *upbringing* and *care*. Moreover, the Japanese word for "education" also holds the connotation of instilling good manners and morals. Through positive interactions with a caregiver or a teacher, children come to learn those important skills they will need for their lifetimes. Such learning can and should be provided in any type of early childhood program. Indeed, there must be more equalities than distinctions between *education* and *care*.

It is clear that Japanese early childhood education has been moving toward an approach that serves children in a more holistic manner—an approach that adds more services to their family and addresses the needs of the community in which they live. Rather than distinguishing yochien, hoikuen, and nintei kodomo-en based on their services and public perceptions, each program should be recognized as providing unique functions that encompass education and care through various services. They are not mutually exclusive; rather, their services differ in some areas because of the interests and needs of the families that choose the program. Yet, all the programs share the same goal: to provide children with education and care that encourage their optimal development. Perhaps in the near future, every Japanese family will be privileged to choose a type of early childhood program that best fits the children's as well as the families' unique interests and needs, knowing that the children will be well cared for and educated.

REFERENCES

Benesse Corporation. (2008). Benesse: [*Education information exchange site*]. Retrieved August 6, 2009, from http://benesse.jp/blog/20080804/p2.html.

Decker, C. A., & Decker, J. R. (2005). *Planning and administering early childhood programs* (8th ed.). Upper Saddle River, NJ: Pearson Education.

History. (n.d.). In [*Chapter 1: History of hoikuenxyochien*]. Retrieved July 31, 2009, from www.kadokawa.co.jp/gakukei/.

Kids Information Service. (2005). [*Differences between yochien and hoikuen*]. Retrieved August 8, 2009, from www.ans.co.jp/kis/what02.htm. (In Japanese)

Kosodate shien: Hataraku mama wo motto hagemasou [Child rearing support: Let us support working mothers more]. (2008, August 22). *The Yomiuri Newspaper*, p. 3. (In Japanese)

Ministry of Education. (1989). *Kindergarten education guidelines.* Tokyo: Author.

Ministry of Education, Culture, Sports, Science and Technology. (2008). [*Kindergarten education guidelines*]. Tokyo: Author. (In Japanese)

Ministry of Health, Labour and Welfare. (2009). [*The number of accredited nintei kodomo on April 1st, 2009*]. Retrieved August 10, 2009, from www.mhlw.go.jp/houdou/2009/06/h0624-l.html.

Nozomi Yochien. (n.d.). *Yochien/hoikuen/nintei kodomoen.* Retrieved August 8, 2009, from www.nozomi.ac.jp/hoikuen/tigai.html.

Oda, Y., & Mori, M. (2006). Current challenges of kindergarten (yochien) education in Japan: Toward balancing children's autonomy and teachers' intention. *Childhood Education, 82,* 369–373.

Rengo. (n.d.). *Nintei kodomo en to ha* [*What is nintei kodomo-en*]. Retrieved August 8, 2009, from www.jtuc-rengo.or.jp/Kuirashi/kodomoen/kodomoentoha.html

WebCrews, Inc. (2009). *Tona-shiba:* Ranking/kyouiku kankyou. Retrieved August 1, 2009, from www.tonashiba.com/ranking/pref_education/. (In Japanese)

Challenge Questions

The following questions will increase understanding of the contents of this article:

1. What percentage of Japanese children attend optional early childhood programs?

2. Explain the main differences between *hoikuen* and *yochien* programs.

3. Give three purposes for *hoikuen* formation. What historical events in Japan prompted these transitions?

4. How did post–World War II independence and a stronger Japanese economy influence curricular changes in early childhood programs?

5. What recent trends led to the development of the *nintei kodomo-en,* and how is it more comprehensive than either a *hoikuen* or a *yochien*?

6. In what ways could Japan provide even more comprehensive support and service systems to young children and their families? What recent trends show that this support is necessary?

9 In Japan, New Nationalism Takes Hold

Robert Marquand *Staff writer of* The Christian Science Monitor

The country's post–World War II Pacificism is being challenged by a more assertive, patriotic attitude.

Learning Objectives

After reading this article, you will more clearly understand the following:

- Generational differences
- Pacifism vs. nationalism
- Japanese tradition
- Post-war identity
- Racism and ethnic stereotyping
- Media influence
- Domestic restructuring

On a pleasant November morning, some 300 Japanese executives paid $150 each to hear a lanky math professor named Masahiko Fujiwara give a secular sermon on restoring Japan's greatness. Mr. Fujiwara spoke quietly, without notes, for 80 minutes. His message, a sort of spiritual nationalism, rang loudly, though: Japan has lost its "glorious purity," its samurai spirit, its traditional sense of beauty, because of habits instilled by the United States after the war. "We are slaves to the Americans," he said.

Fujiwara's remedy is for Japan to recover its emotional strength. He says that Japan "can help save the world"—but its youths are lost in a fog of laxity and don't love Japan enough.

Fujiwara represents the milder side of an assertive discourse rising gradually but powerfully here. What direction it will take in this vibrant and complex society remains unclear. But as a new generation seeks to shed the remnants of what is commonly called the "American occupation" legacy, a range of speech and ideas previously frowned on or ignored, is showing up sharply in mainstream culture.

"We came because Fujiwara is one of few who speaks the truth to our politicians," says Hirofumi Kato, vice president of a family business who attended the talk. Those not there can buy Fujiwara's "Dignity of a Nation," a bestseller at more than 2 million copies this year, that describes how Western concepts like freedom and equality are inappropriate for Japan and don't really work in the US.

■ CARTOONS, MAGAZINES FUEL MESSAGE

The new nationalist sentiment is seen in popular magazines that use provocative language to advocate a more militaristic Japan, question the legitimacy of the Tokyo war-crimes trials, and often cast racist aspersions on China and Korea. Magazines include "Voice," "Bungei-shunju," "Shokun," "Seiron," and "Sapio," among others that are widely available. Sapio issues this fall have detailed how China will soon invade Japan and advocate nuclear weapons for Taiwan and Japan. The Dec. 27 issue details which members of the US Congress "love and hate Japan," including those described by political scientist Takahiko Soejima as helping "US companies take over Japanese banks at cheap prices."

Popular *manga* cartoons, another example, are a vivid entry point for school children and young adult males who read them on the trains. In recent years, *manga* have begun to include stronger and more-open ethnic hate messages. "The 100 Crimes of China," for example, is one in a recent series put out by publisher Yushinsha, with a kicker noting that China is the "world's most evil country." One recent *manga* is titled, "Why We Should Hate South Korea." Drawings are graphic and depict non-Japanese in unflattering ethnic stereotypes.

New programs are emerging, like the weekly Asahi talk show hosted by Beat Takeshi, that have thrown staid political expression into satire for Japanese viewers. There's a higher profile set of "conspiracy theories" that get repeated on TV, including those by writer Hideyuki Sekioka, author of "The Japan That Cannot Say 'No.'" Mr. Sekioka says the US manipulates Japan into adopting weak policies and has a "master plan" to control Japanese business. TV Asahi broadcasts programs detailing various US manipulations, including the idea that the CIA sent the Beatles to Japan in 1966 to dissipate an anti-US mood and "emasculate" Japanese youths.

The rise of this rhetoric is often denied here. Yet by last summer, Yoshinori Katori, then-Foreign Ministry spokesman, acknowledged that nationalism, most often on the right, had become a "new phenomenon."

The Japan of 2006 has quietly adopted a tone very different from the milder pacifism of it postwar identity.

Earlier this month, Prime Minister Shinzo Abe engineered two historic changes—transforming the postwar Defense Agency into a full-scale Defense Ministry, and ushering in a law requiring patriotic education in schools. The new law requires teachers to evaluate student levels of patriotism and eagerness to learn traditions. The Asahi Shimbun warns that this may "force students to vie to be patriotic in the classroom."

"A nationalistic reawakening from Japan's old pacifist identity, is leading to a domestic restructuring of Japan," says Alexander Mansourov, Asia specialist at the Pacific Center for Security Studies in Honolulu. "Along with a new defense ministry, a new national security council, and new intelligence agency, there's debate over whether to go nuclear, a debate on pre-emptive strikes on North Korea."

■ PACIFIST SENSIBILITY STILL STRONG

The new nationalism is not coming as an especially fire-breathing exercise. Japan remains quite cosmopolitan; mildness and politeness are valued. Many Japanese don't notice the stronger messages, or are not interested in politics.

The majority retain a pacifist sensibility. There's little hint of a mass emotional patriotism seen in Japan under Emperor Hirohito. The trend may get redirected as part of a healthy rediscovery of pride.

"I see a Japan that, after the 1990s, is becoming more confident," says one American corporate headhunter who has lived here for two decades.

Still, the extent of change in Japan's discourse can be measured by the number of moderates who say that they have little ground to stand on today. Former Koizumi presidential adviser Yukio Okamoto, a moderate conservative, argues that the "middle or moderate ground" is disappearing. Mr. Okamoto says that on many subjects—membership in the UN Security Council, culpability in World War II—he finds himself without a voice. "Every time I open my mouth to say something, I am bashed by either the left or the right," he says. Recent TV appearances by the granddaughter of Hideki Tojo, a World War II leader who was later executed for war crimes, describing him as a fine fellow, also concern Okamoto, who says that, though not an exact parallel, it would be inconceivable to imagine a granddaughter of Hitler going on German TV.

Most of the current domination of media is by the harder right. Former finance minister Eisuke Sakakibara says, "The sense of nationalism is rising here. I feel threatened . . . any liberal does. We worry about a loss of freedom of speech. [In the US,] the right has not taken complete control in the media, but we are not the US."

The new tone is coupled with the rise of China, fears associated with North Korea, perennial questions of identity—and comes as America, Japan's main ally and security guarantor, is bogged down in Iraq. It was given some license by the repeated visits to the Yasukuni war-memorial shrine by former Prime Minister Junichiro Koizumi. Those angered much of Asia, where they were seen as implicit support of a view that Japan's 20th-century war was justified.

Prime Minister Abe has eased that anger by not visiting the shrine, instead visiting Beijing to promote common points, like trade. But many experts see that decision as tactical.

Radical media, too, are thriving. The magazine "Will," for example, ran a discussion between the ultranationalist governor of Tokyo, Shintaro Ishihara, and Fujiwara, the author. Mr. Ishihara, who won 80 percent of the Tokyo vote in 2005, calls World War II "a splendid war." Fujiwara says Japan must replace its logic-based culture with an emotion-based culture; he pushes to eliminate the teaching of English in schools. Photos in "Will" this year depicted fascist author Yukio Mishima standing atop the high command in 1970 in a military uniform, minutes before he committed ritual suicide. Mr. Mishima's private army had just failed to take control of the building. [**Editor's note:** *The original version incorrectly described Mr. Mishima's death.*]

It's the "mainstreaming" of such material that raises some eyebrows. Yoshinori Kobayashi, a popular far-right cartoonist, now appears regularly on mainstream talk shows. Ishihara recently interviewed Sekioka in "*Bungeishunju,*" a literary magazine akin to the Atlantic Monthly. Ishihara wonders why Japan lacks the spiritual strength to stand up to the Americans.

Behind such views is a shared vision: a return to pure virtues found in medieval Japan. The Tom Cruise film "The Last Samurai" captures some of this. "What we need is a return to the inherent religion and culture of Japan . . . of our ancestors in the middle ages," argues Sekioka.

Japan's education bill is designed to teach such virtues. Prime Minister Abe's new book, "Toward a Beautiful Country," hearkens to the ideas of love of homeland.

The idealized samurai code was given best expression by a Japanese Christian named Inazo Nitobe. His book, "Bushido: The Soul of Japan," was written in English and translated back into Japanese after World War II. It prizes sympathy for the weak and hatred of cowardice—and has been a gold mine for present-day nationalists.

Critics say Japan must confront its wartime past. Much of its pacifist identity emerges from the view that it was a war victim, as epitomized by Hiroshima and Nagasaki. That story, reinforced by textbooks that downplay or deny Japan's role in invading Korea and Manchuria, rang loudly in the 1960s, '70s, and '80s, and was ignored as the economy boomed in the late 1980s. But it had received a boost from tales of Japanese abducted by North Korea. Prime Minister Abe, who has been instrumental in promoting the abductee issue, has of late been trying to mediate between extreme nationalism while still advocating more patriotism.

Challenge Questions

The following questions will increase understanding of the contents of this article:

1. Summarize Mr. Fujiwara's position on renewed nationalism.
2. Give three examples of how nationalism is being advocated in Japan.
3. Why is Japanese pacifism still the attitude of the majority despite efforts to revive nationalism?
4. Why are liberal Japanese worried about a rise in nationalism?
5. What internal and external threats are amplifying the desire for renewed nationalism?
6. What values do nationalists hope to re-instill in Japanese culture?
7. Why do some Japanese consider themselves victims rather than aggressors in World War II?

10 A Second Wind from the Golden Triangle

The Economist

Laos and the Drugs Trade.

Learning Objectives

After reading this article, you will more clearly understand the following:

- Opium production in Laos
- Laos' war on illegal drugs
- International influence
- Global partnerships
- Local economy
- Government action

The new prime minister of Laos, Thongsing Thammavong, has taken the country's drugs problem into his own hands with good Communist brio. At an event co-sponsored by the government and the UN Office on Drugs and Crime (UNODC) in late June Mr Thongsing, wearing a business suit and wielding a giant torch, helped put fire to an enormous stash of seized opium, heroin and cannabis. Three weeks later the prime minister reinforced his message by concluding a co-operation agreement with Myanmar, Laos's big neighbour to the north-west, on the prosecution of drug trafficking. Although official policy in the Lao People's Democratic Republic is usually kept opaque, it is easy to see that the government, led by a man eager to make an impression—is gearing up for a new stage in the war against drug producers and traffickers. In recent years it had shifted to the back foot; the production of both opium and methamphetamine is on the rise.

Laos was long regarded as one side of the Golden Triangle, which was responsible for producing over half of the world's opium as recently as the 1990s. At one point smoking crude opium had become a macabre tourist attraction for foreign visitors slumming it in northern Laos. Facing pressure from America and the UN, the Laotian government, together with its counterparts in Myanmar and Thailand, conducted a wildly successful eradication programme in the late 1990s that saw poppy cultivation plummet. From the 27,000 hectares (over 40,000 football pitches) that were under cultivation in 1998, within eight years the Laotian government had brought the total crop yield close to nil. Close enough that it was able to declare the country opium-free by early 2006. Contemporary reports suggested that it was no Potemkin clean-up job—towns such as Vang Vieng reinvented themselves as destinations for a different type of visitor.

The government's efforts at repressing production were augmented happily by a simultaneous explosion in poppy growth in Afghanistan. The Taliban had seized Kabul in 1996 but it wasn't until 1999 that Afghanistan's opium producers really hit their stride. That year Afghanistan's market overtook Myanmar as the world's largest and began dictating prices worldwide (to ignore the remarkable blip of 2001). By 2006 it was growing seven times the amount of Myanmar, Thailand and Laos combined. A subsequent glut sent global prices plummeting, aiding the eradication efforts of the South-East Asian governments.

It now appears that the Laotian government became complacent almost immediately upon declaring victory. Opium production has grown every year since 2007, and in 2010 the area under cultivation leapt by 58% year-on-year, according to a recent UNODC report. The government has proved its ability to locate and destroy poppy fields, but its dedication to disbursing aid—such as might motivate the erstwhile growers to pursue other livelihoods—is more questionable. UNODC believes that less than 10% of the villages declared opium-free have received funds promised for growing alternative crops. The effects of this failure were exacerbated by the global financial crisis. Weaker demand led to a fall in farm-gate prices for legal crops, while higher input costs raised prices for household goods. As standards of living declined, the reasons to return to poppies grew stronger.

Unhelpfully, the spot price of opium has also continued to rise. As expected, a reduction in local cultivation pushed up domestic prices, from around $250 per kilogram in 2002 to almost five times that in 2008. However, even as local supply began to rise again, the price continued to increase, reaching $1,670 per kilo in 2010. Why this is happening is unclear. One theory is that the drug remains in short supply locally because traffickers have opened new supply routes, taking advantage of new road links to China; the size of its import market is almost totally unknown.

Meanwhile, hidden among the stash burned by Mr Thongsing were 1.2m tablets of methamphetamine, known in Thai or Lao as *yaba*. Production of *yaba* in hidden factories in the Golden Triangle rocketed while opium production shrank in the early 2000s, and it has now supplanted opium as the consumer's drug of choice in Laos. It finds a ready market in the growing cities as well as in the countryside.

According to UNODC, the number of *yaba* pills seized in Laos is rising sharply, from 1.3m in 2007 to 2.3m in 2009. But it is hard to know whether this has anything to do with Mr Thongsing's new campaign, which for once addresses *yaba* on a par with opium. The ease and speed with which *yaba* factories can be assembled and relocated, combined with Laos' porous borders, makes it a cinch to evade the police. Unlike poppy plots, meth labs are not easily spotted by helicopter surveillance. So it is difficult to determine whether police are making inroads or whether factories are simply scaling up production. Nor is it possible to tell if seized pills originated in Laos or only indicted midway along their journey to markets in Europe, America and elsewhere in Asia.

Mr Thongsing's very public involvement in the drug war indicates that the Laotian government is readying itself for another crackdown. The speed with which it set about destroying poppy fields a decade ago indicates that it will be a formidable foe. Its dormancy since then however seems to have given the drug industry a chance to evolve and wise up. Whether it does the country any good or not, Mr Thongsing should have plenty of torch-brandishing ahead of him.

Challenge Questions

The following questions will increase understanding of the contents of this article:

1. What visual reminder did the new Prime Minister of Laos use to show his dedication to the war on illegal drugs?
2. How did external pressure influence the Laotian government's drug policies?
3. Describe the effects of the government crackdown on Laos' local poppy cultivation.
4. Why did the Laotian government fail to maintain its "opium-free" status?
5. How quickly did opium production rebound in Laos?
6. What makes government crackdowns on methamphetamine (yaba) factories more difficult than those on opium production?

11 Distinct Mix Holds On in a Corner of China

Andrew Jacobs

Long before words like "multiculturalism" and "fusion cuisine" entered the modern lexicon, Aida de Jesus and her forebears were mashing up food, language and DNA from far-flung corners of the globe.

A 95-year-old chef whose ancestry is drawn from Goa, Malacca and other former outposts of the Portuguese empire, Senhora de Jesus, as she prefers to be called, grew up celebrating Christmas and Chinese New Year with meals that relied on Portuguese sausage, bok choy and galinha cafreal, a chicken dish with an African pedigree. She spoke Portuguese at school, Cantonese on the street and a lively Creole known as Patuá with "the girls."

"We Macanese are always mixing it up," Senhora de Jesus said with a giggle, speaking in English, as she sat in the restaurant her family has run for decades. "We are very adaptable."

But these days the Macanese—as this former Portuguese colony's mixed-race residents are called—are swimming against a demographic tide that threatens to subsume their rich cultural cocktail. Always outnumbered by the Chinese migrants and Portuguese traders who crammed into this densely settled speck in the Pearl River Delta, the Macanese who stayed after Beijing took back the territory in 1999 are decidedly in the minority. Fewer than 10,000 Macanese reside here; by contrast, Macao's population of 500,000 is about 95 percent Chinese and rising.

"There are probably more Macanese living in California and Canada than Macao," said Miguel de Senna Fernandes, a lawyer and playwright whose father, something of a local cultural institution, chronicled the lives of ordinary Macanese in a series of novels. "Now that we are part of China, we are facing a very absorbing, overpowering force."

Not that Mr. Fernandes is giving up. In addition to organizing social events through his group, the Macaenses

Association, he has also emerged as the Don Quixote of Patuá, which is listed by Unesco as an endangered language. He helped publish a dictionary of Patuá expressions, and for the past 18 years he has staged an annual play that revives what local people call "doci papiaçam," or sweet speech, a stew of archaic Portuguese, Malay and Singhalese spiced with English, Dutch and Japanese, and more recently, a large helping of Cantonese.

Mr. Fernandes, 50, traces his fascination with Patuá to his grandmother, who would slip into it when gossiping with friends during "chá gordo," or fat tea, a typically Macanese interpretation of English high tea whose overabundance of Malaysian noodles, codfish fritters and custard tarts explains the fat.

"Drawn by their laughter, I would hide in the corner and later ask my grandmother about expressions I'd never heard before," he said. More often than not they were unsuitable for an 8-year-old's ears but his grandmother would oblige with sanitized translations, followed by an admonishment to stick to studying proper Portuguese.

"The old-timers considered Patuá broken or bad Portuguese," he said, "but since then I've been hooked."

The language is among the last of the Creoles that once flourished in the constellation of ports that made up Portugal's Asian and African holdings. Unlike British colonizers who maintained some distance from their subjects in Hong Kong, just an hour's ferry trip from Macao, the Portuguese frequently married local women who then converted to Catholicism.

Alan Baxter, a linguist at the University of Macao and an expert on Portuguese-based Creoles, said the roots of Patuá extend back to the 16th century, when Portuguese traders and their camp followers did business with Africans, Indians and Malays, then sailed onward to other colonies in the empire.

"Imagine if you went somewhere new and were deprived of knowledge of a local language and merely picked up the useful bits you heard to get yourself fed," he said, explaining its evolution.

The Cantonese contributions to Patuá came much later, starting in the late 19th century, after the walls dividing Macao's Portuguese and Chinese quarters were torn down and the two groups began to mingle.

These days Macanese give their laundry to a "mainato"—from the southern Indian language Malayalam—and they address their beloveds as "amo chai," a mix of the Portuguese "amor" and the Cantonese expression for "little one." Verbs are unconjugated, nouns are repeated to suggest the plural and words are sometimes assembled in a way that mimics the structure of classic Chinese idioms.

Early on this language served the mixed-blood Macanese well, fostering their role as a bridge between Macao's Portuguese rulers and its predominantly Chinese inhabitants. More recently, after they began sending their children to Portuguese schools, the Macanese became indispensable as managers and bureaucrats. By the time China took over administration of the enclave after more than 400 years of Portuguese rule, the Macanese dominated the territory's civil service.

Although most visitors these days are quickly sucked into Macao's casinos—among them The Venetian, one of the world's largest—those who wander the city's narrow cobblestone streets are struck by the effortless coexistence of Orient and Occident. Incense-suffused Buddhist temples, pastel Baroque churches, Portuguese bakeries and dried shark fin dispensaries are crammed together without complaint.

That same intermingling plays out in the lives of the Macanese, many of whom are devoted Catholics but give their children small red envelopes of cash on the Lunar New Year. Come Mid-Autumn Festival, another Chinese holiday, they take to the streets with rabbit-shaped lanterns.

"Many of us have been educated in Europe, but no Macanese would dare move to a new house without consulting a feng shui expert," said Carlos Marreiros, an architect who designed the Macao pavilion at the 2010 Shanghai World Expo. "I'm a Christian but I also believe God is a big ocean and all the rivers of religion are running to meet him."

In the years leading up to the transfer to China, thousands of apprehensive Macanese left, with many settling in Portugal. But over the past decade, as Beijing stayed true to its promise to give Macao 50 years of relative autonomy, the emigration has slowed and a small but steady number have returned.

One irresistible draw has been breakneck economic growth, mostly spurred by gambling and construction, which last year helped drive 20 percent growth in the economy. Fueled by players from the mainland, Macao's gambling revenue is now quadruple that of the Las Vegas Strip.

The impact on local people has been mixed. A law that bars nonresidents from working as croupiers and dealers has helped deliver well-paid jobs but it has also lured prized teachers out of the classroom. The draw has also been irresistible to young people, a growing number of whom are dropping out of high school or skipping college to head straight to the casino floor.

All that prosperity has brought other downsides as well: frenzied real estate speculation is pricing local people out of the housing market. The sleepy Macao that many once held dear is increasingly subsumed by the horn-honking and manic rhythms commonly associated with Hong Kong.

"Everything is happening very fast: construction is fast, business is fast and everyone is more stressed," said José Sales Marques, 55, the enclave's last Portuguese mayor, who now works to promote better ties between Macao and Europe. "Prosperity is wonderful, but at the end of the day all that money can't buy you a culture and an identity."

Filomeno Jorge is determined to keep alive one strand of that identity. Every Wednesday he rustles up the seven other members of his band, Tuna Macaense, to run through a startlingly diverse repertory that includes Portuguese fados, Cantonese ballads and Filipino pop songs. The mainstays, however, are vintage Patuá, some dating from 1935, when the band was first established by José dos Santos Ferreira, a poet and lyricist widely credited with bringing cultural legitimacy to the Macanese dialect.

At one time, Tuna Macaense had three dozen members and the band was known for making unannounced visits at weddings and birthday parties. "They would travel on foot through the streets because Macao is so small," said Mr. Jorge, a security manager at the MGM Macau who joined the band 25 years ago. "We can't do that now because there is too much traffic."

Although Tuna Macaense is blessed with frequent gigs, Mr. Jorge, 54, is increasingly preoccupied with finding new blood for the band, a quest that has so far been unsuccessful.

"All of us in the band are over 50," he said. "After we die, our music will die, and I can't let that happen."

Challenge Questions

The following questions will increase understanding of the contents of this article:

1. Approximately what percentage of Macao's population is Macanese?

2. What is Patuá, and how has it historically served the mixed-blood Macanese?

3. Give two examples of how different cultures in Macao have intermingled.

4. What has been the main source of Macao's economic growth, and how has this affected the local community?

5. In your opinion, what steps can Macao take to maintain prosperity while still preserving its unique culture and identity?

12 Myanmar's Young Artists and Activists

Joshua Hammer

In the country formerly known as Burma, these free thinkers are a force in the struggle for democracy.

Learning Objectives

After reading this article, you will more clearly understand the following:

- Political artists
- Aung San Suu Kyi
- Government censorship
- Human rights
- Military dictatorship
- HIV/AIDS awareness

The New Zero Gallery and Art Studio looks out over a scruffy street of coconut palms, noodle stalls and cybercafés in Yangon (Rangoon), the capital of Myanmar, the Southeast Asian country formerly known as Burma. The two-story space is filled with easels, dripping brushes and half-finished canvases covered with swirls of paint. A framed photograph of Aung San Suu Kyi, the Burmese opposition leader and Nobel Peace Prize laureate who was released from seven years of house arrest this past November, provides the only hint of the gallery's political sympathies.

An assistant with spiky, dyed orange hair leads me upstairs to a loft space, where half a dozen young men and women are smoking and drinking coffee. They tell me they're planning an "underground" performance for the coming week. Yangon's tiny avant-garde community has been putting on secret exhibitions in spaces hidden throughout this decrepit city—in violation of the censorship laws that require every piece of art to be vetted for subversive content by a panel of "experts."

"We have to be extremely cautious," says Zoncy, a diminutive 24-year-old woman who paints at the studio. "We are always aware of the danger of spies."

Because their work is not considered overtly political, Zoncy and a few other New Zero artists have been allowed to travel abroad. In the past two years, she has visited Thailand, Japan and Indonesia on artistic fellowships—and come away with an exhilarating sense of freedom that has permeated her art. On a computer, she shows me videos she made for a recent government-sanctioned exhibition. One shows a young boy playing cymbals on a sidewalk beside a plastic doll's decapitated head. "One censor said [the head] might be seen as symbolizing Aung San Suu Kyi and demanded that I blot out the image of the head," Zoncy said. (She decided to withdraw the video.) Another video consists of a montage of dogs, cats, gerbils and other animals pacing around in cages. The symbolism is hard to miss. "They did not allow this to be presented at all," she says.

The founder and director of the New Zero Gallery is a ponytailed man named Ay Ko, who is dressed on this day in jeans, sandals and a University of California football T-shirt. Ay Ko, 47, spent four years in a Myanmar prison following a student uprising in August 1988. After he was released, he turned to making political art—challenging the regime in subtle ways, communicating his defiance to a small group of like-minded artists, students and political progressives. "We are always walking on a tightrope here," he told me in painstaking English. "The government is looking at us all the time. We [celebrate] the open mind, we organize the young generation, and they don't like it." Many of Ay Ko's friends and colleagues, as well as two siblings, have left Myanmar. "I don't want to live in an abroad country," he says. "My history is here."

Myanmar's history has been turbulent and bloody. This tropical nation, a former British colony, has long worn two faces. Tourists encounter a land of lush jungles, golden pagodas and monasteries where nearly every Burmese is obliged to spend part of one year in serene contemplation. At the same time, the nation is one of the world's most repressive and isolated states; since a military coup in 1962, it has been ruled by a cabal of generals who have ruthlessly stamped out dissent. Government troops, according to witnesses, shot and killed thousands of students and other protesters during the 1988 rebellion; since then, the generals have intermittently shuttered universities, imprisoned thousands of people because of their political beliefs and activity, and imposed some of the harshest censorship laws in the world.

In 1990, the regime refused to accept the results of national elections won by the National League for Democracy (NLD) Party led by Aung San Suu Kyi—the charismatic daughter of Aung San, a nationalist who negotiated Myanmar's independence from Britain after World War II. He was killed at age 32 in 1947, by a hit squad loyal to a political rival. Anticipating the victory of Suu Kyi's party, the junta had placed her under house arrest in 1989; she would remain in detention for 15 of the next 21 years. In response, the United States and Europe imposed economic sanctions that included freezing the regime's assets abroad and blocking nearly all foreign investment. Cut off from the West, Myanmar—the military regime changed the name in 1989, though the U.S. State Department and others continue to call it Burma—fell into isolation and decrepitude: today, it is the second-poorest nation in Asia, after Afghanistan, with a per capita income of $469 a year. (China has partnered with the regime to exploit the country's natural gas, teak forests and jade deposits, but the money has mostly benefited the military elite and their cronies.)

The younger generation has been particularly hard hit, what with the imprisonment and killing of students and the collapse of the education system. Then, in September 2007, soldiers shot and beat hundreds of young Buddhist monks and students marching for democracy in Yangon—quelling what was called the Saffron Revolution. Scenes of the violence were captured on cellphone video cameras and quickly beamed around the world. "The Burmese people deserve better. They deserve to be able to live in freedom, just as everyone does," then Secretary of State Condoleezza Rice said in late September of that year, speaking at the United Nations. "The brutality of this regime is well known."

Now a new generation of Burmese is testing the limits of government repression, experimenting with new ways of defying the dictatorship. The pro-democracy movement has taken on many forms. Rap musicians and artists slip allusions to drugs, politics and sex past Myanmar's censors. Last year, a subversive art network known as Generation Wave, whose 50 members are all under age 30, used street art, hip-hop music and poetry to express their dissatisfaction with the regime. Members smuggled underground-music CDs into the country and created graffiti insulting Gen. Than Shwe, the country's 78-year-old dictator, and calling for Suu Kyi's release. Half the Generation Wave membership was jailed as a result. Young bloggers, deep underground, are providing reportage to anti-regime publications and Web sites, such as *Irrawaddy Weekly* and *Mizzima News,* put out by Burmese exiles. The junta has banned these outlets and tries to block access to them inside the country.

Young activists have also called attention to the dictatorship's lack of response to human suffering. According to the British-based human rights group Burma Campaign, the Burmese government abandoned victims of the devastating 2008 cyclone that killed more than 138,000 people and has allowed thousands to go untreated for HIV and AIDS. (Although more than 50 international relief organizations work in Myanmar, foreign donors tend to be chary with humanitarian aid, fearing that it will end up lining the pockets of the generals.) Activists have distributed food and supplies to cyclone victims and the destitute and opened Myanmar's only private HIV-AIDS facility, 379 Gayha (*Gayha* means shelter house; the street number is 379). The government has repeatedly tried to shut the clinic down but has backed off in the face of neighborhood protests and occasional international press attention.

It's not quite a youth revolution, as some have dubbed it—more like a sustained protest carried out by a growing number of courageous individuals. "Our country has the second-worst dictatorship in the world, after North Korea," said Thxa Soe, 30, a London-educated Burmese rapper who has gained a large following. "We can't sit around and silently accept things as they are."

Some in Myanmar believe they now have the best chance for reform in decades. This past November, the country held its first election since 1990, a carefully scripted affair that grafted a civilian facade onto the military dictatorship. The regime-sponsored party captured 78 percent of the vote, thus guaranteeing itself near-absolute power for another five years. Many Western diplomats denounced the result as a farce. But six days later, The Lady, as her millions of supporters call Suu Kyi, was set free. "They presumed she was a spent force, that all of those years of being in confinement had reduced her aura," says a Western diplomat in Yangon. Instead, Suu Kyi quickly buoyed her supporters with a pledge to resume the struggle for democracy, and exhorted the "younger generation" to lead the way. Myanmar's youth, she told me in an interview at her party headquarters this past December, holds the key to transforming the country. "There are new openings, and people's perceptions have changed," she said. "People will no longer submit and accept everything the [regime says] as the truth."

I first visited Myanmar during a post-college backpacking trip through Asia in 1980. On a hot and humid night, I took a taxi from the airport through total darkness to downtown Yangon, a slum of decaying British-colonial buildings and vintage automobiles rumbling down potholed roads. Even limited television broadcasts in Myanmar were still a year away. The country felt like a vast time warp, entirely shut off from Western influence.

Thirty years later, when I returned to the country—traveling on a tourist visa—I found that Myanmar has joined the modern world. Chinese businessmen and other Asian investors have poured money into hotels, restaurants and other real estate. Down the road from my faux-colonial hotel, the Savoy, I passed sushi bars, trattorias and a Starbucks knockoff where young Burmese fire text messages to one another over bran muffins and latte macchiatos. Despite efforts by the regime to restrict Internet use (and shut it down completely in times of crisis), young people crowd the city's many cybercafés, trading information over Facebook, watching YouTube and reading about their country on a host of political Web sites. Satellite dishes have sprouted like mushrooms from the rooftop of nearly every apartment building; for customers unable or unwilling to pay fees, the dishes can be bought in the markets of Yangon and Mandalay and installed with a small bribe. "As long as you watch in your own home, nobody bothers you," I was told by my translator, a 40-year-old former student activist I'll call Win Win, an avid watcher of the Democratic Voice of Burma, a satellite TV channel produced by Burmese exiles in Norway, as well as the BBC and Voice of America. Win Win and his friends pass around pirated DVDs of documentaries such as *Burma VJ,* an Academy Award–nominated account of the 2007 protests, and CDs of subversive rock music recorded in secret studios in Myanmar.

After a few days in Yangon, I flew to Mandalay, Myanmar's second-largest city, to see a live performance by J-Me, one of the country's most popular rap musicians and the star attraction at a promotional event for *Now*, a fashion and culture magazine. Five hundred young Burmese, many wearing "I Love *Now*" T-shirts, packed a Mandalay hotel ballroom festooned with yellow bunting and illuminated by strobe lights.

Hotel employees were handing out copies of the *Myanmar Times*, a largely apolitical English-language weekly filled with bland headlines: "Prominent Monk Helps Upgrade Toilets at Monasteries," "Election Turnout Higher Than in 1990." In a sign of the slightly more liberal times, the paper did carry a photograph inside of Suu Kyi, embracing her younger son, Kim Aris, 33, at Myanmar's Yangon International Airport in late November—their first meeting in ten years. Suu Kyi was married to British academic Michael Aris, who died of cancer in 1999; he failed to gain permission to visit his wife during his final days. The couple's older son, Alexander Aris, 37, lives in England.

At the hotel, a dozen Burmese fashion models ambled down a catwalk before J-Me leapt onto the stage wearing sunglasses and a black leather jacket. The tousle-haired 25-year-old rapped in Burmese about love, sex and ambition. In one song, he described "a young guy in downtown Rangoon" who "wants to be somebody. He's reading English language magazines, looking inside, pasting the photos on his wall of the heroes he wants to be."

The son of a half-Irish mother and a Burmese father, J-Me avoids criticizing the regime directly. "I got nothing on my joint that spits against anyone," the baby-faced rapper told me, falling into hip-hop vernacular. "I'm not lying, I'm real. I rap about self-awareness, partying, going out, spending money, the youth that's struggling to come up and be successful in the game." He said his songs reflect the concerns of Myanmar's younger generation. "Maybe some kids are patriotic, saying, 'Aung San Suu Kyi is out of jail, let's go down and see her.' But mostly they're thinking about getting out of Burma, going to school abroad."

Not every rapper treads as carefully as J-Me. Thxa Soe needles the regime from a recording studio in a dilapidated apartment block in Yangon. "I know you're lying, I know you're smiling, but your smile is lying," he says in one song. In another, titled "Buddha Doesn't Like Your Behavior," he warns: "If you behave like that, it's gonna come back to you one day." When I caught up with him, he was rehearsing for a Christmas Day concert with J-Me and a dozen other musicians and preparing for another battle with the censors. "I have a history of politics, that's why they watch me and ban so many things," the chunky 30-year-old told me.

Thxa Soe grew up steeped in opposition politics: his father, a member of Suu Kyi's NLD Party, has been repeatedly jailed for participating in protests and calling for political reform. One uncle fled the country in 2006; a cousin was arrested during student protests in the 1990s and was put in prison for five years. "He was tortured, he has brain damage, and he can't work," Thxa Soe said. His musical awakening came in the early 1990s, when a friend in Myanmar's merchant marine smuggled him cassettes of Vanilla Ice and M.C. Hammer. Later, his father installed a satellite dish on their roof; Thxa Soe spent hours a day glued to MTV. During his four years as a student at London's School of Audio Engineering, he says, "I got a feeling about democracy, about freedom of speech." He cut his first album in 2000 and has tangled with censors ever since. Last year, the government banned all 12 tracks on his live-concert album and an accompanying video that took him a year to produce; officials claimed he showed contempt for "traditional Burmese music" by mixing it up with hip-hop.

During a recent trip to New York City, Thxa Soe participated in a benefit concert performed before hundreds of members of the Burmese exile community at a Queens high school. Some of the money raised there went to help HIV/AIDS sufferers in Myanmar.

Thxa Soe isn't the only activist working for that cause. Shortly after Suu Kyi's release from house arrest, I met the organizers of the 379 Gayha AIDS shelter at the NLD Party headquarters one weekday afternoon. Security agents with earpieces and cameras were watching from a tea shop across the street as I pulled up to the office building near the Shwedagon Pagoda, a golden stupa that towers 30 stories over central Yangon and is the most venerated Buddhist shrine in Myanmar. The large, ground-floor space was bustling with volunteers in their 20s and 30s, journalists, human-rights activists and other international visitors, and people from Myanmar's rural countryside who had come seeking food and other donations. Posters taped on the walls depicted Suu Kyi superimposed over a map of Myanmar and images of Che Guevara and her father.

Over a lunch of rice and spicy beef delivered by pushcart, Phyu Phyu Thin, 40, the founder of the HIV/AIDS shelter, told me about its origins. In 2002, concerned by the lack of treatment facilities and retroviral drugs outside Yangon and Mandalay, Suu Kyi recruited 20 NLD neighborhood youth leaders to raise awareness of HIV/AIDS. Estimates suggest that at least a quarter million Burmese are living with HIV.

Even in Yangon, there is only one hospital with an HIV/AIDS treatment facility. Eventually, Phyu Phyu Thin established a center in the capital where rural patients could stay. She raised funds, gathered building materials and constructed a two-story wooden building next door to her house. Today, a large room, crammed wall to wall with pallets, provides shelter to 90 HIV-infected men, women and children from the countryside. Some patients receive a course of retroviral drugs provided by international aid organizations and, if they improve sufficiently, are sent home with medication and monitored by local volunteers. At 379 Gayha, says Phyu Phyu Thin, patients "get love, care and kindness."

In trying to close the shelter, the government has used a law that requires people staying as houseguests anywhere in Myanmar to obtain permits and report their presence to local authorities. The permits must be renewed every seven days. "Even if my parents come for a visit, I have to inform," Yar Zar, the 30-year-old deputy director of the shelter, told me. In November, a day after Suu Kyi visited the shelter, officials refused to renew the permits of the 120 patients at the facility, including some close to death, and ordered them to vacate the premises. "The authorities were jealous of Aung San Suu Kyi," says Phyu Phyu Thin. She and other NLD youth leaders sprang into action—reaching out to foreign journalists, rallying Burmese artists, writers and neighborhood leaders. "Everybody came out to encourage the patients," Phyu Phyu Thin told me. After a week or so, the authorities backed down. "It was a small victory for us," she says, smiling.

Ma Ei is perhaps the most creative and daring of the avant-garde artists. To visit her in Yangon, I walked up seven dingy flights of stairs to a tiny apartment where I found a waif-like woman of 32 sorting through a dozen large canvases. Ma Ei's unlikely journey began one day in 2008, she told me, after she was obliged to submit canvases from her first exhibit—five colorful abstract oil paintings—to the censorship board. "It made me angry," she said in the halting English she picked up watching American movies on pirated DVDs. "This was my own work, my own feelings, so why should I need permission to show them? Then the anger just started to come out in my work."

Since then, Ma Ei has mounted some 20 exhibitions in Yangon galleries—invariably sneaking messages about repression, environmental despoliation, gender prejudice and poverty into her work. "I am a good liar," she boasted, laughing. "And the censors are too stupid to understand my art." Ma Ei set out for me a series of disturbing photographic self-portraits printed on large canvases, including one that portrays her cradling her own decapitated head. Another work, part of an exhibit called "What Is My Next Life?" showed Ma Ei trapped in a giant spider's web. The censors questioned her about it. "I told them it was about Buddhism, and about the whole world being a prison. They let it go." Her most recent show, "Women for Sale," consisted of a dozen large photographs showing her own body tightly swaddled in layers and layers of plastic wrap, a critique, she said, of Myanmar's male-dominated society. "My message is, 'I am a woman, and I am treated here like a commodity.' Women in Burma are stuck at the second level, far below men."

Ma Ei's closest encounter with the government involved an artwork that, she says, had no political content whatsoever: abstract swirls of black, red and blue that, at a distance, looked vaguely like the number eight. Censors accused her of alluding to the notorious pro-democracy uprising that erupted on August 8, 1988, and went on for five weeks. "It was unintentional," she says. "Finally they said that it was OK, but I had to argue with them." She has come to expect confrontation, she says. "I am one of the only artists in Burma who dares to show my feelings to the people."

Suu Kyi told me that pressure for freedom of expression is growing by the day. Sitting in her office in downtown Yangon, she expressed delight at the proliferation of Web sites such as Facebook, as well as at the bloggers, mobile phone cameras, satellite TV channels and other engines of information exchange that have multiplied since she was placed back under house arrest in 2003, after a one-year release. "With all this new information, there will be more differences of opinion, and I think more and more people are expressing these differences," she said. "This is the kind of change that cannot be turned back, cannot be stemmed, and if you try to put up a barrier, people will go around it."

From *Smithsonian*, March 2011. Copyright © 2011 by Joshua Hammer. Reprinted by permission of the author.

Challenge Questions

The following questions will increase understanding of the contents of this article:

1. How have artists tested the limits of government oppression in Myanmar?
2. In what ways has Myanmar "joined the modern world," and what effect might it have on Burmese society?
3. Aside from politics, what other causes are Burmese artists promoting?
4. What examples of government censorship are given in the article?
5. In your opinion, how can art promote change, and why would the Burmese government be so quick to censor it?

13 The Korean Peninsula on the Verge

Charles K. Armstrong

"It has become apparent that a policy of isolation, sanctions, lack of dialogue, and 'strategic patience' has not worked to weaken North Korea or alter its behavior, much less bring the regime down."

Learning Objectives

After reading this article, you will more clearly understand the following:

- International relations
- Lee Myung-bak
- Sunshine Policy
- Kim Jong-un
- Chinese influence

The only thing less productive than dealing with North Korea is not dealing with North Korea. Six-party talks—established in 2003 to resolve the North Korean nuclear issue through multilateral dialogue among North and South Korea, the United States, China, Japan, and Russia—have been stalled since 2008. In the intervening years, what little trust that had developed between Pyongyang and Seoul has all but evaporated, and tensions on the Korean peninsula have become explosive. South Korean investment in the North has dried up. North Korea successfully tested a nuclear device in October 2009. And violence along the disputed maritime boundary between the two Koreas in the spring and fall of 2010 brought the peninsula closer to open conflict than at any time in decades.

Finally, on July 28, 2011—by coincidence or planning, a day after the 58th anniversary of the Korean War armistice—North Korea and the United States began direct talks in New York, after almost four years of silence. This was preceded a week before by a meeting between the foreign ministers of North and South Korea at the Regional Forum of the Association of Southeast Asian Nations (ASEAN) in Bali, Indonesia. At long last, inter-Korean diplomacy is showing signs of thaw.

◼ STRATEGIC PATIENCE

The Barack Obama administration has referred to its approach toward North Korea as "strategic patience." One might also call this a policy of doing nothing while outsourcing North Korea policy to a particularly hawkish government in Seoul. In some ways, the United States and the Republic of Korea (ROK) have shown surprising commonality in their approach to Pyongyang. Just as a Republican US administration and the progressive

government of Roh Moo-hyun came together rather unexpectedly in a policy of engagement with the North during President George W. Bush's second term, so Obama's Democratic administration and the conservative government of Lee Myung-bak seemed to reach a de facto agreement on a policy of sanctions and hostility toward Pyongyang.

The problem is that this approach has not worked: It is based on a misreading of the North Korean regime and has only made the situation more dangerous. Hopes for "running down the clock"—while sanctions, isolation, and political instability lead to regime change in North Korea—are misplaced. The regime of Kim Jong-il and of his father before him, Kim Il-sung, has, for all its many faults, shown a remarkable knack for survival over more than six decades. It is not yet clear what the face of new leadership will be in Pyongyang, or even if a genuine leadership transition is under way. But predictions of power struggle and instability in the North Korean leadership have proved wrong over the last 40 years, and regime change from below, while conceivable, does not appear likely any time soon.

Crucially, China will give North Korea the political and economic support it needs to stay afloat, whatever Beijing's reservations about the regime and its nuclear ambitions. China's greatest fear is instability in the North, not nuclear weapons. In this regard Beijing has staked a position opposite to that of Washington and Seoul.

More than any previous South Korean administration, the Lee government has aligned itself fully with the US priority on the threat of North Korea's nuclear program. South Korea has called repeatedly for North Korea's denuclearization as a prerequisite for diplomatic engagement and economic cooperation, an approach that had little success during Bush's first term as president. South Korea's economic isolation of the North has not hurt the North Korean economy so much as increased its dependence on China.

Indeed, the United States and South Korea are increasingly at odds with China on Korean peninsular issues, pushing North Korea further into Beijing's embrace. Kim's recent visits to China reinforce the sense of renewed closeness between the Democratic People's Republic of Korea (DPRK) and the People's Republic of China, while South Korea's outspoken alignment with Japan and the United States signifies—perhaps—a new cold war dynamic in Northeast Asia. The difference this time, however, is that the two sides (ROK-Japan-US vs. DPRK-China-Russia) are much more economically interdependent than in the heyday of cold war hostility. South Korea in particular takes a great risk by aligning with the United States and alienating China.

■ NO COLLAPSE YET

Visiting North Korea in June 2011, I certainly did not get the impression that the country was in severe economic distress, much less on the verge of collapse. If anything, Pyongyang and its environs, as well as the southwestern countryside I traveled through, appeared better off than several years ago. The capital city had more traffic, more lights (including newly installed traffic signals), and more hustle and bustle than I had seen in my previous two visits to the DPRK. The Pyongyang central market, rumored to have closed as part of a government clampdown on market activities a few years back, was thriving. Small stalls and outdoor markets appeared frequently along country roads.

To be sure, an ongoing energy shortage was very apparent outside the capital: Hardly any farm vehicles were to be seen, cars on the highways were few and far between, and slow-moving "smoker trucks" (retrofitted wood-burning vehicles) were as common as the gasoline-powered variety. Still, farmers appeared reasonably healthy, and city folk looked as affluent as could be expected in North Korea's clean but Spartan capital.

Granted, I was not able to visit the remote eastern and northeastern areas of the country, which were hardest hit by an economic implosion in the 1990s and may currently be facing a new threat of famine. In fact, the carefully cultivated appearance of affluence in Pyongyang and the surrounding regions directly contradicts reports of a food emergency in North Korea by international nongovernmental organizations.

In February 2011, five US-based aid organizations (Mercy Corps, World Vision, Samaritan's Purse, Christian Friends of Korea, and Global Resource Services) traveled to three central and western provinces; based on their assessment of rising malnutrition and food shortages, the consortium recommended immediate emergency food assistance. In March, a United Nations interagency food security assessment also appealed for emergency food aid to the DPRK. The international community's reaction to these assessments has been mixed. The European Union promised $14.5 million in food aid to be delivered by August. South Korea has disputed the food assessments and refused aid altogether. The United States sat on the fence, expressing concern about the suffering of ordinary North Koreans but reluctant to give aid that might be diverted and misused by the regime.

Even if North Korea is doing reasonably well economically, it is from a very low baseline: By the most generous estimates, North Korea is far poorer than most regions of China, and incomparably worse off than South Korea. Politically, relative affluence is a double-edged sword for the DPRK. Substantial movement toward a more open market economy would inevitably expose North Korean citizens to information about the outside world, and might thereby call into question the propaganda image of North Korea as a place where people have "nothing to envy" and the South as a brutalized colony of the United States.

From the point of view of the Pyongyang regime, the ideal scenario would be improvement in material conditions with continued political and information control by the ruling Workers' Party. So far, North Korea's steps toward economic reform have been cautious and tentative. Far-reaching reform and a genuine opening of the economy and society could be much more dangerous to the regime than sanctions and isolation.

■ FIVE LOST YEARS

Some of South Korea's current conservative leaders have referred to the previous two left-of-center administrations as Korea's "ten lost years." When it comes to dealing with the North, Lee's time in office might well end up as the Korean peninsula's "five lost years." The momentum gained in inter-Korean relations since the inauguration of Kim Dae-jung's "sunshine policy" in 1998, despite frequent stumbles and occasional crises, had reached a peak in early 2007. The six-party agreement of February 13, 2007, called for the DPRK to shut down and abandon its Yongbyon nuclear reactor, invite back International Atomic Energy Agency inspectors, and fully reveal the extent of its nuclear program. In exchange, the United States and Japan would move toward normalization of ties with the DPRK, and they and other countries would offer energy and humanitarian assistance to North Korea.

In October 2007, North Korea promised that it would shut down its nuclear facilities in Yongbyon and "provide a complete and correct declaration of all its nuclear programs in accordance with the February 13 agreement" by the end of the year. Furthermore, Pyongyang reaffirmed its promise not to transfer nuclear materials, technology, or know-how. The United States and Japan, for their part, reaffirmed their commitments to move toward normalization of relations with the DPRK. North Korea would also receive the equivalent of up to 1 million tons of heavy fuel oil—twice as much as in a 1994 agreement—in an arrangement to be worked out by a working group on economy and energy cooperation. North Korea continued to hand over key documents on its nuclear program in the first half of 2008, and took steps to shut down its Yongbyon facilities. It looked like North Korea would live up to its pledges after all.

At the time the February 13 agreement was being finalized in 2007, South Korean President Roh Moo-hyun met Kim in Pyongyang in early October for the second inter-Korean summit. (The first had been held seven years earlier.) The summit had originally been scheduled for late August, but North Korea had requested a postponement due to severe flooding in the North that summer. Roh was determined, even desperate, to hold a summit meeting before the December 2007 presidential election. Roh himself could not run for reelection, but he hoped the summit would give a boost to his hand-picked successor, former Unification Minister Chung Dong-young.

Kim and Roh on October 4 signed an eight-point agreement that outlined a wide range of cooperative activities. North-South "cooperation" has meant, of course, South Korean aid to and investment in the North; critics accused Roh of giving away the store to North Korea and getting nothing in return. Still, by the end of the Kim-Roh decade, the South Korean economic presence in the North was significant, especially with two major Hyundai projects: a tourism complex in the Kumgang Mountains in the east, and above all the vast Kaesong industrial zone complex in the west, just 20 kilometers above the demilitarized zone. The October 4, 2007, agreement suggested an expansion and deepening of South-North economic cooperation, including possibly a new South Korean investment zone in the Haeju area.

Washington at the time was also engaged. During Bush's first term in office, his administration's almost visceral rejection of everything associated with President Bill Clinton's policies—sometimes criticized as "ABC" ("anything but Clinton")—had prominently included a repudiation of the previous administration's engagement with the DPRK. The differences between the United States and South Korea over diplomacy with the North that emerged after Bush came to office grew even more pronounced under Roh. Indeed, North Korea was the main cause of public friction between South Korea and the United States in the first five years of Bush's presidency. However, after Bush was reelected in 2004, his administration pursued a more active policy of engagement with the DPRK, despite much criticism from hard-line former members of the Bush team.

For some 15 months before Lee's inauguration in February 2008, the United States and South Korea were generally in sync in their approach to the North. Then their positions became almost exactly the reverse of the early Bush-Roh years, with South Korea advocating a hard line toward the North and the United States pushing for engagement.

In June 2007 Lee's Grand National Party (GNP), long hawkish on the North, had revised its North Korea policy to favor engagement over pressure, little different from the position of the two "liberal" presidents, Kim Dae-jung and Roh. Thus the GNP managed to appropriate the most important political asset held by Chung, the progressive candidate, in the presidential election: the Roh government's success in engaging North Korea.

Nevertheless, Lee ran for president promising to be tougher on Pyongyang and to link more closely inter-Korean economic cooperation to progress in North Korea's denuclearization process. And indeed, in his first few weeks in office Lee seemed to take a page from the playbook of Bush's first term. Criticizing his predecessors' engagement policy toward North Korea as "unilateral appeasement," just as Bush had done with regard to Clinton, Lee emphasized North Korea's complete compliance with the denuclearization agreement as a precondition for future inter-Korean cooperation and, in particular, large-scale investment—such as the development of the Haeju-West Sea area promised by Roh at the October 2007 summit in Pyongyang. Lee also promised not to shy away from criticizing North Korea on human rights. His government's initial position could be considered the equivalent of the Bush administration's "ABC"—perhaps "ABR," "anything but Roh."

■ THE HARD-LINERS PREVAIL

There was an element of self-contradiction in Lee's approach to the North, which in its early stages gave the impression of being more ad hoc than a conscious policy. On the one hand, Lee had to demonstrate his toughness on Pyongyang to please his conservative support base. On the other hand, given his former company Hyundai's record as South Korea's largest corporate investor in the North, Lee would seem particularly well-positioned to continue and deepen the South's economic penetration of North Korea.

One might have thought that a long-term strategy of maintaining South Korean influence in the North and pulling North Korea more fully into the orbit of Southern capital calls for more economic engagement, not less. But as it turned out, Lee's initial promises of conditional engagement with the North were greeted with hostility in Pyongyang—and belligerent rhetoric from the North in turn hardened conservative responses in the South, creating an escalating series of hard-line words and actions on both sides. In the first two years of the Lee presidency, North-South relations sank to their lowest level in over a decade.

Under Lee, South Korea has explicitly pursued a policy of "conditional engagement," as opposed to what Lee and other conservatives consider the previous two administrations' naïve and dangerous unconditionality. During his presidential campaign in 2007, Lee announced a plan of "denuclearization, openness, and 3,000" for North Korea, meaning that the South would help raise the per capita GDP of the DPRK to $3,000 per annum if the North gave up nuclear weapons and opened its society and economy. Once in office, Lee restated his policy as a "grand bargain," in which Seoul would offer North Korea economic assistance and security guarantees in exchange for the North's denuclearization and other concessions. The Pyongyang leadership reacted angrily to both the substance and the perceived arrogance of South Korea's new approach.

From the spring of 2008, after an initial period of relative neutrality in their references to the new South Korean president, the DPRK media began attacking Lee with a gusto not seen since the days of the South Korean military dictatorship, calling him a traitor, a pro-American, and an enemy of unification. In April 2008, Pyongyang suspended North-South dialogue and demanded that Lee honor the inter-Korean summits of 2000 and

2007. In effect North Korea asked the South to continue the "sunshine policy" of Kim Dae-jung and Roh, something that the Lee administration's reflexive "all but Roh" instincts could not likely accommodate.

But just as it appeared that Lee might moderate his North Korea policy under domestic and foreign (that is, American) pressure, a South Korean tourist was shot dead at the Mt. Kumgang resort, leading the South to suspend the Mt. Kumgang tourism program. South Korea's demand for an apology was dismissed out of hand by Pyongyang. Lee's call for a resumption of inter-Korean dialogue and economic cooperation in his Liberation Day address on August 15, 2008, elicited no interested response from the North.

North-South Korean relations deteriorated further in 2009, as North Korea escalated its threats and provocations. In January, the DPRK threatened to "nullify" all agreements for reducing conflict between Seoul and Pyongyang; in March, North Korea used the occasion of the first high-level North-South talks to condemn ROK-US military exercises; in April, North Korea fired a series of long-range missiles, eliciting condemnation from the United States and South Korea and bringing most North-South economic exchanges to a standstill.

Pyongyang reserved its harshest condemnation for Seoul's decision to join the US-led Proliferation Security Initiative, an undertaking started by the Bush administration in an effort to block trafficking in nuclear weapons materials. Seoul's decision was announced in May shortly after North Korea conducted a nuclear test. Pyongyang called the decision a "declaration of war" against the DPRK and announced that the Korean War armistice was therefore no longer valid.

North-South relations reached a new level of crisis with the sinking of the South Korean navy ship *Cheonan* on March 26, 2010. The ship, carrying 104 naval personnel, sank following an explosion close to Baengnyeong Island on the west coast, near the disputed maritime boundary between North and South Korea. An international investigation team led by South Korea concluded that a torpedo fired by a North Korean submarine had caused the sinking.

The United States supported the claim, though China, Russia, and perhaps one-third of the South Korean population held serious reservations about the investigation's conclusions. The Obama administration expressed no doubts about North Korea's guilt, and pushed for new international sanctions against Pyongyang while at the same time staging massive military exercises with South Korea, including the largest peacetime naval exercises ever conducted in the seas around the Korean peninsula.

> **North Korea is consumed these days with political transition, and this may entail a reorientation of its relations with the outside world.**

North Korea, for its part, vehemently denied responsibility for the sinking of the *Cheonan* and sought permission to undertake its own investigation of the incident. South Korea refused to allow it. Why North Korea would undertake such a risky attack, possibly triggering all-out war with the South, remained a mystery. North Korea threatened a "sacred war" against outside forces in the face of the US-ROK military exercises. But by the end of the summer, the two sides appeared to have pulled back from the brink.

Then, in November 2010, North Korea fired artillery at Yeonpyeong Island, located on the South Korean side of the maritime boundary called the Northern Limit Line—a boundary drawn unilaterally by the United Nations Command at the end of the Korean War, and which the North has long refused to recognize. South Korea had been engaged in military exercises in those waters, and the North claimed its artillery barrage had been in response to South Korean shelling of North Korean territory.

Four South Koreans were killed and nineteen injured, in the most serious exchange of fire between the two Koreas since the 1953 armistice. The Korean War, frozen in place for decades, looked poised to break out into hot war again. Ultimately, the conflict did not escalate out of control, but President Lee called for a "massive response" should another such incident occur.

COLD WAR MENTALITY

There has long been a cold war-retro look to the Lee administration's East Asian and trans-Pacific relationships. In some respects Lee is less a neoliberal, or even a neoconservative, than a paleoconservative throwback to the heyday of South Korea's authoritarian past. Regionally, Lee's political base is the predominantly conservative North Gyeongsang province in the southeastern part of the country, the home territory of South Korea's military presidents Park Chung-hee and Chun Doo-hwan, who ruled successively from 1961 to 1988 (except for a brief democratic interregnum after Park's assassination in 1979).

Demographically, Lee's base skews toward the above-50 and especially the above-60 age group, a generation shaped more by the Korean War and the fear of communism than by the struggle for democracy in the 1980s and 1990s. Many of Lee's supporters, including key members of his government, make no secret of their distaste for the liberal and "soft-on-communism" tendencies of post-democratization South Korea, and their desire (at least to some extent) to turn the clock back.

The government has clamped down on the press, which has become decidedly more conservative. The power and influence of NGOs, which had thrived under Presidents Kim and Roh (1998–2008), have been curtailed. And South Korea's "Truth and Reconciliation Commission," established in 2005 to investigate human rights violations under previous authoritarian regimes, has been systematically weakened.

This is not to say that democracy as such is threatened in South Korea, or that military-authoritarian rule has any chance of making a comeback in the foreseeable future. But under Lee, South Korea has veered sharply to the right after a decade of center-left administrations. Whether this rightward shift continues, or whether the pendulum is set to shift back to the progressive camp, is a major question for next year's National Assembly and presidential elections.

The ROK under Lee has reaffirmed its alliance with the United States and deepened its economic, political, and military linkages with Japan. But there can be no turning back the cold war clock with China, which is the largest trading partner for both South Korea and Japan.

China has positioned itself to gain the most from the current North-South impasse. On the one hand, economic ties with the South are deeper and more extensive than ever. On the other hand, China's political and economic links with the North have greatly expanded. Since the collapse of the Soviet Union, China had been by far the largest supplier of aid to the DPRK. Although detailed information is hard to come by, it seems that China's business investment in North Korea and the purchase of North Korea's raw materials account for a large portion of North Korea's foreign exchange. In short, China has become North Korea's economic lifeline, and the DPRK has become a Chinese dependency as never before.

Sino-North Korean ties are also growing stronger on the political front. Kim Jong-il, not a leader known to make frequent overseas trips, visited China three times in just over a year between 2010 and 2011. On at least one of these trips, Kim was accompanied by his youngest son and heir apparent, Kim Jong-un. China, which was initially somewhat reluctant to recognize Kim Jong-il's own succession to his father in the 1970s, has effectively given its blessing to the third-generation succession that appears to be under way

A policy of pressure on North Korea is fatally undermined by China's lack of cooperation.

It has become increasingly clear that China will do whatever it deems necessary to keep the Kim regime afloat and maintain a friendly buffer state on its eastern rim, as it faces an array of more or less pro-American states on its other borders (with the exceptions of Myanmar and Russia). As has often been noted, China has greater leverage over North Korea than does any other country. But Beijing is unlikely to use that leverage forcefully so long as its interests are served by the status quo. A policy of pressure on North Korea, favored by South Korea, Japan, and the United States, is fatally undermined by China's lack of cooperation.

It is in this context that the Obama policy of "strategic patience" looked to be giving way to action in the summer of 2011. Secretary of State Hillary Clinton called for dialogue between Pyongyang and Seoul at the ASEAN Regional Forum in Bali, and the two foreign ministers met on the sidelines of the conference on July 22, the first such meeting since 2008.

Vice Foreign Minister Kim Kye-kwan, head of the North America division in Pyongyang's foreign ministry and the North's longtime chief nuclear negotiator, was upbeat about meeting with Obama's special envoy for North Korean affairs, Ambassador Stephen W. Bosworth, and the possibility of negotiating a peace treaty to replace the Korean War armistice. Washington, for its part, called the meetings "exploratory talks" and tried to downplay expectations. Differences between the two sides remain substantial, and reaching an agreement will be further complicated by changes facing several of the principal countries involved in the six-party process in 2012.

■ TOWARD 2012

Next year will mark significant political transitions in North and South Korea, as well as the United States. Both the United States and South Korea face general and presidential elections, and in both countries the results may be close. In particular, the change of leadership in South Korea—where presidents are limited to one five-year term—could alter relations with the North considerably.

Change of leadership in South Korea—where presidents are limited to one five-year term—could alter relations with the North considerably.

The future of conservative rule in South Korea is by no means assured. With unemployment and especially underemployment high—among South Korea's youth especially—dissatisfaction with the Lee government is widespread. A number of leading political analysts in Seoul expect substantial gains by the opposition Democratic Party (DP) in the National Assembly elections scheduled for April. Should the DP unite with other opposition parties to field a viable presidential candidate, the GNP in December could lose the presidency as well.

Park Geun-hye, the daughter of the former president Park Chung-hee, narrowly lost to Lee in the 2007 presidential primary and is the most likely candidate for the GNP in 2012. But even if she were to win, she might very well face a majority opposition in South Korea's unicameral legislature. In any case, Park—whose father signed the first North-South communiqué in 1972 and who as a National Assembly member visited Pyongyang and met with Kim Jong-il in 2002—would likely have a much more pro-engagement policy than the Lee administration.

North Korea, for its part, continues to build up Kim Jong-un as the next leader of the country. At the moment, the propaganda regarding the son is fairly muted: North

Korean media reports duly note his military genius as a newly appointed "four-star general" and record his "on-the-spot guidance" at military sites and factories, following in the footsteps of his father and grandfather. But unlike Kim Jong-il and Kim Il-sung, whose names and images have been ubiquitous throughout North Korea, Kim Jong-un does not appear in the large-scale posters and signs that cover the urban and rural landscapes. Indeed, the only public, outdoor references to Kim Jong-un I saw on my visit to the DPRK were in code, always accompanied by references to his father and grandfather: The Great Leader, the Revered General, and the Four-Star General.

There can be little doubt that the grandson's star is on the rise, but now that Kim Jong-il's health appears to be stable, following a stroke in 2008, the regime is building up the younger son's power base and public image gradually, as it prepares for a formal leadership transition to take place sometime in the future.

Meanwhile, North Korea has announced that it will declare itself a "powerful and prosperous country" (*Kangsong Daeguk*) in 2012, the one-hundredth anniversary of Kim Il-sung's birth. Exactly how this status will be determined is unclear. But preparations are under way for major celebrations surrounding Kim Il-sung's birthday on April 15, and construction in Pyongyang—of new housing, monuments, roads, and public buildings, as well as the long-delayed completion of a 105-story hotel—has hit a pace not seen in over two decades.

North Korea is consumed these days with internal political (and perhaps economic) transition, and this may entail a reorientation of its relations with the outside world as well. This, at least in part, lies behind its recent overtures toward the United States.

■ NEWS OF THE WORLD

There is no existing example of a powerful and prosperous country, or even a moderately affluent one, that is as isolated from the world beyond its borders as is the DPRK. But North Korea cannot keep the world out forever, and its citizens are already much more connected to the outside than was the case a decade ago.

Mobile phones are common in Pyongyang, and although North Korean cell phones cannot reach outside the country, Chinese and South Korean cell phones circulate widely, especially in the border regions. With very few exceptions, North Koreans have no access to the internet; an intranet, not linked to the world wide web, connects computers in the country. This too cannot but change as North Korea develops its information technology, as the regime has stated repeatedly it wants to do. And tens of thousands of North Koreans who have visited China—legally or illegally—have brought news of the outside world to uncounted numbers of their countrymen through personal contact.

It may be futile to try to predict the future of North Korea, and hence that of the Korean peninsula as a whole and the relations among the countries with a vested interest in the place. But it has become apparent that a policy of isolation, sanctions, lack of dialogue, and "strategic patience" has not worked to weaken North Korea or alter its behavior, much less bring the regime down. Any change in North Korea must come from within. And at present the North seems on the verge of change—as is, in a different way, the South. A neo-cold war has been avoided, at least for now, and the Korean peninsula is once again moving forward.

Challenge Questions

The following questions will increase understanding of the contents of this article:

1. What is China's main concern regarding North Korea?

2. How is relative affluence in North Korea both good and bad for the nation?

3. What were the terms of the agreement signed between North Korea and other countries on February 13, 2007?

4. On what position did South Korean leader Lee Myung-bak campaign during the election? What incident in 2008 prevented Lee from moderating his policy?

5. What was the U.S. response to the sinking of the South Korean naval ship on March 2, 2008?

6. What accounts for a large portion of North Korean foreign exchange?

7. In your opinion, how do the elections in South Korea and the United States affect relations with North Korea?

14 Taiwan Jet Deal Aids Ally without Provoking Rival China: View

Bloomberg BusinessWeek

Learning Objectives

After reading this article, you will more clearly understand the following:

- Defense upgrades
- Arms sales and diplomacy
- U.S.-China ties
- Military imbalances
- Taiwan's self-governance

The 100th anniversary marking the fall of China's last imperial dynasty upped tensions in the Taiwan Strait, with Chinese President Hu Jintao calling for "reunification through peaceful means" and his Taiwanese counterpart, Ma Ying-jeou, responding that he was just fine with the status quo.

The last thing needed is outside provocation. So we applaud the Obama administration's compromise decision last month to go ahead with a $5.85 billion deal to upgrade Taiwan's existing fleet of F-16 fighter jets.

Taiwan had wanted the U.S. to sell it 66 newer versions of the aircraft, and leaders of both parties in Congress pressured the administration to do so. China objects to any arms sale to Taiwan. The U.S. is obligated by a 1979 law to supply weapons to Taiwan, whose government broke from mainland China in 1949 after the Communist Party came to power.

The campaign for selling newer aircraft to Taiwan was led by Democratic Senator Robert Menendez of New Jersey and Republican Senator John Cornyn of Texas, where Lockheed Martin Corp. manufactures the F-16. A study commissioned by Lockheed estimated that selling Taiwan new planes would have been worth $8.7 billion and created about 16,000 jobs at a time when the U.S desperately needs to reduce unemployment. The assembly line might close now that the sale is off the table.

■ NEEDLESS PROVOCATION

Yet selling new planes to Taiwan might have needlessly provoked China, a vital U.S. trade partner. Although China vies with the U.S. for influence in the region, the two nations also cooperate (sometimes fitfully) on a number of military and diplomatic issues, including keeping North Korea in check. China temporarily severed military ties with the U.S. after a round of arms sales to Taiwan in 2010 and has said it may do so again.

Taiwan, which publicly expressed gratitude for the upgrade deal, has to be privately disappointed. It has 400 combat aircraft, including 145 F-16s, 56 French-made Daussault Mirages and 60 F-5s. This fleet is no match for the 2,300 combat aircraft that China possesses. This imbalance doesn't take account of 1,100 missiles aimed across the 100-mile-wide Taiwan Strait at the island that China considers a province temporarily occupied by a renegade regime.

The U.S. commitment to ensuring that the island is adequately armed makes diplomatic sense. Taiwan's deterrent capability goes hand-in-hand with a thaw in relations between Taipei and Beijing. Direct air service between China and Taiwan started in 2008, and educational and cultural exchanges are increasing. Taiwan President Ma has renewed calls for China to respect the island's self-governance.

■ JET CRASHES

Under the agreement, Taiwan's A/B versions of the F-16 will be overhauled and given most of the capabilities of more advanced C/D models of the plane, including all-weather navigation and offensive firepower. The upgrade may have one distinct advantage: The Obama administration says it can be done more quickly than building new planes.

But the F-16 retrofitting deal leaves one issue unresolved. Many of Taiwan's planes have outlived their useful life. Most troubling: Two Taiwanese F-5s crashed last month killing three people. An investigation is under way to determine the cause, but crashes of combat aircraft are often traced to mechanical failures. Taiwan temporarily grounded its F-5s, planes from the Vietnam War era, and repeated its request to buy new F-16s.

If Taiwan can prove—perhaps with assistance from U.S. military analysts—that the wear and tear on its fighter jets has rendered too many of them obsolete or unsafe, the Obama administration should reconsider selling the newer aircraft. Placating China at the expense of an ally's ability to defend itself isn't a trade-off worth making.

Challenge Questions

The following questions will increase understanding of the contents of this article:

1. How did Taiwan respond to China's call for peaceful reunification?
2. What was the main reason the United States decided not to sell new jets to Taiwan?
3. What are the pros and cons of the U.S. decision to upgrade Taiwan's existing jet fighters rather than to sell the country new combat aircraft?
4. Explain the military imbalance between China and Taiwan. What is Taiwan's biggest concern about only upgrading the jet fighters?
5. How does China view Taiwan? Do you think this is an internationally accepted view? Why or why not?
6. Give two examples of recent positive developments between China and Taiwan.

15 The Thai Rice Bowl May Get a Little Skimpier

Alan Bjerga and Supunnabul Suwannakij

The world's biggest rice exporter faces stiffer competition.

"Cutting the supply may lead to food shortages."

Learning Objectives

After reading this article, you will more clearly understand the following:

- Rice exports
- Sustainability
- Global food supply
- Production costs and benefits
- Policy making
- Regional connectivity

Many Thais revere Me Posop, the rice goddess who guards humankind and rewards good stewards of her grain. Me Posop has been kind to Thailand in recent decades. While its neighbors Vietnam, Myanmar, Cambodia, and Laos struggled through war, Marxist-Leninism, and authoritarian rule, Thailand prospered from its new factories and booming rice exports. The nation surpassed Myanmar as the world's top rice shipper in 1965: Last year 9 million tons of Thai rice were exported around the world. Thailand, like the Saudis in oil, became the key producer, the country that could always moderate global prices with its abundant reserves. This year, while corn and wheat prices have reached new highs, ample stockpiles of Thai rice have driven rice prices down.

Now the Thai government is proposing a major change in strategy for its rice growers, who feel hard pressed by low prices, an assault of pests, and the presence of low-cost competition from emerging rivals. The government seems ready to abandon Thailand's position as the world top rice exporter—a serious decision, considering the mounting anxiety over the size and stability of the global food supply.

Thai farmers are certainly worried about their business. In the rice paddies near Ayutthaya, a former Siamese capital that 17th century emissaries from Louis XIV compared with Paris in its wealth and importance, Payao Ruangpueng must battle an infestation of rice planthoppers that are munching their way through the paddies. That's not all. "We're suffering from a rice price slump, crop damage, and lower-than-expected production," she says, standing on the edge of a rain-soaked paddy. "Production costs are higher than income. We can't afford to continue planting."

In March the Thai government stated its intention to eliminate a third planting this year to improve rice quality and to combat the hopper, which dies if deprived of rice plants for 25 days. The plan may eventually reduce annual exports by 2 million metric tons, or about 20 percent of Thailand's shipments.

Thai officials say they want the industry to focus on fancier grades of rice that fetch higher prices. While Thai rice shipments have increased 33 percent in the past decade, Vietnamese exports are up 70 percent in the same period to 6 million tons, according to the U.S. Agriculture Dept. Cambodia and even Myanmar are also emerging as global rice powers, says Pramote Vanichanont, honorary president of the Thai Rice Mills Assn. and a member of the National Rice Policy Committee. Thailand, following the classic curve of development, has priced itself out of much of its own market, he says. Land prices have shot up, as well as the cost of tractors and the wages of farmhands.

> **With costs of land, tractors, and farmhand wages rising, Thailand has priced itself out of much of its market**

The government also plans to turn the country into the warehouse, finance, and marketing hub of Southeast Asia's rice trade. The Agricultural Futures Exchange of Thailand, the nation's government-backed rice and rubber bourse, is rolling out a new futures contract on Apr. 29 intended to be a regional benchmark for standard quality rice.

This long-term strategy may not be good for global food needs. The U.N. expects world food demand to rise 70 percent by 2050, and its Food and Agriculture Organization in February urged Thailand and its neighbors to grow more rice. Reductions in Thailand's production may end up hurting poor consumers in Africa and elsewhere while doing little for Thai prices, says Kiattisak Kanlayasirivat, a director at the Thai office of trading company Novel Commodities. "I doubt whether it is a good policy, as cutting the supply may lead to food shortages," says Kanlayasirivat, whose firm trades about $600 million of rice a year.

The Vietnamese may not even have the resources needed to replace major cuts in Thai production. "I personally think that Vietnam doesn't need to become No. 1 in rice exports," says Nguyen Van Bo, president of the Vietnamese Academy of Agricultural Science. "To export a lot, Vietnam will have to exploit a lot of land, use a lot of fertilizers. That could cause degradation oil natural resources."

From *Bloomberg BusinessWeek,* April 11–17, 2011. Copyright © Bloomberg 2011 by Bloomberg BusinessWeek. Reprinted by permission of Bloomberg LP via YGS Group.

Challenge Questions

The following questions will increase understanding of the contents of this article:

1. Who is Me Posop, and why do many Thais revere her?
2. What difficulties are causing the Thai government to propose a major change for rice growers?
3. How would eliminating a third planting affect Thailand, and what methods are proposed to make up for the lost crops?
4. How does the "classic curve of development" support the Thai government's strategy for reducing rice production?
5. What are plausible effects on other countries if Thailand adopts its long-term strategy, and thereby loses its position as the world's top rice exporter?

16 The Vietnam Case

Angie Ngoc Tran

Workers versus the Global Supply Chain.

Learning Objectives

After reading this article, you will more clearly understand the following:

- Migrant workers and labor migration
- Foreign direct investment
- Minimum wage issues
- Multinational companies
- Supply chain
- Codes of conduct

Vietnam—a socialist country integrating into the global capitalist system—can serve as a useful case study on labor and globalization because it reflects larger global trends. These include domestic and international labor migration with the increased mobility of capital, and the complex role of the government and recruitment companies in all types of migration. Workers—both inside and outside of Vietnam—have used their agency to fight for their rights and human dignity in this environment. Human rights advocates have forced open a small space in the multi-level global supply chain for the establishment of codes of conduct toward workers, and ethical consumers and investors try to speak and act on behalf of migrant workers worldwide. This article will discuss the causes and implications of these changes in Vietnam, as well as how they relate to a broader global framework of labor.

■ DOMESTIC MIGRATION AND WORKERS' AGENCY

Domestic labor migration is no longer a one-way workers' movement from rural to urban areas to work in manufacturing hubs. It also includes workers who return to their hometowns, villages, or provinces, as mobile global capital expands and sets up factories in poor provinces to take advantage of even lower wages. This circular rural-urban-rural labor migration has occurred in Vietnam, China, and other countries.

In Vietnam, as of 2009, about 47.7 million people worked in all economic sectors, while 6.85 million worked in manufacturing, second only to agriculture and forestry. Most manufacturing workers were in textile/apparel and leather shoe factories in export processing zones (EPZs) and industrial zones (IZs); they earned very low wages, on average less than US$100 per month. Over 60 percent of the workforce in 140 EPZs and IZs nationwide are young females, around 20–35 years old, from poor provinces in the northern and central regions of Vietnam. Most domestic migrants found work through both informal channels (families and friends) and a formal recruitment process in which they had to pay (mostly state-owned) recruitment companies around US$50 to secure their jobs. Most foreign-owned suppliers are from Taiwan, Japan, South Korea, and Hong Kong. Recent trends show that many suppliers have established factories in rural areas to get access to even cheaper labor in provinces such as Tra Vinh, Long An, Thanh Hoa, and Quang Ngai, far from the global cities such as Ho Chi Minh City and Hanoi, which now have developed labor shortages.

However, workers are not victims: they have agency and have been rising up against both management and governments to fight for their rights and human dignity. This is not exclusive to Vietnam: there have been protests in China, Bangladesh, Cambodia, Thailand, and Mexico. In each country, workers develop strategies and tactics to fight for their rights and dignity that reflect the political and economic conditions of the factories' locations. In Vietnam, workers have developed strategies and tactics that reflect Vietnam's socialist past and market-oriented present.

> "In each country, workers develop strategies and tactics to fight for their rights and dignity that reflect the political and economic conditions of the factories' locations."

The common reasons for strikes in those countries have concerned labor rights (pay, work hours, working conditions, labor contracts, overtime compensation, daily productivity targets, meals, fines as disciplinary action, apprenticeship period) and non-wage benefits (social security, health and unemployment insurance, paid leaves, meals at work). The Vietnamese workers invoked the bonus and benefits they used to receive under the socialist system, such as the 13th month pay (usually given around the Vietnamese Lunar New Year), and demanded to be treated with dignity and respect. They have also exposed physical and verbal abuses by foreign experts, managers, and owners.

The minimum wage strike waves that started in the middle of the past decade galvanized the collective action of hundreds of thousands of workers in EPZs and IZs in Ho Chi Minh City. In 2006, workers in three foreign-owned shoe factories in a Ho Chi Minh City export processing zone sparked a series of minimum wage strikes. Up until then, the minimum wage in the foreign direct investment (FDI) sector had been frozen for seven years (1999–2005) at less than US$40 per month. While on the surface these strikes were staged against FDI owners, in reality, by demanding higher minimum wages, these strikes were actually against the state policy. In 2007, workers demanded raises in wages to compensate for spiraling inflation, which led to the Prime Minister's decision to adjust wages for inflation, effective in January 2008. Again, in 2008, workers went on strike to expose FDI factories that refused to implement the inflation-adjustment decision or did not implement the adjustment properly.

But the minimum wage strike waves have had some larger implications for workers in all sectors in Vietnam. First, higher minimum wages are not only adjusted for inflation, but also tied to correspondingly higher social benefits, such as social insurance and health benefits. The 2006 minimum wage strike resulted in the establishment of an automatic annual increase in the minimum wage that adjusts for inflation: every November, the state announces the minimum wage increases (for both FDI and domestic sectors) to take effect in January of the following year. Second, the minimum wage strikes have forced the state to bridge the disparity between the minimum wage levels in the FDI sector and those in the state and domestic private sectors (which are lower than the FDI rates), which is expected to reach parity by 2012. Moreover, as foreign-invested factories expanded to poor provinces outside of big cities to take advantage of even lower minimum wage requirements, strikes spread beyond Ho Chi Minh City and Hanoi. In short, workers' agency in Vietnam shows that when workers took ownership of their rights and responsibilities, they effectively secured positive responses from the labor newspapers, the labor unions, and the Ministry of Labor.

■ LABOR MIGRATION, NEWSPAPERS, BROKERAGE STATES

International labor migration has become another option for workers, in which the nation states play a complex and active role. In socialist Vietnam, the state has become a "labor brokerage state," sending hundreds of thousands of workers overseas, and benefiting from these workers who must pay to work outside of Vietnam. This global practice of inter-governmental agreements to send and receive migrant workers is seen in many countries: Vietnam, the Philippines, Indonesia, and even the United States, such as the 1942–1964 Bracero Program that sent millions of Mexican men to the United States

to work temporary jobs in agriculture. While the duration of the contracts varied in these cases, they shared some common features: management controlled and considered workers as disposable commodities and sent them home at the end of their contracts, paying no attention to their well-being and working conditions. Both home (labor sending) and host (labor receiving) governments have strong shared interests with management to sustain transnational labor migration. In Vietnam, the state meets host countries' demands for cheap, temporary, and compliant workers who will fulfill the terms of the contract and return to Vietnam at the end of their contracts. While some host countries do care for workers' well-being, overseas migrant workers do not have the right to organize and most are not represented by labor unions.

> "The Vietnamese state and their recruitment companies have benefited more from Vietnamese labor migration: both domestic and international."

However, the Vietnamese case is different from others due to the socialist pro-labor legacy and the inherent contradictions in the state's self-proclaimed "market economy with socialist orientation." The media, especially the labor newspapers, while well ensconced within the state and labor union structures, have used their connections and knowledge within this system to expose migrant labor violations perpetrated by capital (both foreign and domestic) and some state recruitment companies. As early as 2004, they alerted the public to fraudulent activities and irresponsible behavior of many state recruitment companies when state bureaucracies and people's committees had turned a blind eye. But the media also faces constraints because they are part of the state structure, which can compromise and censor journalists' reportage. At the same time, while the Vietnamese General Confederation of Labor (Vietnam's only legal labor union) has made considerable efforts to overcome their structural and capacity weaknesses to represent workers within Vietnam's borders, they have played no role in representing and protecting Vietnamese migrant workers overseas thus far.

Exporting labor is not new in Vietnam. Since 2000, the Vietnamese state has been brokering its workers through inter-governmental agreements: over 500,000 Vietnamese workers had paid to work in over 40 countries. When still a part of the former Soviet Bloc from 1980 until its disintegration in 1989, the Vietnamese state exported over 250,000 Vietnamese workers to Eastern Europe, primarily to the former Soviet Union, in order to repay the debts and also benefit from their remittances to relatives back home. After the fall of the former Soviet Bloc, the Vietnamese state redirected exports and monopolized the authority to control and regulate workers to work in other countries. In 2000, Prime Minister

Phan Van Khai echoed support for this strategy, saying that, "Exporting labor is a very important and major strategy because it helps solve the unemployment problem, increase foreign exchange for the country. . . . We must consider labor export as an important and long-term strategy . . ."

Export labor has expanded tremendously since Vietnam joined the World Trade Organization in 2007. The state-sanctioned recruitment companies—most owned by state ministries, state corporations, local state governments, and social-political mass organizations, such as the labor unions—grew to 165 companies as of 2011. Vietnamese migrant workers had been sent to work in over 40 countries; topping the list are Taiwan, Japan, South Korea, and Malaysia. Nations of the Middle East, including United Arab Emirates, Saudi Arabia, Bahrain, and Qatar, follow, as well as some countries in the former Soviet Bloc.

WHO BENEFITS FROM EXPORT LABOR?

The Vietnamese state and their recruitment companies have benefited most from Vietnamese labor migration: both domestic and international. Over 80% of total costs charged to Vietnamese migrants working overseas goes to processing fees accumulated by recruitment companies (in both home and host countries), interests on loans disbursed by state banks, and fees to send remittances home. The economy in general benefits from labor remittances to about US$1.7 billion (annual average between 2005 and 2008) from these fees and from money sent home.

The Vietnamese state has used labor export as a social policy to reduce poverty, but the results are mixed at best. In 2009, the government started a 12-year poverty reduction campaign (for the period between 2009 and 2020) in the 62 poorest rural districts in the northern and central regions by encouraging ethnic minorities (53 groups throughout Vietnam) to work abroad. They offered low-interest loans from the state-owned banks and other subsidies on vocational training, room and board while studying, free health check-ups, and passports. Since 2009, about 4,500 people of different ethnic minorities in these poor districts have left to work overseas; most can only afford to go to Malaysia for low-skilled and low-paid jobs (on average about less than US$300 per month), not Korea and Japan, which require higher fees, more education, and language skills, and in turn pay higher wages. A strategy many ethnic workers use is to pool their wages and take turns sending remittances home to repay the debts they incurred from borrowing to work in Malaysia. While this prompt repayment saves them some interest costs, long-term benefits for these workers and their families remain to be seen. In reality, parents of many ethnic workers are already in debt—mostly due to poor health, illnesses, and natural disasters—so as soon as they receive remittances from their sons and daughters, they must repay these existing

debts, and have very little left to invest in any productive ventures. The vicious cycle of poverty and indebtedness thus continues for these people, who have very little financial cushion for any mishap. Preliminary evidence shows that this process exacerbates the gap between the rich and the poor in local communities and villages and provides no alternatives to escape poverty.

Moreover, there are a lot of uncertainties and unforeseen risks in host countries. In Malaysia, the most common destination for the poorest migrant workers, these adversities include not knowing where they are being sent to work, since they are vulnerable at the hands of Malaysian outsourcing companies. They have also included inadequate official response from Vietnamese recruitment companies and embassy officials to their concerns and passports being impounded by their bosses. And of course, there exist the constant fears of gang violence and robbery. Most rely on bonding and social networking with fellow Vietnamese migrants to support each other.

> "While both female and male Vietnamese workers have to pay the same fees to the recruitment and outsourcing companies, women face more demeaning entry regulations and fears while they work in Malaysia . . ."

While both female and male Vietnamese workers have to pay the same fees to the recruitment and outsourcing companies, women face more demeaning entry regulations and fears while they work in Malaysia, consistent with other scholars' findings. Upon arrival in Malaysia, most women workers had to undergo blood and urine tests to determine whether they were pregnant. Moreover, there were cases of abortion in which these women workers went to back-alley doctors to abort their pregnancies with poor safety conditions.

Overall, the Vietnamese migrant workers in Malaysia have responded to those uncertainties and forms of exploitation on the factory floor in creative ways. They slowed down the speed-up process by saving the completed products for another day. They protested to demand safety in the hostels. They even reached out to religious support groups, Malaysian legal assistance, labor unions, and nongovernmental organizations for help. However, these types of assistance are not systemic, and therefore do not spread the benefits to all Vietnamese migrant workers there.

GLOBAL SUPPLY CHAIN AND CODES OF CONDUCT

Workers in Vietnam and other countries that manufacture for the world operate in the multi-level global supply chain. Understanding how this global production works would reveal the responsibilities of all stakeholders and what we can do as ethical consumers and investors to ensure decent working and living conditions for workers worldwide.

As corporate social responsibility initiatives have spread globally, the Vietnam case can shed light on its practices and the need to reclaim and restore the intent of the codes of conduct to protect all workers and the environment worldwide. Most corporate social responsibility studies, including International Labor Organization (ILO)-sponsored ones like Better Work Vietnam, do not analyze the multi-level subcontracting in the global supply chain, which is key to understanding the effects of uneven power relations in this chain. Multinationals (MNCs) or brands dictate all the terms of production and income distribution (including piece-rates to pay workers) and transfer labor responsibilities to their suppliers, who oversee manufacturing in developing countries like Vietnam. Most brands promise consumers that they will terminate contracts with suppliers who fail to enforce the codes of conduct (CoCs) which are developed from the core ILO conventions that the MNCs promise on their websites. But the brands' actions have never been transparent for public scrutiny: it is very difficult to monitor the relationships between the brands and their suppliers. Even when some suppliers in Vietnam are found to be non-compliant, there is no concrete action from the brands to uphold their social responsibilities to workers as promised to their consumers.

At the top of the chain are the corporate buyers/brands (MNCs) that place orders with their suppliers (mostly owned and managed by Taiwanese and South Koreans in Vietnam), who subcontract to small and medium-sized Vietnamese factories to meet just-in-time delivery schedules. These Vietnamese factories are legal in that they register with die Vietnamese government and pay taxes, but they may not be licensed directly by the foreign brands that produce for them (therefore brand-unlicensed). Their existence may be unknown to the MNCs, or the MNCs may prefer to close their eyes to their existence. As such, they are not subjected to codes of conduct scrutiny and are thus not monitored.

With the MNCs dictating all the terms, their suppliers and domestic subcontractors compete fiercely to drive down the piece rates in order to win contracts in this race to the bottom. Ultimately, the brands win: they secure the lowest contract prices and fastest turnaround delivery time. The distribution of earnings is grossly unequal. On average, for a pair of shoes for which a consumer pays US$100, a multinational corporation would pay a supplier a freight-on-board price of US$16, out of which US$5 goes towards (mostly imported) raw materials, US$2 for local management (in a domestic subcontractor), and US$2 for local labor, leaving US$7 for the supplier's remaining expenses and profit. In 2008, the brand took a massive 84 percent of the sales price, while workers received only 2 percent. Statistics for 2010 show that workers received even less: only one percent.

Moreover, in Vietnam, the brands are very savvy. They base their subcontracted prices on a firm legal ground: the government-set minimum wages, to which they add bonuses such as perfect attendance and bus fares to go home, are not tied to long-term benefits such as social security and health and unemployment insurance. Such is the collusion of capital and the state: the state encourages investment by keeping labor costs low, and capital points to state-mandated minimum wage laws as its defense for poor wages.

■ GLOBAL LEVERAGE AND ETHICAL DIMENSIONS

What can be done to protect migrant workers both domestically and internationally, with a good knowledge of the global supply chain structure and the interests of key stakeholders? The general public should hold MNCs and their suppliers accountable, and not allow them to pay lip service to consumers and shirk their responsibilities.

First, exertion of some existing leverage could be used to protect migrant workers worldwide. Nation states need reminders to be held accountable to a declaration that they signed in 2007: the ASEAN Declaration on the Protection and Promotion of the Rights of Migrant Workers, signed by ten heads of state, including Vietnam and Malaysia, in Cebu, the Philippines, in January 2007. This document is based on the Universal Declaration of Human Rights adopted by the General Assembly in 1948, as well as other appropriate international instruments adopted by all the ASEAN member countries to safeguard the human rights and fundamental freedoms of individuals. It clearly stipulates obligations of both sending and receiving states to promote the full potential and dignity of migrant workers in a climate of freedom, equity, and stability in accordance with the laws and regulations of ASEAN countries. As one of the general principles, it further states that for humanitarian reasons, both sending and receiving states shall "cooperate to resolve the cases of migrant workers who, through no fault of their own, have subsequently become undocumented."

Second, empowering global labor standards, articulated in most brands' codes of conduct, can be achieved by pressuring the brands to be transparent and have concrete actions to comply with their CoCs. Specifically, ethically conscious consumers can vote with their feet if the MNCs do not make transparent the violations of their suppliers or continue to subcontract with the violators. Also, consumers can pressure the brands to raise the contract prices the brands pay their suppliers in order to raise wages for workers, with an understanding that the brands would receive a small reduction in their lion's share of the profits. The brands should share the monitoring costs with their suppliers and subcontractors in Vietnam, as well as the costs to fix non-compliance issues, notably working hours, overtime compensation, and occupational safety and health.

Finally, as conscious consumers and investors, people can reclaim the real intent of CSR. Investors can influence the decisions of MNCs through stockholder meetings, or pressure their investment companies to uphold labor and environmental standards. With the increasingly aging population, we can put pressure on ways in which pension plans are invested by demanding that MNCs do ethical business and uphold their CoCs. In short, simple direct action can improve the circumstances of millions of workers globally, who deserve to live enriching and meaningful lives.

Challenge Questions

The following questions will increase understanding of the contents of this article:

1. How have minimum wage strikes in Vietnam affected foreign direct investment and state policy in Vietnam?

2. How do both host countries and home countries benefit from labor migration, and in what way is the Vietnam case different?

3. What risks do Vietnamese migrant workers face in their host countries?

4. How do multinational companies (MNCs) avoid codes of conduct scrutiny through their suppliers?

5. What example is given of collusion between state and capital in Vietnam?

6. In what ways can the general public hold MNCs and their suppliers accountable to protect migrant workers domestically and abroad?

Internet References

GENERAL SITES

CNN Online Page
www.cnn.com
A U.S. 24-hour video news channel. News is updated every few hours.

C-SPAN ONLINE
www.c-span.org
See international and archived C-SPAN (Cable-Satellite Public Affairs Network) programs, which broadcast unedited federal government and public affairs proceedings.

NationMaster.com
www.nationmaster.com/countries
A central data source that graphically compares nations by generating maps and graphs of statistics from the *CIA World Factbook,* UN, OECD, and others.

The New York Times
topics.nytimes.com/topics/news/international/countriesandterritories/
View the newspaper's recent and archived news online organized by country.

United Nations
www.un.org
Offical site of the United Nations with reports on international programs in Asia financed by the UN. Available in English, Arabic, Chinese, French, Russian, and Spanish.

U.S. Central Intelligence Agency Home Page
www.cia.gov
This site includes publications of the CIA, current *World Factbook,* and maps.

U.S. Department of State Home Page
www.state.gov/
This website is organized by topic categories.

World Bank Group
www.worldbank.org
Find news (i.e., press releases, summary of new projects, speeches), publications, topics in development, countries and regions on this website. It links to other financial organizations. This site is available in English, Chinese, Spanish, and French, and 13 other languages.

World Health Organization (WHO)
www.who.int
Maintained by WHO's headquarters in Geneva, Switzerland, it is possible to use Excite search engine to conduct keyword searches from here, in 6 languages.

World-Newspapers.com
www.world-newspapers.com
A central resource for English language newspapers, magazines and news sites by country and region.

World Trade Organization
www.wto.org
The website's topics include legal frameworks, trade and environmental policies, recent agreements, etc. Available in English, Spanish, and French.

WWW Virtual Library Database
www.vlib.org
Easy search for country-specific sites that provide news, government, and other information is possible from this site.

JAPAN

Daily Yomiuri Online
www.yomiuri.co.jp/dy
Online edition of the *Daily Yomiuri* newspaper in Japan reporting top news of the day, including Japanese political, social, and general news, reports from foreign news services, and overseas English-language newspapers.

Japan Ministry of Foreign Affairs
www.mofa.go.jp
"What's New" lists events, policy statements, press releases on this website. The Foreign Policy section has speeches, archives, and information available in English and Japanese.

Japan Policy Research Institute (JPRI)
www.jpri.org
Find the latest JPRI network publications, news, projects, and events related to Japan and the Pacific Rim.

The Asahi Shimbun AJW (Asahi Japan Watch)
ajw.asahi.com/
News lineup for Japan and Asia, and covering the Great East Japan Earthquake.

The Japan Times Online
www.japantimes.co.jp
This daily online newspaper published in Japan is offered in English and contains late-breaking news.

THE PACIFIC RIM

BBC News–Asia
www.bbc.co.uk/news/world/asia/
News on Asian business by the British Broadcasting Corporation (BBC); up-to-date country profiles available by scrolling to the bottom.

Inside China Today
www.einnews.com/china/
The European Information Network provides digital news about China and is searchable by country, topic, and editor's picks. Requires membership to access data.

Smithsonian.com
www.smithsonianmag.com/people-places/asia-pacific/
Read articles on Asia and the Pacific under headings such as History, People, Science, and Art.

U.S. Agency for International Development (USAID)
www.usaid.gov/locations/asia/countries/
Contains articles and brief reports of current challenges on some Pacific Rim countries, plus extensive descriptions of U.S. assistance programs.

See individual country reports for country-specific websites.

The United States
(United States of America)

GEOGRAPHY

Area in Square Miles (Kilometers): 3,794,100 (9,826,675) (about half the size of Russia)

Capital (Population): Greater Washington, DC (4.4 million)

Environmental Concerns: air and water pollution; limited freshwater resources, desertification; loss of habitat; waste disposal; acid rain

Geographical Features: vast central plain, mountains in the west, hills and low mountains in the east; rugged mountains and broad river valleys in Alaska; volcanic topography in Hawaii

Climate: mostly temperate, but ranging from tropical to arctic

PEOPLE

Population

Total: 313,232,044

Annual Growth Rate: 0.963%

Rural/Urban Population Ratio: 18/82

Major Languages: English; Pacific Islander; many others

Ethnic Makeup: 80% white; 13% black; 4% Asian; 3% Amerindian and others

Religions: 51% Protestant; 24% Roman Catholic; 2% Mormon; 2% Jewish; 4% others; 17% none or unaffiliated

Health

Life Expectancy at Birth: 76 years (male); 81 years (female)

Infant Mortality: 6/1,000 live births

Physicians Available: 1/417 people

HIV/AIDS Rate in Adults: 0.6%

Education

Adult Literacy Rate: 99% (official)

Compulsory (Ages): 7–16; free

COMMUNICATION

Telephones: 141 million main lines (2009)

Telephones: 255 million cellular

Internet Users: 245 million (2010)

Internet Penetration (% of pop): 77%

TRANSPORTATION

Roadways in Miles (Kilometers): 4,042,768 (6,506,204)

Railroads in Miles (Kilometers): 139,679 (224,792)

Usable Airfields: 15,079

GOVERNMENT

Type: federal republic

Jndependence Date: July 4, 1776

Head of State/Government: President Barack H. Obama is both head of state and head of government

Political Parties: Democratic Party; Republican Party; Green Party; Libertarian Party

Suffrage: universal at 18

MILITARY

Military Expenditures (% of GDP): 4.06%

Current Disputes: boundary and territorial disputes with Canada, the Bahamas, Haiti and other countries; water-sharing disputes with Mexico; "war on terrorism"

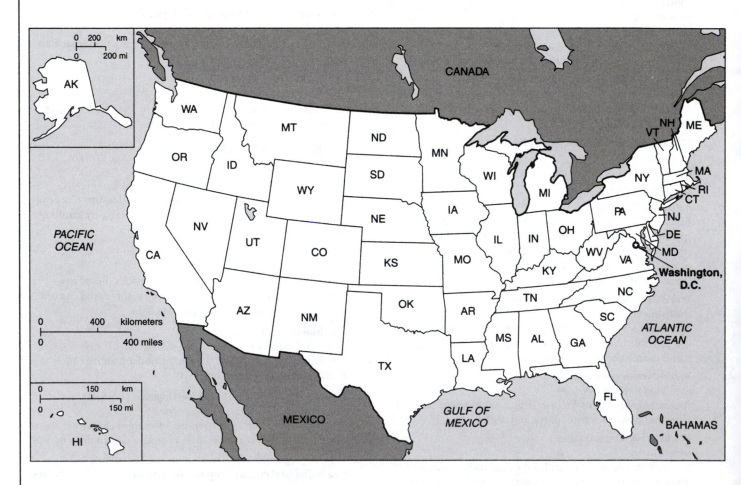

ECONOMY

Per Capita Income/GDP: $47,200/$14.66 trillion
GDP Growth Rate: 2.8%
Inflation Rate: 1.4%
Unemployment Rate: 9.6% (2010)
Labor Force by Occupation: 79% services; 20% industry; 1% agriculture
Population Below Poverty Line: 15%

Natural Resources: many minerals and metals; petroleum; natural gas; timber; arable land
Agriculture: food grains; feed crops; fruits and vegetables; oil-bearing crops; livestock; dairy products
Industry: diversified in both capital and consumer-goods industries
Exports: $1.27 trillion (primary partners Canada, Mexico, China)
Imports: $1.903 trillion (primary partners, China, Canada, Mexico)

Canada

GEOGRAPHY
Area in Square Miles (Kilometers): 3,854,083 (9,984,670) (slightly larger than the United States)
Capital (Population): Ottawa (1.2 million)
Environmental Concerns: air and water pollution; acid rain; industrial damage to agriculture and forest productivity
Geographical Features: permafrost in the north; mountains in the west; central plains; lowlands in the southeast
Climate: varies from temperate to arctic

PEOPLE
Population
Total: 34,030,589
Annual Growth Rate: 0.79%
Rural/Urban Population Ratio: 19/81
Major Languages: English and French
Ethnic Makeup: 28% British Isles origin; 23% French origin; 15% other European; 6% others; 2% indigenous; 26% mixed
Religions: 43% Roman Catholic; 23% Protestant; 18% others; unaffiliated 16%

Health
Life Expectancy at Birth: 79 years (male); 84 years (female)
Infant Mortality: 5/1,000 live births
Physicians Available: 1/498 people *HIV/AIDS Rate in Adults:* 0.3%

Education
Adult Literacy Rate: 99%
Compulsory (Ages): primary school

COMMUNICATION
Telephones: 18.3 million main lines (2009)
Telephones: 23.1 million cellular (2009)
Internet Users: 27 million (2011)
Internet Penetration (% of pop): 79%

TRANSPORTATION
Roadways in Miles (Kilometers): 647,655 (1,042,300)
Railroads in Miles (Kilometers): 29,926 (46,552)
Usable Airfields: 1,404

GOVERNMENT
Type: confederation with parliamentary democracy
Independence Date: July 1, 1867
Head of State/Government: Queen Elizabeth II; Prime Minister Stephen Harper
Political Parties: Bloc Québécois; Liberal Party; New Democratic Party; Green Party; Conservative Party of Canada
Suffrage: universal at 18

MILITARY
Military Expenditures (% of GDP): 1.1%
Current Disputes: maritime boundary disputes with the United States, and sovereignty dispute with Denmark

ECONOMY
Per Capita Income/GDP: $39,400/$1.33 trillion
GDP Growth Rate: 3.1%
Inflation Rate: 1.6%
Unemployment Rate: 8%

Labor Force by Occupation: 76% services; 13% manufacturing; 6% construction; 2% agriculture and 3% others

Natural Resources: petroleum; natural gas; fish; minerals; timber; wildlife; coal; hydropower

Agriculture: grains; dairy products; tobacco; fruits and vegetables; forest products; fish

Industry: oil production and refining; natural-gas development; fish products; wood and paper products; chemicals; transportation equipment

Exports: $461.8 billion (primary partners United States, United Kingdom, China)

Imports: $436.7 billion (primary partners United States, European Union, Japan)

Glossary

of Terms and Abbreviations

Animism The belief that all objects, including plants, animals, rocks, and other matter, contain spirits. It is one of the oldest religious traditions in the world and undergirds ancestor worship in many societies in the Pacific Rim.

Anti-Fascist People's Freedom League (AFPFL) An anti-Japanese resistance movement and political party that was organized in Burma (Myanmar) by, among others, Aung San, father of Aung San Suu Kyi. It wielded influence from 1946 to 1962.

ANZUS The name of a joint military-security agreement originally among Australia, New Zealand, and the United States. New Zealand cancelled its relationship with the United States under this treaty in 1984 but remains affiliated with Australia directly.

Asia Pacific Economic Cooperation Council (APEC) Organized in 1989 and headquartered in Singapore, APEC provides a forum for heads of governments to meet and discuss economic development for the region. It consists of 21 members and meets annually in different Pacific Rim cities.

Asian Development Bank (ADB) With contributions from some 66 industrialized nations, the ADB provides loans to alleviate poverty in the less developed nations of the Pacific Rim.

Association of Southeast Asian Nations (ASEAN) Established in 1967 to accelerate economic growth among the 10 member nations of Southeast Asia. It is headquartered in Jakarta, Indonesia.

Avian Flu (Bird Flu) A deadly, infectious influenza disease usually found in birds but occasionally migrating to humans. Some strains of the virus are resistant to drugs. The virus has been found in birds and humans throughout the Pacific Rim.

(British) Commonwealth of Nations A 54-member voluntary association of mostly former British colonial nations. Officials meet regularly in member countries to decide issues of common economic, military, and political concern.

Buddhism A religious and ethical philosophy of life that originated in India in the fifth and sixth centuries B.C., partly in reaction to the caste system. Buddhism holds that people's souls are weighed down in misery by false desires. People can escape this life of illusion only by following the precepts of Buddha and attaining a state of enlightenment.

Capitalism An economic system in which productive property is owned by individuals or corporations, rather than by the government. Prices of product are determined largely by market demand, and the greater portion of profits belonging to individuals or corporations is used as they see fit.

Chaebol A Korean term for a large business conglomerate. Similar to the Japanese *keiretsu.*

Chinese Communist Party (CCP) Founded in 1921 by Mao Zedong and others, the CCP became the ruling party of the People's Republic of China in 1949 upon the defeat of the Nationalist Party and the army of Chiang Kai-shek.

Cold War The intense rivalry, short of direct "hot-war" military conflict, between the Soviet Union and the United States, which began at the end of World War II and continued until approximately 1990.

Communism An economic system in which land and businesses are owned collectively by everyone in the society rather than by individuals. Modern communism is founded on the teachings of the German intellectuals Marx and Engels.

Confucianism A system of ethical guidelines for managing one's personal relationships with others and with the state. Confucianism stresses filial piety and obligation to one's superiors. It is based on the teachings of the Chinese intellectuals Confucius and Mencius.

Cultural Revolution A period between 1966 and 1976 in China when, urged on by Mao, students attempted to revive a revolutionary spirit in China. Intellectuals and even Chinese Communist Party leaders who were not zealously Communist were violently attacked or purged from office.

Demilitarized Zone (DMZ) A heavily guarded border zone separating North and South Korea. It is 155 miles (248 km) long and 2.5 miles (4 km) wide.

European Union (EU) An organization of 27 European nations formed by treaty in 1992. Its goal is the establishment of a single economic and political European entity. It is headquartered in Brussels, Belgium.

Extraterritoriality The practice whereby the home country exercises jurisdiction over its diplomats and other citizens living in a foreign country, effectively freeing them from the authority of the host government.

Feudalism A social and economic system of premodern Europe, Japan, China, and other countries, characterized by a strict division of the populace into social classes, an agricultural economy, and governance by lords controlling vast parcels of land and the people thereon.

Greater East Asia Co-Prosperity Sphere The Japanese description of the empire it created in the 1940s by military conquest. It included almost all of the nations in the Asia-Pacific.

Gross Domestic Product (GDP) A statistic describing the entire output of goods and services produced by a country in a year, less income earned on foreign investments.

Hinduism A 5,000-year-old religion, especially of India, that advocates a social caste system but anticipates the eventual merging of all individuals into one universal world soul.

Indochina The name of the colony in Southeast Asia controlled by France and consisting of the countries of Laos, Cambodia, and Vietnam. The colony ceased to exist after 1954, but the term still is often applied to the region.

International Monetary Fund (IMF) An agency of the United Nations whose goal it is to promote freer world trade by assisting nations in economic development.

Islam The religion founded by Mohammed and codified in the Koran. Believers, called Muslims, submit to Allah (Arabic for God) and venerate his name in daily prayer.

Keiretsu A Japanese word for a large business conglomerate.

Khmer Rouge The Communist guerrilla army, led by Pol Pot, that controlled Cambodia in the 1970s and subsequently attempted to overthrow the UN-sanctioned government.

Kuomintang The National People's Party (Nationalists), which, under Chiang Kai-shek, governed China until Mao Zedong's revolution in 1949; it dominated politics in Taiwan until the mid-1990s.

Laogai A Mandarin Chinese word for a prison or concentration camp where political prisoners are kept. It is similar in concept to the Russian word *gulag.*

Liberal Democratic Party (LDP) The conservative party that ruled Japan almost continuously between 1955 and the present. The LDP has guided Japan's rapid economic development.

Martial Law Rule applied to a territory by military or governmental authorities when normal civilian law is at risk of breaking down. Under martial law, residents are usually restricted in their movement and in their exercise of such rights as freedom of speech and of the press.

Meiji Restoration The restoration of the Japanese emperor to his throne in 1868. The period is important as the beginning of the modern era in Japan and the opening of Japan to the West after centuries of isolation.

Monsoons Winds that bring exceptionally heavy rainfall to parts of Southeast Asia and elsewhere. Monsoon rains are essential to the production of rice in Asia.

National League for Democracy An opposition party in Myanmar that won the national elections in 1990. Despite its overwhelming popularity, the military refused to allow the party's elected leaders to take office and placed its leader, Aung San Suu Kyi, under house arrest.

New Economic Policy (NEP) An economic plan advanced in the 1970s to restructure the Malaysian economy and foster industrialization and ethnic equality.

Newly Industrializing Country (NIC) A designation for those countries of the developing world, particularly Taiwan, South Korea, and other Asian nations, whose economies have undergone rapid growth; sometimes also referred to as newly industrializing economies.

Non-Aligned Movement A loose association of mostly non-Western developing nations, many of which had been colonies of Western powers but during the cold war chose to remain detached from either the United States or Soviet bloc. Initially Indonesia and India, among others, were enthusiastic promoters of the movement.

Opium Wars Conflicts between Britain and China in 1839–1842 and 1856–1866 in which England used China's destruction of opium shipments and other issues as a pretext to attack China and force the government to sign trade agreements.

Pacific Islands Forum An organization established by Australia and other South Pacific nations to provide a forum for discussion of common problems and opportunities in the region. It is headquartered in Suva, Fiji.

Pacific War The name frequently used by the Japanese to refer to that portion of World War II in which they were involved and which took place in Asia and the Pacific.

SARS (Severe Acute Respiratory Syndrome) A pneumonia-like virus that first appeared in China in 2002 and spread rapidly causing hundreds of deaths and adversely affecting the economy of the region.

Shintoism An ancient indigenous religion of Japan that stresses the role of *kami,* or supernatural gods, in the lives of people. For a time during the 1930s, Shinto was the state religion of Japan, and the emperor was honored as its high priest.

Smokestack Industries Heavy industries such as steel mills that are basic to an economy but produce objectionable levels of air, water, or land pollution.

Socialism An economic system in which productive property is owned by the government as are the proceeds from the productive labor. Most socialist systems today are actually mixed economies in which individuals as well as the government own property.

Southeast Asia Treaty Organization (SEATO) A collective-defense treaty signed by the United States and several European and Southeast Asian nations. It was dissolved in 1977.

Subsistence Farming Farming that meets the immediate needs of the farming family but that does not yield a surplus sufficient for export.

Swine Flu An infectious virus, officially called A/H1N1, that affected over 180,000 people globally in its first year and took the lives of nearly 2,000. Most Asian countries have had outbreaks of the virus, and its presence has affected tourism in several countries.

Taoism (pronounced *Daoism*) An ancient religion of China inspired by Lao-tze that stresses the need for mystical contemplation to free one from the desires and sensations of the materialistic and physical world.

Tiananmen Square Massacre The violent suppression by the Chinese Army of a prodemocracy movement that had been organized in Beijing by thousands of Chinese students in 1989 and that had become an international embarrassment to the Chinese regime.

Tsunami Japanese word composed of two Chinese characters meaning "harbor wave." Tsunamis or tidal waves are created when earthquakes, underwater volcanic activity, or undersea mudslides upset ocean stability and produce unusually high and fast-traveling ocean waves.

United Nations (UN) An international organization established immediately after World War II to replace the League of Nations. The organization includes most of the countries of the world and works for international understanding and world peace.

World Health Organization (WHO) Established in 1948 as an advisory and technical-assistance organization to improve the health of peoples around the world.

World Trade Organization (WTO) Successor organization to the General Agreement on Trade and Tariffs (GATT) treaties. WTO attempts to standardize the rules of free trade throughout the world.

Bibliography

SOURCES FOR STATISTICAL REPORTS

U.S. Department of State *Background Notes* (2011)
C.I.A. *World Factbook* (2011)
World Bank *World Development Report and World Development Indicators* (2011)
UN *Population and Vital Statistics Report* (2011)
World Statistics in Brief (2010)
The Statesman's Yearbook Online (2011)
Population Reference Bureau *World Population Data Sheet* (2011)
The World Almanac (2010, 2011)
The Economist Intelligence Unit (2011)
Human Development Reports Country Profiles and International Human Development Indicators (2011)
Internet World Stats Usage and Population Statistics (2011)
O.E.C.D. Statistics (2011)

GENERAL WORKS

Amitav Acharya, *The Quest for Identity: International Relations of Southeast Asia* (New York/Oxford: Oxford University Press, 2000).
Explores the complex relationship between intense feelings of nationalism and the international order post–cold war.

Mark Borthwick, *East Asian Civilizations: A Dialogue in Five Stages* (Cambridge, MA: Harvard University Press, 1988). The development of philosophical and religious thought in China, Korea, Japan, and other regions of East Asia.

Richard Bowring and Peter Kornicki, *Encyclopedia of Japan* (New York: Cambridge University Press, 1993).

Barbara K. Bundy, Stephen D. Burns, and Kimberly V. Weichel, *The Future of the Pacific Rim: Scenarios for Regional Cooperation* (Westport, CT: Praeger, 1994).

Commission on U.S.–Japan Relations for the Twenty-First Century, *Preparing for a Pacific Century: Exploring the Potential for Pacific Basin Cooperation* (Washington, D.C.: November 1991).
Transcription of an international conference on the Pacific with commentary by representatives from the United States, Malaysia, Japan, Thailand, Indonesia, and others.

Susanna Cuyler, *A Companion to Japanese Literature, Culture, and Language* (Highland Park, NJ: B. Rugged, 1992).

Owen G. Norman, ed., *The Emergence of Modern Southeast Asia: A New History* (Honolulu: University of Hawaii Press, 2005).

Paul J. Smith, ed., *Terrorism and Violence in Southeast Asia: Transnational Challenges to States and Regional Stability* (New York: M.E. Sharpe, 2005).

William Theodore de Bary, *East Asian Civilizations: A Dialogue in Five Stages* (Cambridge, MA: Harvard University Press, 1988).
An examination of religions and philosophical thought in several regions of East Asia.

Richard E. Feinberg, ed., *APEC as an Institution: Multilateral Governance in the Asia-Pacific* (Singapore: Institute of Southeast Asian Studies, 2003).
Explores the strategy of the Asia Pacific Economic Cooperation Council and analyzes its accomplishments and failures.

Syed N. Hossain, *Japan: Not in the West* (Boston: Vikas II, 1995).

James W. McGuire, ed., *Rethinking Development in East Asia and Latin America* (Los Angeles: Pacific Council on International Policy, 1997).

Charles E. Morrison, ed., *Asia Pacific Security Outlook 1997* (Honolulu: East-West Center, 1997).

Seijiu Naya and Stephen Browne, eds., *Development Challenges in Asia and the Pacific in the 1990s* (Honolulu: East-West Center, 1991).
A collection of speeches made at the 1990 Symposium on Cooperation in Asia and the Pacific. The articles cover development issues in East, Southeast, and South Asia and the Pacific.

Edwin O. Reischauer and Marius B. Jansen, *The Japanese Today: Change and Continuity* (Cambridge: Belknap Press, 1995).
A description of the basic geography and historical background of Japan.

Leo Suryadinata, ed., *Nationalism and Globalization: East and West* (Singapore: Institute of Southeast Asian Studies, 2000).
Case studies of six Asian countries and six European countries, showing how each has handled the pressures of the modern era.

NATIONAL HISTORIES AND ANALYSES

Australia

Boris Frankel, *From the Prophets Deserts Come: The Struggle to Reshape Australian Political Culture* (New York: Deakin University [St. Mut.], 1994).
Australia's government and political aspects are described in this essay.

Peter Brown and Julian Thomas, eds., *A Win and a Prayer: Scenes from the 2004 Australian Election* (University of New South Wales Press, 2005).
Post-election analysis of the sociology of the campaigning process.

Ann Capling, *All the Way with the USA: Australia, the US, and Free Trade* (University of New South Wales Press, 2004).

Herman J. Hiery, *The Neglected War: The German South Pacific and the Influence of WW I* (Honolulu: University of Hawaii Press, 1995).

David Alistair Kemp, *Society and Electoral Behaviors in Australia: A Study of Three Decades* (St. Lucia: University of Queensland Press, 1978).
Elections, political parties, and social problems in Australia since 1945.

David Meredith and Barrie Dyster, *Australia in the International Economy in the Twentieth Century* (New York: Cambridge University Press, 1990).
Examines the international aspects of Australia's economy.

Brunei

Wendy Hutton, *East Malaysia and Brunei* (Berkeley, CA: Periplus, 1993).

Graham Saunders, *A History of Brunei* (New York: Oxford University Press, 1995).

Nicholas Tarling, *Britain, the Brookes, and Brunei* (Kuala Lumpur: Oxford University Press, 1971).
A history of the sultanate of Brunei and its neighbors.

Cambodia

David P. Chandler, *The Tragedy of Cambodian History, War, and Revolution since 1945* (New Haven, CT: Yale University Press, 1993).
A short history of Cambodia.

Michael W. Doyle, *UN Peacekeeping in Cambodia: UNTAC's Civil Mandate* (Boulder, CO: Lynne Rienner, 1995).
A review of the current status of Cambodia's government and political parties.

Craig Etcheson, *The Rise and Demise of Democratic Kampuchea* (Boulder, CO: Westview Press, 1984).
A history of the rise of the Communist government in Cambodia.

Caroline Hughes, *Dependent Communities: Aid and Politics in Cambodia and East Timor* (New York: Cornell University Press, 2009).

Richard Lunn, *Leaving Year Zero: Stories of Surviving Pol Pot's Cambodia* (Crawley, Western Australia: University of Western Australia Press, 2004).
Stories of six Cambodian refugees who settled in Australia.

William Shawcross, *The Quality of Mercy: Cambodia, Holocaust, and Modern Conscience; with a report from Ethiopia* (New York: Simon & Schuster, 1985).
A report on political atrocities, relief programs, and refugees in Cambodia and Ethiopia.

Usha Welaratna, ed., *Beyond the Killing Fields: Voices of Nine Cambodian Survivors* (Stanford, CA: Stanford University Press, 1993).
A collection of nine narratives by Cambodian refugees in the United States and their adjustments into American society.

China

Julia F. Andrews, *Painters and Politics in the People's Republic of China, 1949–1979* (Berkeley, CA: University of California Press, 1994).
A fascinating presentation of the relationship between politics and art from the beginning of the Communist period until the eve of major liberalization in 1979.

Ma Bo, *Blood Red Sunset* (New York: Viking, 1995).
A compelling autobiographical account by a Red Guard during the Cultural Revolution.

Jung Chang, *Wild Swans: Three Daughters of China* (New York: Simon and Shuster, 1992).
An autobiographical/biographical account that illuminates what China was like for one family for three generations.

James Fallows, *Postcards from Tomorrow Square* (New York: Vintage Books, 2008).

John Gittings, *The Changing Face of China: From Mao to Market* (Oxford: Oxford University Press, 2005).
Review of the path China has taken from communism to partial capitalism.

Kwang-chih Chang, *The Archaeology of China*, 4th ed. (New Haven, CT: Yale University Press, 1986).

———, *Shang Civilization* (New Haven, CT: Yale University Press, 1980).
Two works by an eminent archaeologist on the origins of Chinese civilization.

Nien Cheng, *Life and Death in Shanghai* (New York: Penguin Books, 1988). A view of the Cultural Revolution by one of its victims.

Qing Dai, *Yangtze! Yangtze!* (Toronto: Probe International, 1994).
Collection of documents concerning the debate over building the Three Gorges Dam on the upper Yangtze River in order to harness energy for China.

John King Fairbank, *China: A New History* (Cambridge, MA: Harvard University Press, 1992).
An examination of the motivating forces in China's history that define it as a coherent culture from its earliest recorded history to 1991.

David S. G. Goodman and Beverly Hooper, eds., *China's Quiet Revolution: New Interactions between State and Society* (New York: St. Martin's Press, 1994).
Articles examine the impact of economic reforms since the early 1980s on the social structure and society generally, with focus on changes in wealth, status, power, and newly emerging social forces.

Richard Madsen, *China and the American Dream: A Moral Inquiry* (Berkeley, CA: University of California Press, 1995).
A history on the emotional and unpredictable relationship the United States has had with China from the nineteenth century to the present.

Jim Mann, *Beijing Jeep: A Case Study of Western Business in China* (Boulder, CO: Westview Press, 1997).
A crisp view of what it takes for a Westerner to do business in China.

Suzanne Ogden, *China's Unresolved Issues: Politics, Development, and Culture* (Englewood Cliffs, NJ: Prentice-Hall, 1992).
A complete review of economic and cultural issues in modern China.

Jie Tang, *Managers and Mandarins in Contemporary China* (New York: Routledge, 2005).
An ethnographic study of cultural differences in an international joint venture.

Tyrene White, *China's Longest Campaign: Birth Planning in the PRC 1949–2005* (Ithaca/London: Cornell University Press, 2006).
How China created the world's strictest family planning program.

John Wong and Lu Ding, *China's Economy into the New Century* (London: Imperial College Press, 2002).
A policy-oriented and fact-based analysis of the new Chinese economy.

John Wong and Nan Seok Ling, *China's Emerging New Economy* (London: Imperial College Press, 2000).
This book is intended for readers interested in China's Internet and e-commerce sectors.

Li Zhisui, *The Private Life of Chairman Mao* (New York: Random House, 1994).
Memoirs of Mao's personal physician.

East Timor

Hal Hill and Joao M. Saldanha, eds., *East Timor: Development Challenges for the World's Newest Nation* (ISEAS/Palgrave Publishers, 2001).
A summary of economic information about East Timor and an analysis of the pitfalls ahead as the country attempts to recover from its difficult birth as a new nation.

Damien Kingsbury, *East Timor: The Price of Liberty* (Basingstoke, U.K.: Palgrave MacMillan, 2009).

Richard Tanter, Desmond, and Gerry van Klinken, eds., *Masters of Terror: Indonesia's Military and Violence in East Timor* (Lanham, MD: Rowman & Littlefield Publishers, 2006).
Description of those who masterminded the destruction of East Timor in 1999.

Hong Kong

"Basic Law of Hong Kong Special Administrative Region of the People's Republic of China," *Beijing Review,* Vol. 33, No. 18 (April 30–May 6, 1990), supplement.

Bob Beatty, *Democracy, Asian Values, and Hong Kong: Evaluating Political Elite Beliefs* (Westport, CT: Praeger, 2003).

Ming K. Chan and Gerard A. Postiglione, *The Hong Kong Reader: Passage to Chinese Sovereignty* (Armonk, NY: M. E. Sharpe, 1996).
A collection of articles about the issues facing Hong Kong during the transition to Chinese rule after July 1, 1997.

Stephen Chiu and Tai-Lok Lui Abingdon, *Hong Kong: Becoming a Chinese Global City* (New York: Routledge, 2009).

Jonathan Dimbleby, *The Last Governor* (Doubleday Canada, 1997).

Walter Hatch and Kozo Yamamura, *Asia in Japan's Embrace: Building a Regional Production Alliance* (Cambridge: Cambridge University Press, 1996).
Discusses the future likelihood of Japan building an exclusive trading zone in Asia.

Berry Hsu, ed., *The Common Law in Chinese Context* in the series entitled *Hong Kong becoming China: The Transition to 1997* (Armonk, NY: M. E. Sharpe, Inc., 1992).
An examination of the common law aspects of the "Basic Law," the mini-constitution that will govern Hong Kong after 1997.

Christine Loh, ed., *Building Democracy: Creating Good Government for Hong Kong* (Hong Kong: Hong Kong University Press, 2003).

Benjamin K. P. Leung, ed., *Social Issues in Hong Kong* (New York: Oxford University Press, 1990).
A collection of essays on select issues in Hong Kong, such as aging, poverty, women, pornography, and mental illness.

Jan Morris, *Hong Kong: Epilogue to an Empire* (New York: Vintage, 1997).
A detailed portrait of Hong Kong that gives the reader the sense of actually being on the scene in a vibrant Hong Kong.

Mark Roberti, *The Fall of Hong Kong: China's Triumph and Britain's Betrayal* (New York: John Wiley & Sons, Inc., 1994).
An account on the decisions Britain and China made about Hong Kong's fate since the early 1980s.

Frank Welsh, *A Borrowed Place: The History of Hong Kong* (New York: Kodansha International, 1996).
A presentation on Hong Kong's history from the time of the British East India Company in the eighteenth century through the Opium Wars of the nineteenth century to the present.

Indonesia

Amarendra Bhattacharya and Mari Pangestu, *Indonesia: Development, Transformation, and Public Policy* (Washington, D.C.: World Bank, 1993).
An examination of Indonesia's economic policy.

Frederica M. Bunge, *Indonesia: A Country Study* (Washington, D.C.: U.S. Government, 1983).
An excellent review of the outlines of Indonesian history and culture, including politics and national security.

Freek Colombijn and J. Thomas Lindblad, eds., *Roots of Violence in Indonesia* (Singapore: ISEAS/KITLV Press, 2002).
The authors explain the violence in Aceh, East Timore, and other places by exploring the long history of Indonesia.

Helen Creese, *Women of the Kakawin World* (New York: M.E. Sharpe, 2004).
Sexuality of women in two regions of Indonesia.

East Asia Institute, *Indonesia in Transition* (New York: Columbia University, 2000).
An analysis of the revolution in Indonesian politics since the overthrow of Suharto.

Philip J. Eldridge, *Non-government Organizations and Political Participation in Indonesia* (New York: Oxford University Press, 1995).
An examination of Indonesia's nongovernment agencies (NGOs).

Audrey R. Kahin, ed., *Regional Dynamics of the Indonesian Revolution: Unity from Diversity* (Honolulu: University of Hawaii Press, 1985).

A history of Indonesia since the end of World War II, with separate chapters on selected islands.

Hamish McConald, *Suharto's Indonesia* (Australia: The Dominion Press, 1980).
The story of the rise of Suharto and the manner in which he controlled the political and military life of the country, beginning in 1965.

Tamara Nasir, *Indonesia Rising: Islam, Democracy and the Rise of Indonesia as a Major Power* (St. Louis: Select Publishing, 2009).

Susan Rodgers, ed., *Telling Lives, Telling Histories: Autobiography and Historical Immigration in Modern Indonesia* (Berkeley, CA: University of California Press, 1995).
Reviews the history of Indonesia's immigration.

David Wigg, *In a Class of Their Own: A Look at the Campaign against Female Illiteracy* (Washington, D.C.: World Bank, 1994).
Looks at the work that is being done by various groups to advance women's literacy in Indonesia.

Japan

David Arase, *Buying Power: The Political Economy of Japan's Foreign Aid* (Boulder CO: Lynne Rienner Publishers, Inc., 1995).
An attempt to explain the complexities of Japan foreign-aid programs.

Michael Barnhart, *Japan and the World since 1868* (New York: Routledge, Chapman, and Hall, 1994).
An essay that addresses commerce in Japan from 1868 to the present.

Marjorie Wall Bingham and Susan Hill Gross, *Women in Japan* (Minnesota: Glenhurst Publications, Inc., 1987).
An historical review of Japanese women's roles in Japan.

Roger W. Bowen, *Japan's Dysfunctional Democracy: The Liberal Democratic Party and Structural Corruption* (New York: M. E. Sharpe, 2003).
Explores the Japanese public's weakness in the face of major government corruption.

John Clammer, *Difference and Modernity: Social Theory and Contemporary Japanese Society* (New York: Routledge, Chapman, and Hall, 1995).

Dean W. Collinwood, "Japan," in Michael Sodaro, ed., *Comparative Politics* (New York: McGraw-Hill, 2000).
An analysis of Japan's government structure and history, electoral process, and some of the issues and pressure points affecting Japanese government.

Dennis J. Encarnation, *Rivals beyond Trade: America versus Japan in Global Competition* (Ithaca NY: Cornell University Press, 1993).
Explains how the economic rivalry that was once bilateral has turned into an intense global competition.

Mark Gauthier, *Making It in Japan* (Upland, PA: Diane Publishers, 1994).
An examination of how success can be attained in Japan's marketplace.

Shu Hagiwara, *Origins: The Creative Spark behind Japan's Best Product Designs* (Tokyo: Kodansha International, 2007).

Walter Hatch and Kozo Yamamura, *Asia in Japan's Embrace: Building a Regional Production Alliance* (Cambridge: Cambridge University Press, 1996).
Discusses the future likelihood of Japan building an exclusive trading zone in Asia.

Paul Herbig, *Innovation Japanese Style: A Cultural and Historical Perspective* (Glenview, IL: Greenwood, 1995).
A review of the implications for international competition.

Ronald J. Hrebenar, *Japan's New Party System* (Boulder, CO: Westview Press, 2000).
An analysis of the political structure in Japan since the end of complete LDP dominance.

Masae Kato, *Women's Rights? The Politics of Eugenic Abortion in Modern Japan* (Amsterdam: Amsterdam University Press, 2009).

Harold R. Kerbo and John McKinstry, *Who Rules Japan? The Inner-Circle of Economic and Political Power* (Glenview, IL: Greenwood, 1995).
The effect of Japan's politics on its economy is evaluated in this essay.

Hiroshi Komai, *Migrant Workers in Japan* (New York: Routledge, Chapman, and Hall, 1994).
An examination of the migrant labor supply in Japan.

Makoto Kumazawa, *Portraits of the Japanese Workplace: Labor Movements, Workers, and Managers* (Boulder, CO: Westview Press, 1996).
Translated into English from Japanese, the book includes reviews of the workplace lifestyle of bankers, women, steel workers, and others.

Solomon B. Levine and Koji Taira, eds., *Japan's External Economic Relations: Japanese Perspectives,* special issue of *The Annals of the American Academy of Political and Social Science* (January 1991).
An excellent overview of the origin and future of Japan's economic relations with the rest of the world, especially Asia.

John Lie, *Multi-Ethnic Japan* (Cambridge, MA: Harvard University Press, 2001).
The existence of Ainu, Koreans, Chinese, Taiwanese, Burakumin, and Okinawans challenges the widely held belief that Japan is a monoethnic society.

Gavan McCormack, *Target North Korea: Pushing North Korea to the Brink of Nuclear Catastrophe* (New York: Nation Books, 2004).

E. Wayne Nafziger, *Learning from the Japanese: Japan's Pre-War Development and the Third World* (Armonk, NY: M. E. Sharpe, 1995).
Presents Japan as a model of "guided capitalism," and what it did by way of policies designed to promote and accelerate development.

Nippon Steel Corporation, *Nippon: The Land and Its People* (Japan: Gakuseisha Publishing, 1984).
An overview of modern Japan in both English and Japanese.

Asahi Shimbun, *Japan Almanac 1998* (Tokyo: Asahi Shimbun Publishing Company, 1997).
Charts, maps, statistical data about Japan in both English and Japanese.

Patrick Smith, *Japan: A Reinterpretation* (New York: Pantheon Books, 1997).
A discussion of the rapidly changing Japanese national character.

Yoichiro Sato and Satu Limaye, *Japan in a Dynamic Asia: Coping with the New Security Challenges* (Lexington Books, 2006).
The political and economic power struggles Japan is facing in the new Asia.

Korea: North and South Korea

Chai-Sik Chung, *A Korean Confucian Encounter with the Modern World* (Berkeley, CA: IEAS, 1995).
Korea's history and the effectiveness of Confucianism are addressed.

Donald Clark et al., *U.S.–Korean Relations* (Farmingdale, NY: Regina Books, 1995).
A review on the history of Korea's relationship with the United States.

James Cotton, *Politics and Policy in the New Korean State: From Rah Tae-Woo to Kim Young-Sam* (New York: St. Martin's, 1995).
The power and influence of politics in Korea are examined.

James Hoare, *North Korea* (New York: Oxford University Press, 1995).
An essay that addresses commerce in Japan between 1868 and the present.

Dae-Jung Kim, *Mass Participatory Economy: Korea's Road to World Economic Power* (Landham, MD: University Press of America, 1995).

Korean Overseas Information Service, *A Handbook of Korea* (Seoul: Seoul International Publishing House, 1987).
A description of modern South Korea, including social welfare, foreign relations, and culture. The early history of the entire Korean Peninsula is also discussed.

———, *Korean Arts and Culture* (Seoul: Seoul International Publishing House, 1986).
A beautifully illustrated introduction to the rich cultural life of modern South Korea.

Callus A. MacDonald, *Korea: The War before Vietnam* (New York: The Free Press, 1986).
A detailed account of the military events in Korea between 1950 and 1953, including a careful analysis of the U.S. decision to send troops to the peninsula.

Christopher J. Sigur, ed., *Continuity and Change in Contemporary Korea* (New York: Carnegie Ethics and International Affairs, 1994).
A review of the numerous stages of change that Korea has experienced.

Joseph A. B. Winder, ed., *Korea's Economy 1999* (Washington, D.C.: Korea Economic Institute, 1999).
A review of the economic impact of the Asian financial crisis on South Korea.

Yonhap News Agency, ed., *North Korea Handbook* (New York: M. E. Sharpe, 2002).
A comprehensive guide to North Korea's military, economy, and culture.

Laos

Sucheng Chan, ed., *Hmong: Means Free Life in Laos and America* (Philadelphia: Temple University Press, 1994).

Arthur J. Dommen, *Laos: Keystone of Indochina* (Boulder, CO: Westview Press, 1985).
A short history and review of current events in Laos.

Grant Evans, ed., *Laos: Culture and Society* (Silkworm Books, 2000).
A comprehensive social and cultural analysis of the nation and people of Laos.

Joel M. Halpern, *The Natural Economy of Laos* (Christiansburg, VA: Dalley Book Service, 1990).

———, *Government, Politics, and South Structures of Laos: Study of Traditions and Innovations* (Christiansburg, VA: Dalley Book Service, 1990).

Viliam Phraxayavong, *History of Aid to Laos* (Chiang Mai, Thailand: Mekong Press, 2009).

Vatthana Pholsena, *Post-War Laos: The Politics of Culture, History, Identity* (Ithaca: Cornell University Press, 2006).

Macau

Charles Ralph Boxer, *The Portuguese Seaborne Empire, 1415–1825* (New York: A. A. Knopf, 1969).
A history of Portugal's colonies, including Macau.

W. G. Clarence-Smith, *The Third Portuguese Empire, 1825–1975* (Manchester: Manchester University Press, 1985).
A history of Portugal's colonies, including Macau.

Cathryn Clayton, *Sovereignty on the Edge: Macau and the Question of Chineseness* (Cambridge, MA: Harvard University Press, 2009).

Malaysia

Mohammed Ariff, *The Malaysian Economy: Pacific Connections* (New York: Oxford University Press, 1991).
The report on Malaysia examines Malaysia's development and its vulnerability in world trade.

Richard Clutterbuck, *Conflict and Violence in Singapore and Malaysia, 1945–1983* (Boulder, CO: Westview Press, 1985).
  The Communist challenge to the stability of Singapore and Malaysia in the early years of their independence from Great Britain is presented.
Virginia Hooker and Norani Othman, eds., *Malaysia: Islam, Society and Politics* (Singapore: Institute of Southeast Asian Studies, 2003).
  Collection of essays written by sociologist Clive Kessler on modern Malaysian religion and society.
Noboru Ishikawa, *Between Frontiers: Nation and Identity in a Southeast Asian Border Zone,* (Copenhagen: NIAS Press, 2009).
K. S. Jomo, ed., *Japan and Malaysian Development: In the Shadow of the Rising Sun* (New York: Routledge, 1995).
  A review of the relationship between Japan and Malaysia's economy.
Gordon Means, *Malaysian Politics: The Second Generation* (New York: Oxford University Press, 1991).
R. S. Milne, *Malaysia: Tradition, Modernity, and Islam* (Boulder, CO: Westview Press, 1986).
  A general overview of the nature of modern Malaysian society.
Chris Nyland et al., eds., *Malaysian Business in the New Era* (Cheltenham, UK: Edward Elgar Publishing, 2001).
A. B. Shamsul, *From British to Bumiputera Rule: Local Politics and Rural Development in Peninsular Malaysia* (Singapore: Institute of Southeast Asian Studies, 2004).
  Originally published in 1986, this book nevertheless provides keen insight into the mindset of the Malay people as they deal with domestic and international pressures.

Myanmar (Burma)

Michael Aung-Thwin, *Pagan: The Origins of Modern Burma* (Honolulu: University of Hawaii Press, 1985).
  A treatment of the religious and political ideology of the Burmese people and the effect of ideology on the economy and politics of the modern state.
Aye Kyaw, *The Voice of Young Burma* (Ithaca, NY: Cornell SE Asia, 1993).
  The political history of Burma is presented in this report.
Chi-Shad Liang, *Burma's Foreign Relations: Neutralism in Theory and Practice* (Glenview, IL: Greenwood, 1990).
Mya Maung, *The Burma Road to Poverty* (Glenview, IL: Greenwood, 1991).
David L. Steinberg, *Burma: The State of Myanmar* (Washington, D.C.: Georgetown University Press, 2002).
  Issues of authority and legitimacy in Myanmar since the coup of 1988.
Myat Thein, *Economic Development of Myanmar* (Singapore: Institute of Southeast Asian Studies, 2004).
  A compilation of studies on the economy of Myanmar from 1948 to 2000.
Justin Wintle, *Perfect Hostage: Aung San Suu Kyi, Burma and the Generals* (New York: Skyhorse Publishing, 2007).

New Zealand

Bev James and Kay Saville-Smith, *Gender, Culture, and Power: Challenging New Zealand's Gendered Culture* (New York: Oxford University Press, 1995).
Patrick Massey, *New Zealand: Market Liberalization in a Developed Economy* (New York: St. Martin, 1995).
  Analyzes New Zealand's market-oriented reform programs since the Labour government came into power in 1984.
Stephen Rainbow, *Green Politics* (New York: Oxford University Press, 1994).
  A review of current New Zealand politics.
Geoffrey W. Rice, *The Oxford History of New Zealand* (New York: Oxford University Press, 1993).

Philippa Mein Smith, *A Concise History of New Zealand* (New York: Cambridge University Press, 2005).

Papua New Guinea

Robert J. Gordon and Mervyn J. Meggitt, *Law and Order in the New Guinea Highlands: Encounters with Enga* (Hanover, NH: University Press of New England, 1985).
  Tribal law and warfare in Papua New Guinea.
David Hyndman, *Ancestral Rainforests and the Mountain of Gold: Indigenous Peoples and Mining in New Guinea* (Boulder, CO: Westview Press, 1994).
Bruce W. Knauft, *South Coast New Guinea Cultures: History, Comparison, Dialectic* (New York: Cambridge University Press, 1993).

The Philippines

Frederica M. Bunge, ed., *Philippines: A Country Study* (Washington, D.C.: U.S. Government, 1984).
  Description and analysis of the economic, security, political, and social systems of the Philippines, including maps, statistical charts, and reproduction of important documents.
  An extensive bibliography is included.
Manual B. Dy, *Values in Philippine Culture and Education* (Washington, D.C.: Council for Research in Values and Philosophy, 1994).
James F. Eder and Robert L. Youngblood, eds., *Patterns of Power and Politics in the Philippines: Implications for Development* (Tempe, AZ: ASU Program, SE Asian, 1994).
  A review of the impact of politics and its power over development in the Philippines.

Singapore

Lai A. Eng, *Meanings of Multiethnicity: A Case Study of Ethnicity and Ethnic Relations in Singapore* (New York: Oxford University Press, 1995).
Paul Leppert, *Doing Business with Singapore* (Fremont, CA: Jain Publishing, 1995).
  Singapore's economic status is examined in this report.
Hafiz Mirza, *Multinationals and the Growth of the Singapore Economy* (New York: St. Martin's Press, 1986).
  An essay on foreign companies and their impact on modern Singapore.
Nilavu Mohdx et al., *New Place, Old Ways: Essays on Indian Society and Culture in Modern Singapore* (Columbia, MO: South Asia, 1994).
Lily Rahim, *Singapore in the Malay World* (New York: Routledge, 2009).

South Pacific

C. Beeby and N. Fyfe, "The South Pacific Nuclear Free Zone Treaty," Victoria University of Wellington *Law Review,* Vol. 17, No. 1, pp. 33–51 (February 1987).
  A good review of nuclear issues in the Pacific.
William S. Livingston and William Roger Louis, eds., *Australia, New Zealand, and the Pacific Islands since the First World War* (Austin, TX: University of Texas Press, 1979).
  An assessment of significant historical and political developments in Australia, New Zealand, and the Pacific Islands since 1917.
Leo Suryadinata, *Ethnic Relations and Nation-Building in Southeast: The Case of the Ethnic Chinese* (Singapore: Institute of Southeast Asian Studies, 2004).
  A review of the ethnic tensions that have long been a feature of Southeast Asian life.
Lim Chong Yah, *Southeast Asia: The Long Road Ahead* (London: Imperial College Press, 2001).
  A cross-country discussion of the problems that will be encountered as Southeast Asia develops.

Taiwan

Joel Aberbach et al., eds., *The Role of the State in Taiwan's Development* (Armonk, NY: M. E. Sharpe, 1994). Articles address technology, international trade, state policy toward the development of local industries, and the effect of economic development on society, including women and farmers.

Chou Bih-er, Clark Cal, and Janet Clark, *Women in Taiwan Politics: Overcoming Barriers to Women's Participation in a Modernizing Society* (Boulder, CO: Lynne Rienner, 1990). Examines the political underrepresentation of women in Taiwan and how Chinese culture on the one hand and modernization and development on the other are affecting women's status.

Richard C. Bush, *Untying the Knot: Making Peace in the Taiwan Strait* (Washington, D.C.: Brookings Institution Press, 2005). Analysis of what it would take to solve the China-Taiwan stand-off.

Catherine Farris, Anru Lee, Murray Rubinstein, *Women in the New Taiwan* (New York: M.E. Sharpe, 2002).

Alain Guilloux, *Taiwan Humanitarianism and Global Governance* (New York: Routledge, 2009).

Stevan Harrell and Chun-chieh Huang, eds., *Cultural Change in Postwar Taiwan* (Boulder, CO: Westview Press, 1994). A collection of essays that analyzes the tensions in Taiwan's society as modernization erodes many of its old values and traditions.

Dennis Hickey, *United States–Taiwan Security Ties: From Cold War to beyond Containment* (Westport, CT: Praeger, 1994). Examines U.S.–Taiwan security ties from the cold war to the present and what Taiwan is doing to ensure its own military preparedness.

Chin-chuan Lee, "Sparking a Fire: The Press and the Ferment of Democratic Change in Taiwan," in Chin-chuan Lee, ed., *China's Media, Media China* (Boulder, CO: Westview Press, 1994), pp. 179–193.

Sheng Lijun, *China and Taiwan: Cross-Strait Relations under Chen Shui-bian* (Singapore: Institute of Southeast Asian Studies, 2002). With the electoral defeat of the Kuomintang (KMT), Taiwan's relations with China are undergoing important changes.

Robert M. Marsh, *The Great Transformation: Social Change in Taipei, Taiwan, since the 1960s* (Armonk, NY: M. E. Sharpe, 1996). An investigation of how Taiwan's society has changed since the 1960s when its economic transformation began.

Robert G. Sutter and William R. Johnson, *Taiwan in World Affairs* (Boulder, CO: Westview Press, 1994). Articles give comprehensive coverage of Taiwan's involvement in foreign affairs.

Wei-Bin Zhang, *Taiwan's Modernization* (London: Imperial College Press, 2003). The impact of Confucianism on the modernization of Taiwan and other Confucian-based societies in East Asia.

Thailand

Duncan McCargo, ed., *Rethinking Thailand's Southern Violence* (Honolulu: University of Hawaii Press, 2007).

Medhi Krongkaew, *Thailand's Industrialization and Its Consequences* (New York: St. Martin, 1995). A discussion of events surrounding the development of Thailand since the mid-1980s, with a focus on the nature and characteristics of Thai industrialization.

Ross Prizzia, *Thailand in Transition: The Role of Oppositional Forces* (Honolulu: University of Hawaii Press, 1985). Government management of political opposition in Thailand.

Susan Wells and Steve Van Beek, *A Day in the Life of Thailand* (San Francisco: Collins SF, 1995).

Vietnam

Chris Brazier, *The Price of Peace* (New York: Okfam Pubs. U.K. [St. Mut.], 1992).

Ronald J. Cima, ed., *Vietnam: A Country Study* (Washington, D.C.: U.S. Government, 1989). An overview of modern Vietnam, with emphasis on the origins, values, and lifestyles of the Vietnamese people.

Chris Ellsbury et al., *Vietnam: Perspectives and Performance* (Cedar Falls, IA: Assn. Text Study, 1994). A review of Vietnam's history.

Bernd Greiner, *War Without Fronts: The USA in Vietnam* (New York: Random House, 2009).

Hy Van Luong, *Postwar Vietnam: Dynamics of a Transforming Society* (Lanham, MD: Rowman and Littlefield Publishers, 2003). The Vietnam War is not the most important thing about Vietnam. This book shows how rich are Vietnam's culture and history.

Hy V Luong, ed., *Postwar Vietnam: Dynamics of a Transforming Society* (ISEAS/Rowman & Littlefield Publishers, 2003). The dynamics of economic reform, socioeconomic inequality, and other social conditions are covered in this book, which attempts to go beyond the usual focus on the Vietnam War.

D. R. SarDeSai, *Vietnam: The Struggle for National Identity* (Boulder, CO: Westview Press, 1992). A good treatment of ethnicity in Vietnam and a national history up to the involvement in Cambodia.

PERIODICALS AND CURRENT EVENTS

The Annals of the American Academy of Political and Social Science
c/o Sage Publications, Inc.
2455 Teller Rd.
Newbury Park, CA 91320
Selected issues focus on the Pacific Rim; there is an extensive book-review section. Special issues are as follows: "The Pacific Region: Challenges to Policy and Theory" (September 1989). "China's Foreign Relations" (January 1992). "Japan's External Economic Relations: Japanese Perspectives" (January 1991).

Asian Affairs: An American Review
Helen Dwight Reid Educational Foundation
1319 Eighteenth St., NW
Washington, D.C. 20036-1802
Publishes articles on political, economic, and security policy.

The Asian Wall Street Journal,
Dow Jones & Company, Inc.
A daily business newspaper focusing on Asian markets.

Asia-Pacific Issues
East-West Center
1601 East-West Rd.
Burns Hall, Rm. 1079
Honolulu, HI 96848-1601
Each contains one article on an issue of the day in Asia and the Pacific.

The Asia-Pacific Magazine
Research School of Pacific and Asian Studies
The Australian National University
Canberra ACT 0200, Australia
General coverage of all of Asia and the Pacific, including book reviews and excellent color photographs.

Asia-Pacific Population Journal
Economic and Social Commission for Asia and the Pacific
United Nations Building
Rajdamnern Nok Ave.
Bangkok 10200, Thailand
A quarterly publication of the United Nations.

Australia Report
1601 Massachusetts Ave., NW
Washington, D.C. 20036
A monthly publication of the Embassy of Australia, Public Diplomacy Office, with a focus on U.S. relations.

Canada and Hong Kong Update
Joint Centre for Asia Pacific Studies
Suite 270, York Lanes
York University
4700 Keele St.
North York, Ontario M3J 1P3, Canada
A source of information about Hong Kong emigration.

Courier
The Stanley Foundation
209 Iowa Ave.
Muscatine, IA 52761
Published three times a year, the *Courier* carries interviews of leaders in Asian and other world conflicts.

Current History: A World Affairs Journal
Focuses on one country or region in each issue; the emphasis is on international and domestic politics.

The Economist
25 St. James's St.
London, England
A news magazine with insightful commentary on international issues affecting the Pacific Rim.

Education About Asia
1 Lane Hall
The University of Michigan
Ann Arbor, MI 48109
Published three times a year, it contains useful tips for teachers of Asian studies. The Spring 1998 issue (Vol. 3, No. 1) focuses on teaching the geography of Asia.

Indochina Interchange
Suite 1801
220 West 42nd St.
New York, NY 10036
A publication of the U.S.–Indochina Reconciliation Project. An excellent source of information about assistance programs for Laos, Cambodia, and Vietnam.

Japan Echo
Maruzen Co., Ltd.
P.O. Box 5050
Tokyo 100-3199, Japan
Bimonthly translation of selected articles from the Japanese press on culture, government, environment, and other topics.

Japan Economic Currents
Keizai Koho Center
1900 K Street NW
Suite 1075
Washington, D.C. 20006
A commentary on selected economic and business trends. Available online at www.kkc-usa.org.

The Japan Foundation Newsletter
The Japan Foundation
Park Building
3-6 Kioi-cho
Chiyoda-ku
Tokyo 102, Japan
A quarterly with research reports, book reviews, and announcements of interest to Japan specialists.

Japan Quarterly
Asahi Shimbun
5-3-2 Tsukiji

Chuo-ku
Tokyo 104, Japan
A quarterly journal, in English, covering political, cultural, and sociological aspects of modern Japanese life.

The Japan Times
The Japan Times Ltd.
C.P.O. Box 144
Tokyo 100-91, Japan
Excellent coverage, in English, of news reported in the Japanese press.

The Journal of Asian Studies
Association for Asian Studies
1 Lane Hall
University of Michigan
Ann Arbor, MI 48109
Formerly *The Far Eastern Quarterly;* scholarly articles on Asia, South Asia, and Southeast Asia.

Journal of Japanese Trade & Industry
11th Floor, Fukoku Seimei Bldg., 2-2-2 Uchisaiwai-cho
Chiyoda Ku
Tokyo 100-0011, Japan
A bimonthly publication of the Japan Economic Foundation, with a focus on trade but including articles on Japanese culture and other topics.

Journal of Southeast Asian Studies
Singapore University Press
Formerly the *Journal of Southeast Asian History;* scholarly articles on all aspects of modern Southeast Asia.

Korea Economic Report
Yoido
P.O. Box 963
Seoul 150-609
South Korea
An economic magazine for people doing business in Korea.

The Korea Herald
2-12, 3-ga Hoehyon-dong
Chung-gu
Seoul, South Korea
World news coverage, in English, with focus on events affecting the Korean Peninsula.

The Korea Times
The Korea Times Hankook Ilbo
Seoul, South Korea
Coverage of world news, with emphasis on events affecting Asia and the Korean Peninsula.

Malaysia Industrial Digest
Malaysian Industrial Development Authority (MIDA)
6th Floor
Industrial Promotion Division
Wisma Damansara, Jalan Semantan
50490 Kuala Kumpur, Malaysia
A source of statistics on manufacturing in Malaysia; of interest to those wishing to become more knowledgeable in the business and industry of the Pacific Rim.

The New York Times
229 West 43rd St.
New York, NY 10036
A daily newspaper with excellent coverage of world events.

News From Japan
Embassy of Japan
Japan-U.S. News and Communication
Suite 520
900 17th St., NW
Washington, D.C. 20006
A twice-monthly newsletter with news briefs from the Embassy of Japan on issues affecting Japan–U.S. relations.

Newsweek
444 Madison Ave.
New York, NY 10022
A weekly magazine with news and commentary on national and world events.

The Oriental Economist
380 Lexington Ave.
New York, NY 10168
A monthly review of political and economic news in Japan by Toyo Keizai America Inc.

Pacific Affairs
The University of British Columbia
Vancouver, BC V6T 1W5
Canada
An international journal on Asia and the Pacific, including reviews of recent books about the region.

Pacific Basin Quarterly
c/o Thomas Y. Miracle
1421 Lakeview Dr.
Virginia Beach, VA 23455-4147
Newsletter of the Pacific Basin Center Foundation. Sometimes provides instructor's guides for included articles.

South China Morning Post
Tong Chong Street
Hong Kong
Daily coverage of world news, with emphasis on Hong Kong, China, Taiwan, and other Asian countries.

Time
Time-Life Building
Rockefeller Center
New York, NY 10020
A weekly newsmagazine with news and commentary on national and world events.

U.S. News & World Report
2400 N St., NW
Washington, D.C. 20037
A weekly newsmagazine with news and commentary on national and world events.

The US-Korea Review
950 Third Ave.
New York, NY 10022
Bimonthly magazine reviewing cultural, economic, political, and other activities of The Korea Society.

Vietnam Economic Times
175 Nguyen Thai Hoc
Hanoi, Vietnam
An English-language monthly publication of the Vietnam Economic Association, with articles on business and culture.

The World & I: A Chronicle of Our Changing Era
2800 New York Ave., NE
Washington, D.C. 20002
A monthly review of current events plus excellent articles on various regions of the world.

Index